Element-Doped Functional Carbon-Based Materials

Element-Doped Functional Carbon-Based Materials

Special Issue Editors

Sergio Morales-Torres
Agustín F. Pérez-Cadenas
Francisco Carrasco-Marín

MDPI • Basel • Beijing • Wuhan • Barcelona • Belgrade

Special Issue Editors

Sergio Morales-Torres
University of Granada
Spain

Agustín F. Pérez-Cadenas
University of Granada
Spain

Francisco Carrasco-Marín
University of Granada
Spain

Editorial Office
MDPI
St. Alban-Anlage 66 4052
Basel, Switzerland

This is a reprint of articles from the Special Issue published online in the open access journal *Materials* (ISSN 1996-1944) from 2018 to 2020 (available at: https://www.mdpi.com/journal/materials/special_issues/element_doped_functional_carbon).

For citation purposes, cite each article independently as indicated on the article page online and as indicated below:

LastName, A.A.; LastName, B.B.; LastName, C.C. Article Title. *Journal Name* **Year**, *Article Number*, Page Range.

ISBN 978-3-03928-224-1 (Pbk)
ISBN 978-3-03928-225-8 (PDF)

Contents

About the Special Issue Editors

Sergio Morales-Torres (Ph.D.) has been Associate Researcher at Faculty of Sciences of University of Granada (UGR, Spain) since 2017. He graduated in Chemistry at University of Jaén (Spain) in 2003 and completed a two-year MSc course at the same institution in 2005. By the end of 2009, he obtained his Ph.D. degree by UGR with European Mention and Extraordinary Award. After that, he moved to the Faculty of Engineering at University of Porto (FEUP, Portugal) for seven years, where he became Associate Researcher and developed a project on carbon-based membranes for water desalination and purification. His research interests involve the development of nanostructured materials as heterogeneous catalysts (including photo- and electrocatalysts) and membranes for energy and environmental applications. To date, he co-authored an international patent and 60 publications, including 45 articles in international JCR journals and 9 book chapters. He is regularly involved in the supervision of Ph.D. and MSc students as well as in the organization of scientific conferences and activities for science dissemination. He has also lectured specialized seminars and courses at FEUP and UGR and served as Guest Editor of six Special Issues of international scientific journals.

Agustín F. Pérez-Cadenas has been Full Professor of Inorganic Chemistry at University of Granada since 2018. His degree in Chemistry was obtained from University of Jaén (1997) and Ph.D. in Chemical Sciences from University of Granada (2002). He is currently the Tutor of the Erasmus Program for Chemistry studies at the University of Granada, and board member of the Spanish Carbon Group. He has supervised 7 doctoral theses on different topics, and his research is focused on the design of advanced carbon-based materials and the study of their industrial and environmental applications, through processes of adsorption and catalysis. His main research interests center around the electroreduction of CO_2 and O_2, energy storage, and photodegradation of pollutants. He has participated in international research stays abroad through different competitive Mobility Programs over 30 months. Co-author of a hundred scientific articles published in international JCR journals, and six patents. Prof. Pérez Cadenas is also co-author of four book chapters and more than 130 communications to national and international conferences.

Francisco Carrasco-Marín is Full Professor in Inorganic Chemistry at the University of Granada. He graduated in Chemistry from the University of Granada in 1984, where he also obtained his Ph.D. in Chemical Science in 1988, with a Special Doctoral Award in 1989. He spent postdoctoral stays at the Pennsylvania State University (USA) and Université Claude Bernard (France), among others. He was appointed Associate Professor in 1993 at the University of Jaén and at the University of Granada in 1996, where he was promoted to Full Professor in 2009. He is co-author of more than 140 papers and book chapters and 6 patents and has supervised 12 Ph.D. students. His research has focused on the synthesis, characterization, and applications of different forms of carbon materials; activated carbons; carbon xerogels and aerogels; carbon micro- and nanospheres; N-, S-, O-, B-, and P-doped carbons; and metal-containing nanoparticles. His most relevant contributions have been in the synthesis of high surface area porous carbons and in the modification of the surface chemistry in order to optimize their applications in the areas of catalysis, environmental protection and energy storage. Prof. Carrasco Marín is Vice-President of the Spanish Carbon Group, board member of the Specialized Group of Adsorption (Spanish Royal Societies of Physics and Chemistry), and member of the Spanish Society of Catalysis.

Editorial

Element-Doped Functional Carbon-Based Materials

Sergio Morales-Torres *, Agustín F. Pérez-Cadenas and Francisco Carrasco-Marín

Carbon Materials Research Group, Department of Inorganic Chemistry, Faculty of Sciences, University of Granada, Avenida de Fuente Nueva, s/n, ES18071 Granada, Spain; afperez@ugr.es (A.F.P.-C.); fmarin@ugr.es (F.C.-M.)
* Correspondence: semoto@ugr.es

Received: 18 December 2019; Accepted: 3 January 2020; Published: 11 January 2020

Abstract: Carbon materials are one of the most fascinating materials because of their unique properties and potential use in several applications. They can be obtained from agricultural waste, organic polymers, or by using advanced synthesizing technologies. The carbon family is very wide, it includes classical activated carbons to more advanced types like carbon gels, graphene, and so on. The surface chemistry of these materials is one of the most interesting aspects to be studied. The incorporation of different types of chemical functionalities and/or heteroatoms such as O, N, B, S, or P on the carbon surface enables the modification of the acidic–basic character, hydrophilicity–hydrophobicity, and the electron properties of these materials, which in turn determines the final application. This book collects original research articles focused on the synthesis, properties, and applications of heteroatom-doped functional carbon materials.

Keywords: carbon materials; heteroatoms; doping; surface chemistry; adsorption; catalysis; environmental remediation; energy storage

---

The broad family of carbon materials includes classical activated carbons to carbon nanostructures like carbon gels, carbon nanotubes, fullerenes, graphene, and so on. In general, these materials present different properties and origins, but all of them possess a common characteristic, in other words, the ability to be prepared in many different shapes such as pellets, granular, powders, cloths, fibers, monoliths, foams, coatings, films, and so on. Furthermore, their porous texture and chemical properties can be tailored by physical/thermal and chemical processes, enabling the development of porosity and specific surface area and the incorporation of different chemical functionalities. Both porosity and surface chemistry have a marked influence on their performance in a specific application, either by themselves or in combination with other materials. In fact, carbon materials have demonstrated to be excellent options as adsorbents [1], catalysts [2,3], or catalyst supports [4] when compared to classic materials (e.g., alumina, silica or ceria) as consequence of their high stability in both acidic and alkaline media.

Surface chemistry is the most attractive property of carbon materials, since the chemical groups anchored on the carbon surface may interact with organic molecules, inorganic salts, and metals. The most common heteroatoms are oxygen (O), nitrogen (N), sulfur (S), boron (B), and phosphorus (P). They are often part of functional groups and determine the acidic–basic character and the hydrophilicity–hydrophobicity [5–7]. For instance, oxygen-containing groups such as carboxylic acids, anhydrides, lactones, and phenols have an acidic character, while quinones, pyrones, and chromene are basic groups [8–10]. On the other hand, delocalized π electrons from the basal planes also contribute to the basicity [11], but also to the variation of the electron density. This effect can also be achieved by the incorporation of boron atoms or nitrogen-containing groups (i.e., pyridine and pyrrole), and deficient or additional electrons being provided, respectively. Thus, changes in the chemical properties of carbon materials influence their adsorption behavior and catalytic activity in some reactions [1,4].

This Special Issue deals with the recent advances in heteroatom-doped carbon materials. Different synthesis procedures, characterization techniques, and applications were investigated for these functional materials. The Special Issue collects eleven full-length articles and a short communication.

H. Hamad et al. [12] prepared carbon–phosphorus–titanium composites from cellulose to be used as photocatalysts in the removal of Orange-G dye. They pointed out that the phosphorus-containing groups incorporated in the composites modified their textural properties, crystallinity, and photocatalytic performance. S. Zhang et al. [13] modified biochars obtained from agricultural waste using 3-mercaptopropyltrimethoxysilane epoxy-chloropropane via an ionic-imprinted technique. These materials were active as adsorbents of Cd (II) in an aqueous solution, showing a higher Cd-selectivity in the presence of Co (II), Pb (II), Zn (II), and Cu (II) and a good stability after several adsorption–desorption cycles. A. Elmouwahidi et al. [14] developed carbon materials from waste woods by KOH activation. The surface chemistry was modified by different chemical agents, which incorporated nitrogen- and oxygen-containing groups on the carbon surface. All doped materials, with the exception of that treated with nitric acid, showed good capacitance values and high cyclic stability when used as electrodes for supercapacitors. An alternative method to obtain N-doped carbon materials for the same application was proposed by T. Ai. et al. [15]. This method consisted of the use of a N-containing bio-phenolic resin as a precursor and subsequent activation by a molten-salt method. Carbon materials have also been demonstrated to be efficient electrocatalysts in the oxygen reduction reaction (ORR). A. Abdelwahab et al. [16] studied Co- and Ni-doped carbon xerogels, while N-doped carbon fibers and microspheres synthesized from apricot sap were proposed by R. Kanuragaran et al. [17].

Carbon capture is a growing technology, whose implementation can be achieved by the research of novel materials. R. Wei et al. [18] prepared N-doped carbon materials from resorcinol and formaldehyde after KOH activation and ammonia carbonization. A. A. Alghamdi et al. [19] employed N-doped graphene oxide sheets (N-GOs) obtained from different N-containing polymers and after KOH activation. In general, the CO_2 capture capacity by N-doped materials was enhanced by the increase of the nitrogen content, the surface area, and the micropore volume. E. Rodriguez-Acevedo et al. [20] demonstrated that shallow reservoirs could be effective for carbon capture after injecting nanofluids based on N-rich carbon nanospheres. Finally, the last articles of this Special Issue deal with the development of N-doped graphene films for high sensitivity electrodes [21]; the functionalization of graphene oxides with p-phenylenediamine as a modifier [22]; and the induction of magnetic moments in graphene by introducing sp^3-defects [23].

All the published papers were strictly peer reviewed following the standard review practices for the Materials journal. As the Guest Editors of this Special Issue, we acknowledge all of the authors for their prime contributions and the reviewers for their valuable comments to improve the quality of the papers. Finally, we would like to thank the staff members of Materials, in particular Clark Xu for its kind assistance.

Author Contributions: S.M.-T.: writing—original draft. A.F.P.-C. and F.C.-M.: review & editing. All authors have read and agreed to the published version of the manuscript.

Funding: This work was supported by Junta de Andalucía (grant numbers P12-RNM-2892 and RNM172) and ERDF/Ministry of Science, Innovation and Universities—State Research Agency/_Project ref. RTI2018-099224-B-I00. SMT acknowledges the financial support from the University of Granada (Reincorporación Plan Propio).

Conflicts of Interest: The authors declare no conflict of interest.

References

1. Moreno-Castilla, C. Adsorption of organic molecules from aqueous solutions on carbon materials. *Carbon* **2004**, *42*, 83–94. [CrossRef]
2. Tsuji, K.; Shiraishi, I. Combined desulfurization, denitrification and reduction of air toxics using activated coke: 2. Process applications and performance of activated coke. *Fuel* **1997**, *76*, 555–560. [CrossRef]

3. Morales-Torres, S.; Silva, A.M.T.; Pérez-Cadenas, A.F.; Faria, J.L.; Maldonado-Hódar, F.J.; Figueiredo, J.L.; Carrasco-Marín, F. Wet air oxidation of trinitrophenol with activated carbon catalysts: Effect of textural properties on the mechanism of degradation. *Appl. Catal. B Environ.* **2010**, *100*, 310–317. [CrossRef]

4. Serp, P.; Figueiredo, J.L. *Carbon Materials for Catalysis*; John Wiley & Sons: Hoboken, NJ, USA, 2009.

5. Boehm, H.P. Some aspects of the surface chemistry of carbon blacks and other carbons. *Carbon* **1994**, *32*, 759–769. [CrossRef]

6. Kapteijn, F.; Moulijn, J.A.; Matzner, S.; Boehm, H.P. The development of nitrogen functionality in model chars during gasification in CO_2 and O_2. *Carbon* **1999**, *37*, 1143–1150. [CrossRef]

7. Salame, I.I.; Bandosz, T.J. Surface Chemistry of Activated Carbons: Combining the Results of Temperature-Programmed Desorption, Boehm, and Potentiometric Titrations. *J. Colloid. Interf. Sci.* **2001**, *240*, 252–258. [CrossRef]

8. Figueiredo, J.L.; Pereira, M.F.R.; Freitas, M.M.A.; Órfão, J.J.M. Modification of the surface chemistry of activated carbons. *Carbon* **1999**, *37*, 1379–1389. [CrossRef]

9. Morales-Torres, S.; Silva, T.L.S.; Pastrana-Martínez, L.M.; Brandão, A.T.S.C.; Figueiredo, J.L.; Silva, A.M.T. Modification of the surface chemistry of single- and multi-walled carbon nanotubes by HNO_3 and H_2SO_4 hydrothermal oxidation for application in direct contact membrane distillation. *Phys. Chem. Chem. Phys.* **2014**, *16*, 12237–12250. [CrossRef]

10. Pastrana-Martínez, L.M.; Morales-Torres, S.; Likodimos, V.; Falaras, P.; Figueiredo, J.L.; Faria, J.L.; Silva, A.M.T. Role of oxygen functionalities on the synthesis of photocatalytically active graphene-TiO_2 composites. *Appl. Catal. B Environ.* **2014**, *158–159*, 329–340. [CrossRef]

11. Lopez-Ramon, M.V.; Stoeckli, F.; Moreno-Castilla, C.; Carrasco-Marin, F. On the characterization of acidic and basic surface sites on carbons by various techniques. *Carbon* **1999**, *37*, 1215–1221. [CrossRef]

12. Hamad, H.; Castelo-Quibén, J.; Morales-Torres, S.; Carrasco-Marín, F.; Pérez-Cadenas, A.F.; Maldonado-Hódar, F.J. On the Interactions and Synergism between Phases of Carbon–Phosphorus–Titanium Composites Synthetized from Cellulose for the Removal of the Orange-G Dye. *Materials* **2018**, *11*, 1766. [CrossRef] [PubMed]

13. Zhang, S.; Yang, X.; Liu, L.; Ju, M.; Zheng, K. Adsorption Behavior of Selective Recognition Functionalized Biochar to Cd(II) in Wastewater. *Materials* **2018**, *11*, 299. [CrossRef] [PubMed]

14. Elmouwahidi, A.; Bailón-García, E.; Romero-Cano, L.A.; Zárate-Guzmán, A.I.; Pérez-Cadenas, A.F.; Carrasco-Marín, F. Influence of Surface Chemistry on the Electrochemical Performance of Biomass-Derived Carbon Electrodes for its Use as Supercapacitors. *Materials* **2019**, *12*, 2458. [CrossRef] [PubMed]

15. Ai, T.; Wang, Z.; Zhang, H.; Hong, F.; Yan, X.; Su, X. Novel Synthesis of Nitrogen-Containing Bio-Phenol Resin and Its Molten Salt Activation of Porous Carbon for Supercapacitor Electrode. *Materials* **2019**, *12*, 1986. [CrossRef] [PubMed]

16. Abdelwahab, A.; Carrasco-Marín, F.; Pérez-Cadenas, A.F. Carbon Xerogels Hydrothermally Doped with Bimetal Oxides for Oxygen Reduction Reaction. *Materials* **2019**, *12*, 2446. [CrossRef]

17. Karunagaran, R.; Coghlan, C.; Shearer, C.; Tran, D.; Gulati, K.; Tung, T.T.; Doonan, C.; Losic, D. Green Synthesis of Three-Dimensional Hybrid N-Doped ORR Electro-Catalysts Derived from Apricot Sap. *Materials* **2018**, *11*, 205. [CrossRef]

18. Wei, R.; Dai, X.; Shi, F. Enhanced CO_2 Adsorption on Nitrogen-Doped Carbon Materials by Salt and Base Co-Activation Method. *Materials* **2019**, *12*, 1207. [CrossRef]

19. Alghamdi, A.A.; Alshahrani, A.F.; Khdary, N.H.; Alharthi, F.A.; Alattas, H.A.; Adil, S.F. Enhanced CO_2 Adsorption by Nitrogen-Doped Graphene Oxide Sheets (N-GOs) Prepared by Employing Polymeric Precursors. *Materials* **2018**, *11*, 578. [CrossRef]

20. Rodriguez Acevedo, E.; Cortés, F.B.; Franco, C.A.; Carrasco-Marín, F.; Pérez-Cadenas, A.F.; Fierro, V.; Celzard, A.; Schaefer, S.; Cardona Molina, A. An Enhanced Carbon Capture and Storage Process (e-CCS) Applied to Shallow Reservoirs Using Nanofluids Based on Nitrogen-Rich Carbon Nanospheres. *Materials* **2019**, *12*, 2088. [CrossRef]

21. Bourquard, F.; Bleu, Y.; Loir, A.-S.; Caja-Munoz, B.; Avila, J.; Asensio, M.-C.; Raimondi, G.; Shokouhi, M.; Rassas, I.; Farre, C.; et al. Electroanalytical Performance of Nitrogen-Doped Graphene Films Processed in One Step by Pulsed Laser Deposition Directly Coupled with Thermal Annealing. *Materials* **2019**, *12*, 666. [CrossRef]

22. Sun, H.-J.; Liu, B.; Peng, T.-J.; Zhao, X.-L. Effect of Reaction Temperature on Structure, Appearance and Bonding Type of Functionalized Graphene Oxide Modified P-Phenylene Diamine. *Materials* **2018**, *11*, 647. [CrossRef] [PubMed]
23. Tang, T.; Wu, L.; Gao, S.; He, F.; Li, M.; Wen, J.; Li, X.; Liu, F. Universal Effectiveness of Inducing Magnetic Moments in Graphene by Amino-Type sp^3-Defects. *Materials* **2018**, *11*, 616. [CrossRef] [PubMed]

 *materials*

Article

On the Interactions and Synergism between Phases of Carbon–Phosphorus–Titanium Composites Synthetized from Cellulose for the Removal of the Orange-G Dye

Hesham Hamad [†], Jesica Castelo-Quibén, Sergio Morales-Torres *, Francisco Carrasco-Marín, Agustín F. Pérez-Cadenas and Francisco J. Maldonado-Hódar

Carbon Materials Research Group, Department of Inorganic Chemistry, Faculty of Sciences, University of Granada, Avenida de Fuentenueva, s/n. ES18071 Granada, Spain; heshamaterials@hotmail.com (H.H.); jesicacastelo@ugr.es (J.C.-Q.); fmarin@ugr.es (F.C.-M.); afperez@ugr.es (A.F.P.-C.); fjmaldon@ugr.es (F.J.M.-H.)
* Correspondence: semoto@ugr.es
† Current address: Fabrication Technology Department, Advanced Technology and New Materials Research Institute (ATNMRI), City of Scientific Research and Technology Applications (SRTA-City), New Borg El-Arab City, Alexandria 21934, Egypt

Received: 11 August 2018; Accepted: 15 September 2018; Published: 18 September 2018

Abstract: Carbon–phosphorus–titanium composites (CPT) were synthesized by Ti-impregnation and carbonization of cellulose. Microcrystalline cellulose used as carbon precursor was initially dissolved by phosphoric acid (H_3PO_4) to favor the Ti-dispersion and the simultaneous functionalization of the cellulose chains with phosphorus-containing groups, namely phosphates and polyphosphates. These groups interacted with the Ti-precursor during impregnation and determined the interface transformations during carbonization as a function of the Ti-content and carbonization temperature. Amorphous composites with high surface area and mesoporosity were obtained at low Ti-content (Ti:cellulose ratio = 1) and carbonization temperature (500 °C), while in composites with Ti:cellulose ratio = 12 and 800 °C, Ti-particles reacted with the cellulose groups leading to different Ti-crystalline polyphosphates and a marked loss of the porosity. The efficiency of composites in the removal of the Orange G dye in solution by adsorption and photocatalysis was discussed based on their physicochemical properties. These materials were more active than the benchmark TiO_2 material (Degussa P25), showing a clear synergism between phases.

Keywords: microcrystalline cellulose; chemical functionalization; polyphosphates; synergism; physicochemical properties; Orange G; photocatalysis

---

1. Introduction

Environmental catalysis tries to overcome the increasing pollution generated by a progressively more industrialized society through the search and development of novel materials and treatment technologies. Global warming, exponential growing population, intensive agricultural practices, among others, are the major factors affecting the availability of freshwater resources worldwide [1]. Treatment technologies, desalination and reuse of water intend to mitigate water scarcity. In fact, porous and catalytically active materials are continuously developed to be applied in different treatment processes for the removal of organic pollutants in water by adsorption and/or advanced oxidation processes (AOPs). Among others, heterogeneous photocatalysis has demonstrated to be an excellence treatment technology to remove water pollutants by the action of highly reactive oxygen species (e.g., hydroxyl radicals) generated from a semiconductor. In fact, a wide variety of the materials based on oxides (e.g., TiO_2, ZnO, WO_3), chalcogenides (e.g., ZnS, CdS, $ZnTe$, Bi_2S_3), nitrides (GaN), phosphides (GaP) and carbides (SiC), as well as free metal semiconductors composed by recent nanostructured carbons, such as graphitic carbon nitride (g-C_3N_4) and graphene derivatives, have been applied to different photocatalytic processes [2–8]. Most of these semiconductors present a limited photocatalytic performance due to a slow transportation of photoelectrons, fast photoelectron-hole recombination, a deficient surface that hinders the redox interaction with reactants and even, metal leaching when are irradiated in water. Thus, expensive and complex binary or ternary combinations of these materials are often proposed [9,10].

TiO_2 is the most widely applied semiconductor due to a high photo-activity, low cost, relative low toxicity and good chemical and thermal stability [4,11,12]. However, its performance in the visible range is poor so that different strategies, including non-metal and/or metal doping, dye sensitization, coupling semiconductor and the modification of properties such as crystalline phase, crystallite size and shapes and so on, have been studied to improve its photocatalytic efficiency [13,14]. On the other hand, the handling facilities and the price and suitability of the precursor materials should be taken into consideration in the design and development of novel photocatalysts. For instance, photocatalysts are used as building materials and some amounts of them are added to the concrete for the control of indoor air quality, preventing the accumulation of volatile organic compounds (VOCs) on building surfaces by oxidation [15]. Different types of industrial residues were recently reviewed in order to optimize the final price of the photocatalyst [16]. Thus, Ti-photocatalysts were prepared by calcination of the sludge containing Ti-salts previously used in the flocculation of sewages effluents [17] and by using natural phosphates [18]. An interesting approach is the synthesis of Ti-carbon composites [14] due to a better dispersion of Ti-nanoparticles, a well-developed porosity (enhanced pollutants adsorption) and the band gap narrowing by the synergism between phases. The employ of biomass, in particular cellulose [19], as support or carbon source is a remarkable alternative to prepare Ti–carbon photocatalysts, because it is the cheapest and most abundant biopolymer [20].

In this manuscript, carbon–phosphorus–Ti composites were sustainably developed, characterized and used for the photodegradation of Orange G (OG), a typical dye used in the textile industry. Microcrystalline cellulose (MCC) was used as a carbon precursor because is cheap, environmentally friendly and the most abundant renewable material; TiO_2 was used as semiconductor for the synthesis of nanocomposites. The crystalline structure of MCC required its previous solubilization with an acid treatment before Ti-impregnation, which in turn improved the contact between phases and the dispersion of the active Ti-phase. The influence of the acid pretreatment and the Ti:cellulose ratio on the physicochemical properties of the nanocomposites obtained and on the photocatalytic efficiency of the samples is discussed.

2. Materials and Methods

The synthesis of the cellulose–Ti composites was carried out using a procedure reported elsewhere [21]. Briefly, MCC (from Merck, Darmstadt, Germany) was suspended in distilled water (200 g L^{-1}) and then, it was completely dissolved by adding 10 mL of phosphoric acid (H_3PO_4) under stirring at 50 °C overnight. After that, an appropriated amount of titanium tetra-isopropoxide (TTIP) in heptane was dropped to the previous cellulose solution to obtain different cellulose-Ti composites by changing the corresponding Ti:cellulose mass ratio, namely 1:1, 6:1 or 12:1. The solid suspension formed during the TTIP hydrolysis was aged under continuous stirring at 60 °C for 24 h and then, the composites were filtered, washed with distilled water and acetone and dried at 120 °C in an oven. Finally, the carbon–phosphorus–Ti composites were obtained by carbonization of the corresponding cellulose–Ti composites in a tubular furnace at 500 or 800 °C under 100 cm^3 min^{-1} N_2 flow. All samples were grinded and sieved to a particle size of 100–200 μm before used in photocatalysis and characterization. The samples will be labelled as CPTX-Y indicating the composition (C = cellulose, P = phosphoric acid, T = TTIP impregnation), "X" refers the Ti:cellulose ratio used (i.e., 1, 6 or 12) and "Y" states the carbonization temperature (500 or 800 °C). For instance, CPT6-500 corresponds to the composite prepared in a 6:1 ratio and at 500 °C.

The morphology of the materials was studied by scanning electron microscopy (SEM) using an AURIGA Carl Zeiss SMT microscope (Carl Zeiss AG, Oberkochen, Germany). Energy dispersive X-ray (EDX) microanalysis (Carl Zeiss AG, Oberkochen, Germany) was carried out to determine the composition and homogeneity of the samples. This information was completed by analysing the samples with X-ray photoelectron spectroscopy (XPS) using a Kratos Axis Ultra-DLD (Kratos Analytical Ltd., Kyoto, Japan). Accurate binding energies (± 0.1 eV) were determined regarding to the position of the C_{1s} peak. The residual pressure in the analysis chamber was maintained below 10^{-9} Torr during data acquisition and survey and multiregion spectra were recorded. Each spectral region of interest was scanned several times to obtain good signal-to-noise ratios. The atomic concentrations were calculated from photoelectron peak areas and sensitivity factors provided by the spectrometer manufacturer. The crystallinity of composites were determined by X-ray diffraction (XRD) using a Bruker D8 Advance X-ray diffractometer (BRUKER, Rivas-Vaciamadrid, Spain) (Cu Kα radiation, wavelength (λ) of 1.541 Å).

The carbonization process of composites was studied by thermogravimetric (TG) and differential thermogravimetric (DTG) analyses by heating the sample in nitrogen flow from 50 °C to 900 °C at 20 °C min^{-1} using a Mettler–Toledo TGA/DSC1 thermal balance (Mettler-Toledo International Inc., Greifensee, Switzerland). The TiO_2 content in a given composite was estimated by ubtracting the weight loss obtained with pure TiO_2 under air atmosphere (oxidizing conditions) until constant weight from the weight loss obtained with the composite [22].

Textural characterization of the samples was carried out by N_2 adsorption-desorption at -196 °C with a Quantachrome Autosorb-1 apparatus (Quantachrome Instruments, FL, USA). The apparent surface area (S_{BET}) was determined by applying the Brunauer–Emmett–Teller (BET) equation [23], while the micropore volume (V_{micro}) and the mean micropore width (L_0) were obtained from Dubinin–Radushkevich and Stoeckli equations, respectively [24,25]. The volume of nitrogen adsorbed at a relative pressure of 0.95 (V_{pore}), was also obtained from the adsorption isotherms, which corresponds to the sum of the micro- and mesopore volumes according to Gurvitch's rule [26].

The performance of materials in the photodegradation of the Orange-G (OG) dye in aqueous solutions was studied under UV irradiation. The experiments were performed using a glass photoreactor (8.5×20 cm) equipped with a low-pressure mercury vapor lamp (TNN 15/32, 15 W, Heraeus Headquarters, Hanau, Germany) emitting at 254 nm placed inside an inner quartz tube of 2.5 cm of diameter. The concentration of OG was determined by a UV–vis spectrophotometer (5625 Unicam Ltd., Cambridge, UK). Before catalytic experiments, all materials (800 mg) were saturated with the dye solution (800 mL) in dark to remove the adsorptive contribution. After saturation, the initial dye concentration (C_0) was fitted again to 10 mg L^{-1} in all cases, and then, a UV lamp was turned on, this time being considered t = 0. Samples were taken from the reactor and centrifuged to separate the catalyst particles before analysis by the UV–vis spectrophotometer.

3. Results and Discussion

The closed structure of MCC required a previous solubilization with H_3PO_4 before Ti-impregnation. This acid treatment improved the dispersion of the Ti-active phase on the cellulose support but also, functionalized it simultaneously with different phosphorus-containing groups leading to carbon–phosphorus–Ti composites.

The morphology of the composites was analyzed by SEM (Figure 1). The composites prepared with low and intermediate Ti:cellulose and carbonized at 500 °C, i.e., CPTi1-500 and CPT6-500, presented open structures formed by a network of elongated particles resembling the raw cellulose fibers (Figure 1a,c). These large structures become round shaped particles with increasing the Ti:cellulose ratio up to 12 wt.% (Figure 1e). After carbonization at 800 °C, round-shaped particles are observed in the surface of all samples; the particle size being increased as the Ti-content (Figure 1b,d,e). The particle size determined for the CPT12 composite after carbonization at 500 °C was always smaller than 50 nm, while some particles larger than 300 nm were detected after carbonizing at 800 °C. EDX-microanalysis of all composites showed high contents of C and Ti, but also of phosphorus (Figure 1g for CPT6-500), which was distributed homogeneously on the composite, as confirmed by EDX.

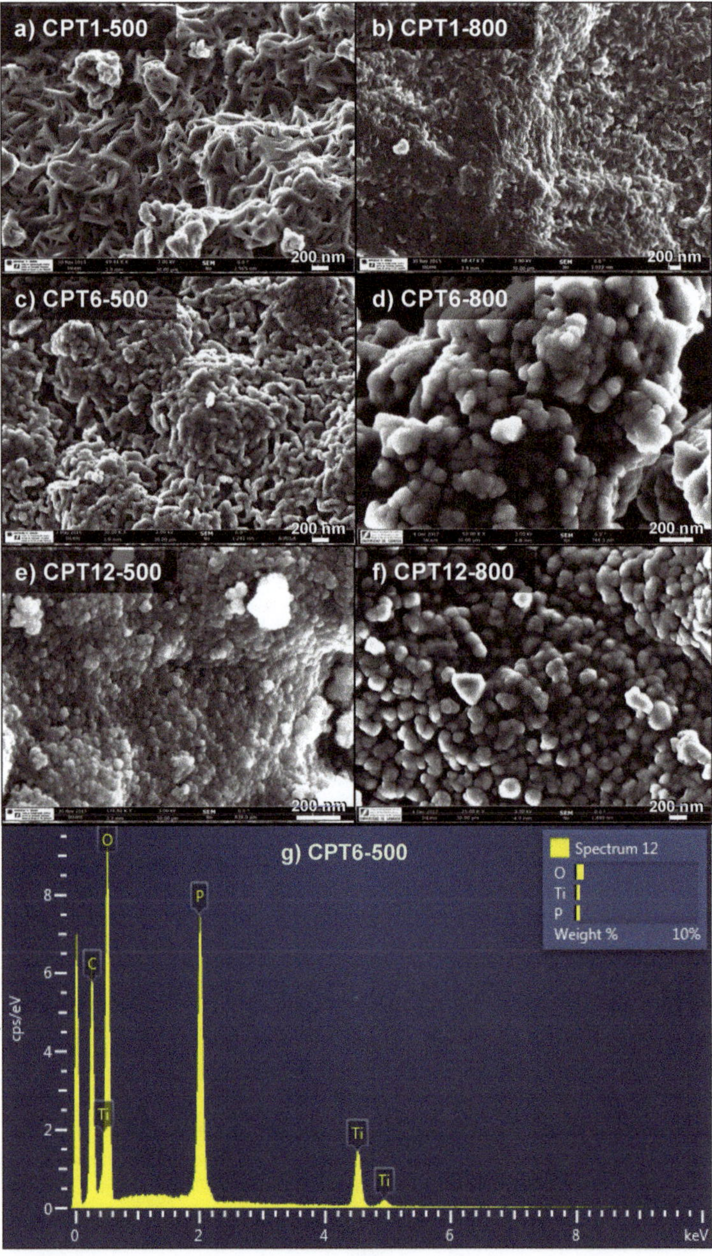

Figure 1. SEM micrographs for the carbon-phosphorus-Ti composites treated at 500 °C (**a,c,e**) and 800 °C (**b,d,f**), as well (**g**) EDX spectrum for the CPT6-500 composite.

Cellulose–phosphate structures formed during the MCC solubilization with H_3PO_4 were reported to be reversible, i.e., they are removed after washed leading to free H_3PO_4 and amorphous cellulose [27]. In our case, although amorphous cellulose was obtained, the phosphorus functionalities were stable not only after being washed but also after carbonization of the composites, as corroborated below by different techniques.

Thus, the stability of the phosphorus-containing groups was confirmed by XPS. As an example, the chemical composition of the CPT6 samples and the variation on the nature of the surface groups with the carbonization temperature are summarized in Table 1. An increase of the carbonization temperature led to the progressive reduction of the samples since the oxygen content decreased (i.e., 42.7 and 36.4% for CPT6-500 and CPT6-800, respectively) due to the thermal decomposition of some oxygen and/or phosphorus functionalities, which were released as CO_x. The deconvolution of the Ti_{2p} region showed for CPT6-500, an only peak placed at $\approx$459.3 eV corresponding to the presence of Ti^{+4}, while the corresponding sample carbonized at 800 °C presented an additional component at $\approx$458.6 eV due to the presence of Ti^{+3} (Table 1).

Table 1. Surface concentration, species percentage and corresponding binding energies (in brackets, eV) obtained for the CTP6 sample obtained at different carbonization temperatures.

Sample	C	O	P	Ti	P_{2p} (%)		Ti_{2p} (%)	
	(wt.%)				C-PO$_3$	C-O-PO$_3$	Ti^{3+}	Ti^{4+}
CPT6-500	22.0	42.7	21.9	13.4	36 (132.9)	64 (133.8)	-	100 (459.3)
CPT6-800	27.3	36.4	22.8	13.5	63 (132.8)	37 (133.8)	48 (458.6)	52 (459.5)

Analogously, a variation of the spectra of the P_{2p} region was observed for the different CPT6 samples. Thus, this region can be deconvoluted in two peaks placed at $\approx$132.8 and $\approx$133.8 eV corresponding to phosphorus linked to carbon (C-PO$_3$) and to pentavalent tetracoordinated phosphorus in phosphates or polyphosphates as (C-O-PO$_3$), respectively [28] (Table 1). In addition, the position of these peaks is shifted to higher binding energies (BE) with increasing the oxidation degree of the P-groups [29,30], while the peak at low BE is favored at high carbonization temperatures.

XRD patterns for the composites treated at 500 °C did not show any peak regardless the Ti:cellulose ratio used, denoting an amorphous character for these samples. Nevertheless, sharp peaks were observed in XRD patters when samples were treated at 800 °C, with different crystalline phases being formed depending on the Ti:cellulose ratio (Figure 2). The TiP_2O_7 crystalline phase (JCPDS 38-1468) was present in all these composites, but also there is a small contribution of $Ti(HPO_4)_2$ (JCPDS 38-334) at low Ti-content, i.e., CPT1-800. On the other hand, when the Ti-content is increased up to 12 wt.% (i.e., CPT12-800), the main crystalline phase was $(TiO)_2P_2O_7$ (JCPDS 39-0207). Thus, richer crystalline Ti-phases are favored when increasing the Ti-content in the composite since the H_3PO_4/cellulose ratio was always maintained. The crystal size obtained by application of the Scherrer equation was 38.9, 53.4 and 57.9 nm for CPT1-800, CPT6-800 and CPT12-800, respectively.

The marked influence of the carbonization temperature on the interactions of Ti-species with the phosphate surface groups was also pointed out by the thermogravimetric analysis of the samples. Thus, TG-DTG profiles obtained during the carbonization process of H_3PO_4-treated cellulose before (i.e., the CP support) and after Ti-impregnation (i.e., the CPT6 composite) are compared in Figure 3a,b, respectively. The support carbonization occurs in three steps denoted by the corresponding minimum in the DTG profile (Figure 3a). The first weight loss occurs at $\approx$120 °C and can be associated with dehydration and drying processes; a second peak at $\approx$240 °C corresponds to the release of CO_x formed by the thermal decomposition of oxygenated surface groups, namely, carboxylic acids that decompose at this temperature range [31]; and the third peak can be due to the reduction of the phosphate surface groups by the organic matrix, causing the gasification of the support, as typically described in the chemical activation process of lignocellulosic materials [32]. In the carbonization of the CPT6 composite (Figure 3b), the Ti-support interactions leads to a certain shifting of the first peaks to slightly higher temperatures compared to the support. This fact should be related with the formation of links between the oxygenated surface groups of cellulose and the Ti-species [19]. However, Ti-species mainly interact with the phosphorus-containing groups, in such a manner that the reduction of these groups by the cellulose matrix at $\approx$750 °C does not occur during the thermal treatment (Figure 3b), this reduction being replaced by the reaction between these groups with the Ti species leading to crystalline Ti-phosphate or polyphosphate compounds, as previously observed by XRD. In fact, the weight loss above 500 °C is clearly negligible.

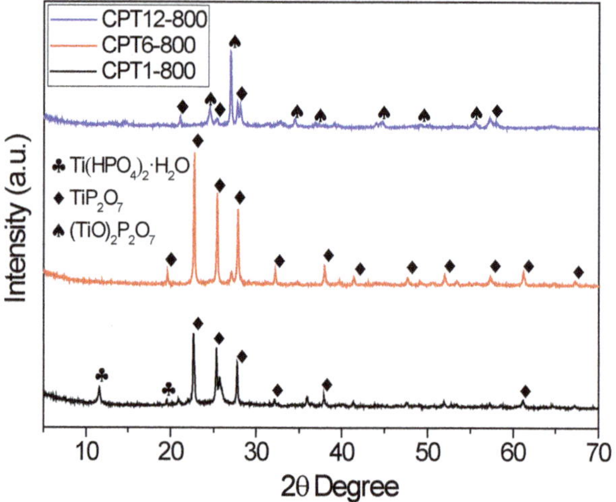

Figure 2. XRD patterns of the different carbon-phosphorus-Ti composites treated at 800 °C.

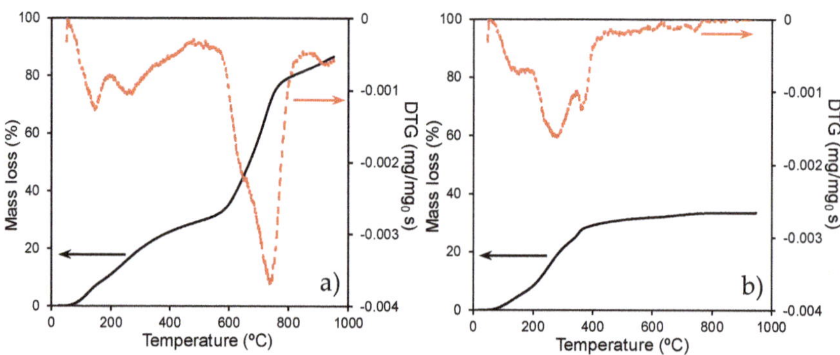

Figure 3. TG and DTG profiles obtained during the carbonization in N_2 flow: (**a**) H_3PO_4-treated cellulose support and (**b**) CPT6 composite.

The morphological and crystalline transformations of the composites previously discussed had a clear effect on their textural properties, which were determined by analyzing the corresponding N_2-adsorption isotherms (Table 2 and Figure 4). In general, the total pore volume (V_{pore}) and the BET surface area (S_{BET}) of the composites decreased as the Ti-content and the carbonization temperature increased, due to the higher porosity of the carbon phase compared to inorganic Ti-phases and the sintering favored under these conditions. The CPT1-500 composite presented the highest surface area (357 m^2 g^{-1}) due to its high micropore volume (0.144 cm^3 g^{-1}) associated with a high adsorbed volume of N_2 at low relative pressure (Figure 4). In general, the isotherms of the composites belong to type-IV or type-II, showing from $P/P_0 > 0.4$ a clear hysteresis cycle due to the presence of mesopores. A similar behavior is also observed for CPT6 and CPTi12 samples with a loss of microporosity but also an enhanced mesoporosity (i.e., higher adsorbed volume of N_2 at high relative pressure), compared to composites prepared with lower Ti-content.

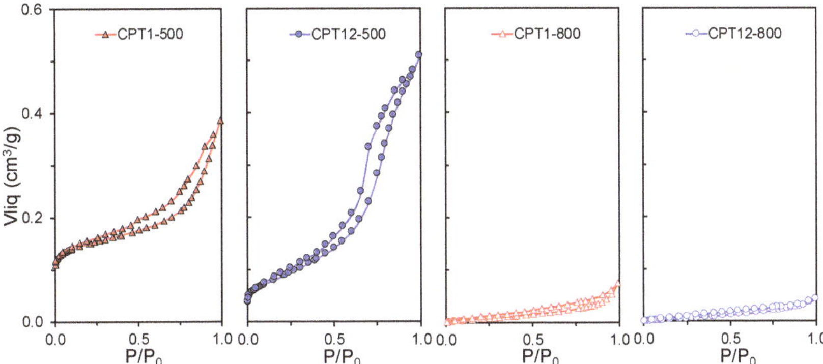

Figure 4. N_2-adsorption isotherms of carbon-phosphorus-Ti composites treated at 500 or 800 °C.

Table 2. Textural properties of selected carbon-phosphorus-Ti composites treated at 500 or 800 °C.

Sample	S_{BET} (m^2 g^{-1})	V_{micro} (cm^3 g^{-1})	V_{pore} (cm^3 g^{-1})
CPT1-500	357	0.144	0.386
CPT6-500	28	0.013	0.160
CPT1-800	9	0.004	0.076
CPT12-500	184	0.073	0.508
CPT6-500	30	0.017	0.239
CPT12-800	5	0.021	0.043

The adsorptive and photocatalytic performance of the carbon–phosphorus–Ti composites were analyzed for the removal of OG (Figure 5a,b, respectively). Firstly, all samples were saturated in dark experiments, which hinders the contribution of the adsorption process to the OG removal in the subsequent photocatalytic experiments. The adsorption capacity and the adsorption rate are not exclusively related to the different porosity of the samples, as observed in Figure 5a. In general, composites with lower Ti-content presented a better adsorptive behavior than those prepared with intermediate and high Ti-contents regardless of the carbonization temperature used. The maximum removal of OG was achieved after 20 min and varied as: CPT1 samples > CPT6 samples > CPT12 samples. This trend could be explained because composites with low Ti:carbon ratio present a larger carbon phase, which has a higher affinity for OG in solution. The CPT1 composites presented the best adsorptive behavior, being the adsorption of both CPT1-500 and CPT1-800 comparable in spite of their different porous characteristics and the crystallinity of their Ti-phases.

In Figure 5b, we show the photocatalytic efficiency obtained for the different carbon–phosphorus–Ti composites and the benchmark TiO_2 material (Degussa P25) for comparison. The complete OG removal was achieved after $\approx$25–35 min or 40–50 min depending on composites treated at 500 or 800 °C, respectively. In general, all composites obtained at low carbonization temperature presented a better efficiency than those treated at 800 °C; in spite of that, all these samples were completely amorphous since no peaks were observed by XRD. The carbon phase retards the crystal growth and the phase transformations of metal oxides in carbon–metallic oxide composites [33,34]. In addition, the presence of phosphorus may influence the TiO_2-crystal structure, obtaining mixtures of TiO_2 amorphous and anatase [35]. In this context, composites obtained at 500 °C could develop a mixture of amorphous and very small TiO_2-anatase nanoparticles (undetectable by XRD), with the latter being responsible for the high activity of composites treated at 500 °C. On the other hand, the photocatalytic efficiency varied as follows: CPT1-500 > CPT6-500 > CPT12-500, i.e., when the Ti-content increased in the composites, which could be related with their lower porosity (Table 2).

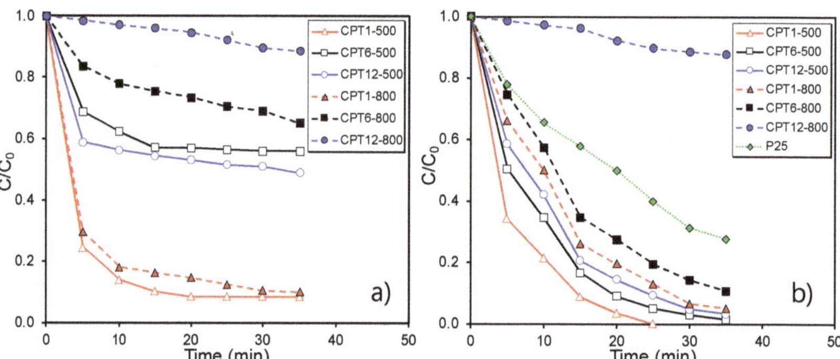

Figure 5. Removal of the Orange-G from water solution by adsorption (**a**) and photocatalytic (**b**) processes using carbon-phosphorus-Ti composites.

Concerning composites treated at 800 °C, the formation of different polyphosphates was pointed out by XRD. The band gap for the titanium pyrophosphate (TiP_2O_7) was estimated to be 3.48 eV [36], with is higher than that for the TiO_2 anatase or rutile phases, i.e., 3.2 and 3.0 eV, respectively. The different nature of polyphosphates can also influence their performance [37,38].

In addition, sintering was favored to this range of temperature, with Ti-particles larger than ≈39 nm being obtained. Otherwise, even the composites carbonized at this temperature, with exception of CPTi12-800, presented a better performance than the benchmark TiO_2 material (Degussa P25). This fact denotes the importance of the carbon phase in Ti-based composites, leading to an enhanced improved performance based on the synergism between both phases. Overall, carbon–phosphorus–titanium composites with low carbon content and carbonization temperature are preferred for the removal of the OG pollutant by photocatalysis because of their enhanced porosity, high dispersion of the active phase (anatase) and strong adsorption capacity (interaction) of OG.

4. Conclusions

The treatment of microcrystalline cellulose with H_3PO_4 leads to a simultaneous functionalization of cellulose chains by incorporating stable phosphorus-containing surface groups, namely, phosphates and polyphosphates. These functionalities interact and react progressively with Ti-species during impregnation and carbonization at high temperatures, with different polyphosphates of titanium being anchored on the carbon phase. The physicochemical properties of these carbon–phosphorus–Ti composites vary according to the Ti-content and carbonization temperature. Thus, the increase of these parameters favors Ti-particle sintering, the formation of Ti-crystalline phases and a marked loss of the porosity. The synergism between phases allows to obtain materials with enhanced photocatalytic efficiency compared to the benchmark TiO_2 material (Degussa P25), in spite of the band gap of polyphosphates being wider than that for anatase/rutile phases. Carbon–phosphorus–Ti composites with anatase TiO_2 nanoparticles and large surface areas seem to be the most active photocatalysts for OG degradation under UV irradiation.

Author Contributions: This manuscript is included in the PhD Thesis developed by H.H. and supervised by F.J.M.-H., A.F.P.-C. and S.M.-T. F.C.-M. and J.C.-Q. contributed to the characterization of the samples by XPS. All authors participated in the writing and revision of the final version of the manuscript.

Funding: This research is supported by the FEDER and Spanish projects CTQ2013-44789-R (MINECO) and P12-RNM-2892 (Junta de Andalucía). H.H. gratefully thanks the support of Erasmus-Mundus (Al-Idrisi II) project for PhD scholarship. S.M.-T. acknowledges the financial support from University of Granada (Reincorporación Plan Propio). J.C.-Q. is grateful to the Junta de Andalucía for her research contract (P12-RNM-2892).

Conflicts of Interest: The authors declare no conflict of interest.

References

1. Silva, T.L.S.; Morales-Torres, S.; Figueiredo, J.L.; Silva, A.M.T. Multi-walled carbon nanotube/PVDF blended membranes with sponge- and finger-like pores for direct contact membrane distillation. *Desalination* **2015**, *357*, 233–245. [CrossRef]
2. Ong, C.B.; Ng, L.Y.; Mohammad, A.W. A review of ZnO nanoparticles as solar photocatalysts: Synthesis, mechanisms and applications. *Renew. Sust. Energ. Rev.* **2018**, *81*, 536–551. [CrossRef]
3. Lee, G.-J.; Wu, J.J. Recent developments in ZnS photocatalysts from synthesis to photocatalytic applications—A review. *Powder Technol.* **2017**, *318*, 8–22. [CrossRef]
4. Pelaez, M.; Nolan, N.T.; Pillai, S.C.; Seery, M.K.; Falaras, P.; Kontos, A.G.; Dunlop, P.S.M.; Hamilton, J.W.J.; Byrne, J.A.; O'Shea, K.; et al. A review on the visible light active titanium dioxide photocatalysts for environmental applications. *Appl. Catal. B* **2012**, *125*, 331–349. [CrossRef]
5. Lima, M.J.; Silva, A.M.T.; Silva, C.G.; Faria, J.L. Graphitic carbon nitride modified by thermal, chemical and mechanical processes as metal-free photocatalyst for the selective synthesis of benzaldehyde from benzyl alcohol. *J. Catal.* **2017**, *353*, 44–53. [CrossRef]
6. Morales-Torres, S.; Pastrana-Martínez, L.M.; Figueiredo, J.L.; Faria, J.L.; Silva, A.M.T. Design of graphene-based TiO$_2$ photocatalysts—A review. *Environ. Sci. Pollut. Res.* **2012**, *19*, 3676–3687. [CrossRef] [PubMed]
7. Pastrana-Martínez, L.M.; Morales-Torres, S.; Likodimos, V.; Figueiredo, J.L.; Faria, J.L.; Falaras, P.; Silva, A.M.T. Advanced nanostructured photocatalysts based on reduced graphene oxide–TiO$_2$ composites for degradation of diphenhydramine pharmaceutical and methyl orange dye. *Appl. Catal. B* **2012**, *123–124*, 241–256. [CrossRef]
8. Liu, X.; Li, Y.; Yang, J.; Wang, B.; Ma, M.; Xu, F.; Sun, R.; Zhang, X. Enhanced Photocatalytic Activity of CdS-Decorated TiO$_2$/Carbon Core-Shell Microspheres Derived from Microcrystalline Cellulose. *Materials* **2016**, *9*, 245. [CrossRef] [PubMed]
9. Du, H.; Liu, Y.-N.; Shen, C.-C.; Xu, A.-W. Nanoheterostructured photocatalysts for improving photocatalytic hydrogen production. *Chin. J. Catal.* **2017**, *38*, 1295–1306. [CrossRef]
10. Bora, L.V.; Mewada, R.K. Visible/solar light active photocatalysts for organic effluent treatment: Fundamentals, mechanisms and parametric review. *Renew. Sust. Energ. Rev.* **2017**, *76*, 1393–1421. [CrossRef]
11. Spasiano, D.; Siciliano, A.; Race, M.; Marotta, R.; Guida, M.; Andreozzi, R.; Pirozzi, F. Biodegradation, ecotoxicity and UV$_{254}$/H$_2$O$_2$ treatment of imidazole, 1-methyl-imidazole and *N,N'*-alkyl-imidazolium chlorides in water. *Water Res.* **2016**, *106*, 450–460. [CrossRef] [PubMed]
12. Zhu, Y.; Wang, H.; Li, X.; Hu, C.; Yang, M.; Qu, J. Characterization of biofilm and corrosion of cast iron pipes in drinking water distribution system with UV/Cl$_2$ disinfection. *Water Res.* **2014**, *60*, 174–181. [CrossRef] [PubMed]
13. Likodimos, V.; Chrysi, A.; Calamiotou, M.; Fernández-Rodríguez, C.; Doña-Rodríguez, J.M.; Dionysiou, D.D.; Falaras, P. Microstructure and charge trapping assessment in highly reactive mixed phase TiO$_2$ photocatalysts. *Appl. Catal. B* **2016**, *192*, 242–252. [CrossRef]
14. Bailón-García, E.; Elmouwahidi, A.; Álvarez, M.A.; Carrasco-Marín, F.; Pérez-Cadenas, A.F.; Maldonado-Hódar, F.J. New carbon xerogel-TiO$_2$ composites with high performance as visible-light photocatalysts for dye mineralization. *Appl. Catal. B* **2017**, *201*, 29–40. [CrossRef]
15. Nath, R.K.; Zain, M.F.M.; Jamil, M. An environment-friendly solution for indoor air purification by using renewable photocatalysts in concrete: A review. *Renew. Sust. Energ. Rev.* **2016**, *62*, 1184–1194. [CrossRef]
16. da Silva, W.L.; dos Santos, J.H.Z. Ecotechnological strategies in the development of alternative photocatalysts. *Curr. Opin. Green Sustain. Chem.* **2017**, *6*, 63–68. [CrossRef]

17. Okour, Y.; Shon, H.K.; Liu, H.; Kim, J.B.; Kim, J.H. Seasonal variation in the properties of titania photocatalysts produced from Ti-salt flocculated bioresource sludge. *Bioresour. Technol.* **2011**, *102*, 5545–5549. [CrossRef] [PubMed]

18. El Bekkali, C.; Bouyarmane, H.; Saoiabi, S.; El Karbane, M.; Rami, A.; Saoiabi, A.; Boujtita, M.; Laghzizil, A. Low-cost composites based on porous titania–apatite surfaces for the removal of patent blue V from water: Effect of chemical structure of dye. *J. Adv. Res.* **2016**, *7*, 1009–1017. [CrossRef] [PubMed]

19. Mohamed, M.A.; Mutalib, M.A.; Hir, Z.A.M.; Zain, M.F.M.; Mohamad, A.B.; Minggu, L.J.; Awang, N.A.; Salleh, W.N.W. An overview on cellulose-based material in tailoring bio-hybrid nanostructured photocatalysts for water treatment and renewable energy applications. *Int. J. Biol. Macromol.* **2017**, *103* (Suppl. C), 1232–1256. [CrossRef] [PubMed]

20. Sun, X.; Wang, K.; Shu, Y.; Zou, F.; Zhang, B.; Sun, G.; Uyama, H.; Wang, X. One-Pot Route towards Active TiO_2 Doped Hierarchically Porous Cellulose: Highly Efficient Photocatalysts for Methylene Blue Degradation. *Materials* **2017**, *10*, 373. [CrossRef] [PubMed]

21. Hamad, H.; Bailón-García, E.; Morales-Torres, S.; Carrasco-Marín, F.; Pérez-Cadenas, A.F.; Maldonado-Hódar, F.J. Physicochemical properties of new cellulose-TiO_2 composites for the removal of water pollutants: Developing specific interactions and performances by cellulose functionalization. *J. Environ. Chem. Eng.* **2018**, *6*, 5032–5041. [CrossRef]

22. Morales-Torres, S.; Pastrana-Martínez, L.M.; Figueiredo, J.L.; Faria, J.L.; Silva, A.M.T. Graphene oxide-P25 photocatalysts for degradation of diphenhydramine pharmaceutical and methyl orange dye. *Appl. Surface Sci.* **2013**, *275*, 361–368. [CrossRef]

23. Brunauer, S.; Emmett, P.H.; Teller, E. Adsorption of Gases in Multimolecular Layers. *J. Am. Chem. Soc.* **1938**, *60*, 309–319. [CrossRef]

24. Stoeckli, F. *Porosity in Carbons. Characterization and Applications*; Arnold: London, UK, 1995.

25. Cazorla-Amorós, D.; Alcañiz-Monge, J.; de la Casa-Lillo, M.A.; Linares-Solano, A. CO_2 As an Adsorptive To Characterize Carbon Molecular Sieves and Activated Carbons. *Langmuir* **1998**, *14*, 4589–4596. [CrossRef]

26. Rouquerol, J.; Rouquerol, F.; Llewellyn, P.; Maurin, G.; Sing, K.S. *Adsorption by Powders and Porous Solids*; Academic Press: London, UK, 1999; pp. 219–228.

27. Zhang, Y.H.P.; Cui, J.; Lynd, L.R.; Kuang, L.R. A Transition from Cellulose Swelling to Cellulose Dissolution by o-Phosphoric Acid: Evidence from Enzymatic Hydrolysis and Supramolecular Structure. *Biomacromolecules* **2006**, *7*, 644–648. [CrossRef] [PubMed]

28. Rosas, J.M.; Bedia, J.; Rodríguez-Mirasol, J.; Cordero, T. HEMP-derived activated carbon fibers by chemical activation with phosphoric acid. *Fuel* **2009**, *88*, 19–26. [CrossRef]

29. Hasegawa, G.; Deguchi, T.; Kanamori, K.; Kobayashi, Y.; Kageyama, H.; Abe, T.; Nakanishi, K. High-Level Doping of Nitrogen, Phosphorus, and Sulfur into Activated Carbon Monoliths and Their Electrochemical Capacitances. *Chem. Mater.* **2015**, *27*, 4703–4712. [CrossRef]

30. Elmouwahidi, A.; Bailón-García, E.; Pérez-Cadenas, A.F.; Maldonado-Hódar, F.J.; Carrasco-Marín, F. Activated carbons from KOH and H_3PO_4-activation of olive residues and its application as supercapacitor electrodes. *Electrochim. Acta* **2017**, *229*, 219–228. [CrossRef]

31. Vivo-Vilches, J.F.; Bailón-García, E.; Pérez-Cadenas, A.F.; Carrasco-Marín, F.; Maldonado-Hódar, F.J. Tailoring the surface chemistry and porosity of activated carbons: Evidence of reorganization and mobility of oxygenated surface groups. *Carbon* **2014**, *68*, 520–530. [CrossRef]

32. Prauchner, M.J.; Rodríguez-Reinoso, F. Chemical versus physical activation of coconut shell: A comparative study. *Microporous Mesoporous Mater.* **2012**, *152* (Suppl. C), 163–171. [CrossRef]

33. Maldonado-Hódar, F.J.; Moreno-Castilla, C.; Rivera-Utrilla, J. Synthesis, pore texture and surface acid–base character of TiO_2/carbon composite xerogels and aerogels and their carbonized derivatives. *Appl. Catal. A* **2000**, *203*, 151–159. [CrossRef]

34. Moreno-Castilla, C.; Maldonado-Hodar, F.J. Synthesis and surface characteristics of silica- and alumina-carbon composite xerogels. *Phys. Chem. Chem. Phys.* **2000**, *2*, 4818–4822. [CrossRef]

35. Kőrösi, L.; Oszkó, A.; Galbács, G.; Richardt, A.; Zöllmer, V.; Dékány, I. Structural properties and photocatalytic behaviour of phosphate-modified nanocrystalline titania films. *Appl. Catal. B* **2007**, *77*, 175–183. [CrossRef]

36. Meng, X.; Hao, M.; Shi, J.; Cao, Z.; He, W.; Gao, Y.; Liu, J.; Li, Z. Novel visible light response Ag_3PO_4/TiP_2O_7 composite photocatalyst with low Ag consumption. *Adv. Powder Technol.* **2017**, *28*, 1047–1053. [CrossRef]

37. Fagan, R.; McCormack, D.E.; Hinder, S.; Pillai, S.C. Improved high temperature stability of anatase TiO_2 photocatalysts by N, F, P co-doping. *Mater. Des.* **2016**, *96*, 44–53. [CrossRef]
38. Yu, H.-F. Phase development and photocatalytic ability of gel-derived P-doped TiO_2. *J. Mater. Res.* **2007**, *22*, 2565–2572. [CrossRef]

 *materials* 

Article

Adsorption Behavior of Selective Recognition Functionalized Biochar to Cd(II) in Wastewater

Shiqiu Zhang [1], Xue Yang [1], Le Liu [1,*], Meiting Ju [1,*] and Kui Zheng [2]

[1] College of Environmental Science and Engineering, Nankai University, Tianjin 300350, China; 1120170189@mail.nankai.edu.cn (S.Z.); 2120170638@mail.nankai.edu.cn (X.Y.)

[2] Analytical and Testing Center, Southwest University of Science and Technology, Mianyang 621010, China; zhengkui@swust.edu.cn

* Correspondence: huanjikejian@163.com (L.L.); nkujumeiting@sohu.com (M.J.)

Received: 9 January 2018; Accepted: 12 February 2018; Published: 14 February 2018

Abstract: Biochar is an excellent absorbent for most heavy metal ions and organic pollutants with high specific surface area, strong aperture structure, high stability, higher cation exchange capacity and rich surface functional groups. To improve the selective adsorption capacity of biochar to designated heavy metal ions, biochar prepared by agricultural waste is modified via Ionic-Imprinted Technique. Fourier transform infrared (FT-IR) spectra analysis and X-ray photoelectron spectroscopy (XPS) analysis of imprinted biochar (IB) indicate that 3-Mercaptopropyltrimethoxysilane is grafted on biochar surface through Si–O–Si bonds. The results of adsorption experiments indicate that the suitable pH range is about 3.0–8.0, the dosage is 2.0 g·L^{-1}, and the adsorption equilibrium is reached within 960 min. In addition, the data match pseudo-second-order kinetic model and Langmuir model well. The computation results of adsorption thermodynamics and stoichiometric displacement theory of adsorption (SDT-A) prove that the adsorption process is spontaneous and endothermic. Finally, IB possesses a higher selectivity adsorption to Cd(II) and a better reuse capacity. The functionalized biochar could solidify designated ions stably.

Keywords: biochar; targeted adsorption; Cd(II); adsorption

1. Introduction

It is well known that farmland, an important section of ecological environment, is closely related to the problems of resource, grain and environment [1–3]. However, with the development of industrialization and urbanization of China, the total area of heavy metal (such as Cd, As, Pb, Hg, Zn, etc.) contaminated soil increases rapidly and is more than 2×10^5 km^2, occupying about 1/5 of the total agricultural area [4]. Meanwhile, the research of the Ministry of Agriculture of the People's Republic of China indicated that the main reasons for the contaminated soil were wastewater irrigation and sludge application [5], which could cause for crop failure up to 1×10^7 t and lead to total economic loss surpassing $2.5 billion. Among all the heavy metal ions, cadmium is the most higher poisonous substance [6], jeopardizing the health of humans and animals through the food chain system of water–soil–plant–animal–human [7]. The migration capacity of cadmium was higher than the other chemical elements from water to humans and animals [8,9]. Once cadmium enters the human body through the respiratory tract or digestive tract, it could destroy the tissues and organs [10]. Consequently, the development of reliable methods for the removal of cadmium from environment and biological samples was particularly significant.

For scavenging contaminants, the methods of biological treatment, adsorption, precipitation, membranes separation and ion exchange had been carried out [11,12]. Among these methods, the effect of biological treatment was slow and precipitation method could cause the new contaminants. For removing cadmium from dilute solution, the adsorption technique possessed of a better applicability than traditional extraction process [13]. In recent years, with the discovery of black earth in Amazon Basin and the development of correlational research [14], biochar has gained much publicity as a new type of environmental functional material. Many studies indicated that biochar has huge application potential in the matter of the reduction of greenhouse gas emissions, agromelioration, and contaminated soil remediation [15–17]. Among these advantages, biochar has a strong capacity to absorb heavy metal ions [1,17], such as Cd(II), Pb(II), Cu(II) etc., and reduce the effectiveness and migration of heavy metal ions in the wastewater. Furthermore, the raw material sources of biochar are extensive, e.g., agricultural wastes (wood, straw, or shell), municipal solid wastes (refuse and sludge), and other organic materials [18]. In addition, biomass solid waste is a big problem. The biomass waste brings not only water, atmospheric and soil pollution, but also the safety problem of humans and animals via the food chain. Hence, the comprehensive utilization of biomass is also a crucial challenge. Moreover, biochar could adsorb contaminants simultaneously which may reduce the contents of beneficial components in the water. Ionic-Imprinted Technique is similar to Molecular-Imprinted Technique [19,20], and can recognize metal ions after imprinting. The effectiveness of the materials in binding metal ions has been attributed to the complexation between the ligand and the metal ions. The specificity of a particular ligand toward target metal ions is the result of a conventional acdi–base interaction between the ligand and the metal ions. Some of these sulfydryl-functionalized sorbents could exhibit specific interactions with soft Lewis acids (e.g., Hg(II), Cd(II), Cu(II), or Ag(I)), and the selectivity of these materials is usually remarkable because many metals have the ability to bind with thiol ligands considering the stereochemical interactions between the ligand and metal ions [21,22]. An efficient adsorption material should consist of a stable and insoluble porous matrix with suitable active groups—typically organic groups interacting with heavy metal ions. In addition, the imprinted polymers could not only possess of better selective adsorption capacity, but also be reused at least 100 times without loss of affinity towards the template ions under acidic and basic conditions, and an elevated temperature [23]. Quartz dispersed on the biochar surface were often accompanied by hydroxyl under hydrolysis. The quartz offered the action sites to the modifier [24,25]. Thus, it is a good idea to make use of the characters to modify biochar.

The objective of this work is to explore the selective recognition performance of a biochar prepared by green waste using 3-Mercaptopropyltrimethoxysilane as the surface conditioning agent and Epoxy-chloropropane as the cross-linking agent via Ionic-imprinted Technique. To investigate the adsorption capacity and selective recognition performance of the selective recognition functionalized biochar (IB), the initial pH of Cd(II) solution, sorbent dosage, adsorption kinetics and adsorption thermodynamics are studied, and analyzed by X-ray diffraction (XRD) patterns, Fourier transform infrared (FT-IR), Zeta potential analysis and X-ray photoelectron spectroscopy (XPS).

2. Experiment

2.1. Materials and Reagents

Biochar (80 mesh, green waste under slow pyrolysis at 600 °C for a retention time of 10 h) was used in this study as the substrate material to prepare the ion-imprinted functionalized sorbent. All chemicals were analytical grade, and the water used in all experiments was deionized water (18.25 MΩ·cm). 3-Mercaptopropyltrimethoxysilane (MPS, Heowns Chemical Factory, Tianjin, China), Epoxy-chloropropane (ECH, Kemiou Chemical Reagent Co. Ltd., Tianjin, China), and CdCl$_2$·2/(5H$_2$O) (Aladdin Chemical Reagent Co. Ltd., Shanghai, China) were used in this study.

2.2. Biochar Modification

Exactly 4.0 g biochar (100 mesh) mixed with 200 mL 6.0 mol·L^{-1} hydrochloric acid were stirred for 8 h, then the solid product was recovered via filtration, and washed by deionized water to pH = 6.0, and dried under vacuum at 70 °C for 8 h. The sample was denoted as activated biochar (AB).

Exactly 1.92 g CdCl$_2$·2/(5H$_2$O) were dissolved in 80 mL alcohol (95%) with stirring and heating at 50 °C, and then 4 mL MPS was injected in and reacted for 2 h. Next, 10.0 g activated biochar was added in the mixed solution, and the mixture was reacted for 20 h at 80 °C under stirring, recovered by filtration and washed 3 times with ethanol. Then, the sample was stirred for 2 h in 100 mL 6 mol·L^{-1} hydrochloric acid, and recovered via filtration, washed by 0.10 mol·L^{-1} NaHCO$_3$ and deionized water up to the eluent pH = 6.0–7.0, dried under vacuum at 80 °C for 12 h, and denoted as imprinted biochar (IB). For comparison, the non-imprinted functionalized biochar was also prepared using an identical procedure, but without the addition of CdCl$_2$·2/(5H$_2$O), and denoted as non-imprinted biochar (NIB). The preparation procedure of IB is shown in Figure 1.

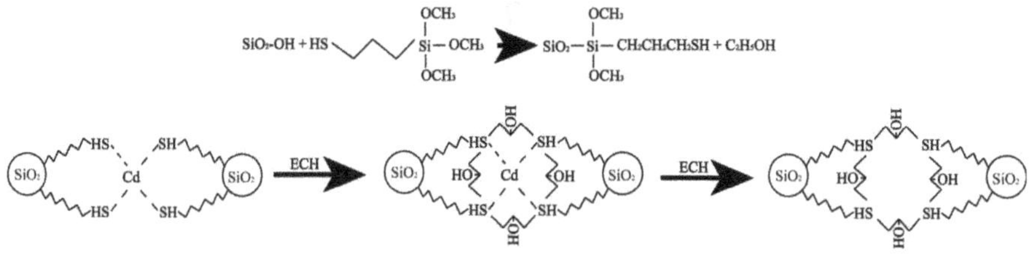

Figure 1. Preparation procedure of IB.

2.3. Adsorption Experiments

The methodology for adsorption experiments is as follows. All adsorption experiments were executed using an air thermostatic shaker (HNYC-2102C, Honour Instrument, Tianjin, China), including the factors of initial pH of Cd(II) solution, sorbent dosage, contact time, initial Cd(II) solution concentration and adsorption temperature. The effect of initial pH of Cd(II) solution was carried out by dispersion of 0.1000 ± 0.0002 g IB and 50.00 mL 0.10 mmol·L^{-1} Cd(II) solution in 100 mL conical flask. The pH of Cd(II) solution was adjusted to the range of about 1.0–14.0 by 0.1 mol·L^{-1} HCl or 0.1 mol·L^{-1} NaOH solutions, and then the mixtures were agitated at 298.15 K for 1440 min. Afterwards, to investigate the effect of sorbent dosage, the IB dosage was changed from 0.5 to 5.0 g·L^{-1} with 50.00 mL 0.10 mmol·L^{-1} Cd(II) solution in 100 mL conical flask which pH about 5.0–6.0. The adsorption kinetics was determined by analyzing adsorption capacity at different time intervals (5–1440 min) with the same sorbent dosage (2.0 g·L^{-1}) and initial Cd(II) concentration (0.1 mmol·L^{-1}, 50.00 mL) in 100 mL conical flask with pH about 5.0–6.0. For adsorption isotherms, different concentrations of Cd(II) solution (0.02–0.10 mmol·L^{-1}, 0.02 mmol·L^{-1} interval, 50.00 mL) were agitated until equilibrium was achieved with the same sorbent dosage (2.0 g·L^{-1}) in 100 mL conical flask with pH about 5.0–6.0. The temperature factor was investigated by determining the adsorption capacity at 298.15 K, 303.15 K, and 308.15 K. In all adsorption experiments, the mixtures were separated by 0.45 μm filter membrane and the Cd(II) concentrations were measured by inductively coupled plasma-atomic emission spectrometry (ICP-AES) (iCAP6500, Thermo fiShher scientific, Franklin, KY, USA).

The selective recognition adsorption experiments of Cu(II), Co(II), Pb(II) and Zn(II) ions with respect to Cd(II) were conducted using IB Biochar-based (BS) and NIB. Next, 0.1000 ± 0.0002 g sorbent were added in 50.00 mL metal ions mixed solution containing 0.10 mmol·L^{-1} Cd(II)/Cu(II), Cd(II)/Zn(II), Cd(II)/Co(II) and Cd(II)/Pb(II) at pH 5.0–6.0 in 100 mL conical flask. After adsorption equilibrium, the concentration of each ion in the remaining solution was measured by ICP-AES. Cd(II) was desorbed by the treatment of with hydrochloric acid. In this section, 0.1000 ± 0.0002 g employed IB with 50.00 mL 6.0 mol·L^{-1} hydrochloric acid in 100 mL conical flask was agitated for different durations (0.5, 1, 2, 4, 8, 12, 16, and 24 h) by a magnetic stirrer. The final Cd(II) concentration in the aqueous phase was measured by ICP-AES. The ratio of desorption was calculated from the amount of Cd(II) adsorbed on IB and the final Cd(II) concentration in the desorption medium. Finally, all experimental results are the averages of thrice repeated experiments.

2.4. Characterizations

XRD measurements were performed using an X-ray Diffractometer (X'Pert PRO, PANalytical, Almelo, Netherlands) with a Cu-Kα (λ = 0.15418 nm) radiation source. A continuous scan mode was used to collect the 2θ scan XRD data from 5° to 70° at the scanning speed of 5°/min; the voltage and current of the source were 40 kV and 40 mA, respectively. The XPS analyses were performed using a Kratos AXIS Ultra XPS system (Shimadzu, Kyoto, Japan) equipped with a monochromatic Al X-ray source at 150 W. Each analysis started with a survey scan from 0 to 1350 eV with a dwell time of 8 s, pass energy of 150 eV at steps of 1 eV with 1 sweep. For the high-resolution analysis, the number of sweeps was increased, the pass energy was lowered to 30 eV at steps of 50 meV, and the dwell time was changed to 0.5 s. FT-IR spectra were collected in the range of 4000–400 cm^{-1} by Spectrum One (Version BM) FT-IR (PerkinElmer, Waltham, MA, USA) spectrometer with 32 scans resolution of 2 cm^{-1}. Approximately 10% (mass fraction) of the solid sample was mixed with spectroscopic grade KBr. The Zeta-potentials of the fresh biochar and ageing biochar were measured using a Zetasizer Nano Zs90 (Malvern Instruments, Worcestershire, UK) at room temperature (25 °C) [26]. They were monitored continuously in terms of the conductivity and pH of the suspension during the measurement. The biochar samples were ground to a size of 2 μm using an agate mill. The suspension was prepared by adding 30 mg of biochars to 50 mL of deionized water. The prepared suspension was conditioned by magnetic stirring for 5 min, during which the pH of the suspension was measured. After settling for 10 min, the supernatant of the dilute fine particle suspension was obtained for zeta-potential measurements. Three measurements of zeta potentials were obtained, and their averages were taken as the results.

2.5. Data Analysis

2.5.1. Adsorption Kinetics Analysis

To investigate the mechanism of Cd(II) adsorption on IB, two kinetic models [17,20,27], pseudo-first-order kinetic model and pseudo-second-order kinetic model, were tested to find the best fitted model for the experimental data.

The pseudo-first-order kinetic equation:

$$\log(q_e - q_t) = \log q_e - \frac{k_1 t}{2.303}$$

where k_1 is constant rate (min^{-1}), and q_e and q_t are Cd(II) adsorption amounts (mg·g^{-1}) at equilibrium t and time t (min), respectively.

The pseudo-second-order kinetic equation:

$$\frac{t}{q_t} = \frac{1}{k_2 q_e^2} + \frac{t}{q_e}$$

where k_2 is constant rate (min^{-1}), and q_e and q_t are the Cd(II) adsorption amount (mg·g^{-1}) at equilibrium t and time t (min), respectively.

2.5.2. Adsorption Thermodynamics Analysis

Langmuir and Freundlich models were used to describe the adsorption process [17,20,27].
Langmuir adsorption isotherms equation:

$$\frac{C_e}{q_e} = \frac{1}{b q_m} + \frac{C_e}{q_m}$$

where C_e is the equilibrium concentration (mg·L^{-1}), q_e is the equilibrium capacity of Cd(II) on the IB, q_m is the monolayer adsorption capacity of the sorbent (mg·L^{-1}), and b is the Langmuir adsorption constant (L·mg^{-1}).

Freundlich adsorption isotherms equation:

$$\log q_e = \log K_f + \frac{1}{n} \log C_e$$

where K_f and n are the Freundlich adsorption constant which indicate the adsorption capacity and intensity, respectively; and q_e is the equilibrium capacity of IB to Cd(II).

The data of adsorption isotherms were used to estimate the thermodynamic parameters, Gibbs free energy change (ΔG^0), Enthalpy change (ΔH^0), and Entropy change (ΔS^0), calculated using the following equations:

$$\Delta G^0 = -RT \times \ln K_d$$

$$\ln K_d = \frac{\Delta S^0}{R} - \frac{\Delta H^0}{RT}$$

$$K_d = \frac{q_e}{C_e}$$

where K_d is the distribution coefficient, T is the temperature (K), and R is the gas constant (8.3145 J·mol^{-1}·K^{-1}).

2.5.3. Stoichiometric Displacement Theory of Adsorption Analysis (SDT-A)

The stoichiometric displacement theory of adsorption equation is as follows.

$$\ln K_d = \beta - \frac{q}{Z} \log C_e$$

where K_d ($K_d = q_e/C_e$) is the partition coefficient of solvent in liquid solid phase, C_e is equilibrium adsorption concentration (mg·L^{-1}), β is a constant that measures the affinity of the solute to the sorbent, Z represents the total moles of the solvent released or adsorbed for 1 mol solute together with its corresponding contact area on the adsorbent surface during the adsorption or desorption process, is q is the reduced molecule number of the solvent.

When the temperature is invariant, β and q/Z are constant, while β and q/Z are linear to $1/T$.

The definition of β is

$$\beta = \frac{k_1}{T} + b_1$$

The definition of q/Z is

$$\frac{q}{Z} = \frac{k_2}{T} + b_2$$

where β and q/Z are obtained by the slope and intercept of the straight line plotting lgK_d versus lgC_e. ΔG_T, ΔH_T, and ΔS_T are calculated as follows.

$$\Delta G_T = -2.303Rk_2 \log C_e - 2.303RTb_2 \log C_e + \Delta G_A, \Delta G_A = -2.303Rk_1 - 2.303RTb_1$$

$$\Delta H_T = 2.303Rk_2 \log C_e \Delta H_A, \Delta H_A = -2.303Rk_1$$

$$\Delta S_T = -2.303Rb_2 \log C_e + \Delta S_A, \Delta S_A = 2.303Rb_1$$

3. Results and Discussion

3.1. Biochar Characterization

3.1.1. XRD Analysis

Figure S1 shows the X-ray diffraction (XRD) patterns of the initial biochar and activated biochar (AB), indicating that the diffraction peaks of the initial biochar matched well with the pattern of standard diffraction peaks of sylvite (*pdf* = No. 41-1476), quartz (*pdf* = No.46-1045), graphite (*pdf* = No. 41-1487), calcite (*pdf* = No. 05-0586) and brushite (*pdf* = No. 09-0077), respectively, and also that the main mineral in AB was quartz (*pdf* = No. 46-1045). The comparison indicates that interfering metallic materials on the carbon surface are removed and the silicon content is relatively enhanced.

3.1.2. Zeta Potential Analysis

Figure S2 presents the zeta-potentials of the activated biochar and imprinted biochar. It indicates that the point of zero charges (PZCs) of the activated biochar and imprinted biochar are absent and may both locate at around pH < 2.0. The decrease in the zeta-potentials of imprinted biochar could be attributed to the specific functional groups onto the initial biochar surface, which is distributed to the –OH groups and –SH groups on the IB surface.

3.1.3. FT-IR Analysis

The FT-IR spectra of MPS, AB and IB are shown in Figure S3. Compared with AB, the spectral features of MPS in IB are obtained [21,28]. The bands at 2923 cm^{-1} and 2853 cm^{-1} reflect –CH$_2$ stretching vibration. A broad peak is noted at 1095 cm^{-1}, due to the Si–O–Si, which shifted from 1048 cm^{-1} and indicates MPS interacted on the Si–O site on the surface of activated biochar. The absorption band at 2152 cm^{-1} is assigned to S–H vibrations of sulfydryl group.

3.1.4. XPS Analysis

Table 1 presents the binding energies and relative contents of major elements on biochar surface. For C1s, the content on the initial biochar surface is 53.75%, and it increases after activating and imprinting, which is attributed to that the Mg, K, etc. ions in elution. For Si2p, after the activated biochar is treated by the MPS, the chemical shift of Si2p increases from 102.69 eV to 103.72 eV, indicating that MPS reacted on the surface of the activated biochar with Si–O forming Si–O–Si bends. For S2p, the results indicate that the content of S on the activated biochar surface is low and the bending energy is 164.41 eV. After imprinting, the content of S increases and the bending energy of S2p decreases to 163.01 eV; however, the bending energy shifts to 163.89 eV and the content increases to 3.03% after elution by hydrochloric acid, presenting that MPS acts on the surface of the activated biochar and the –SH group acts with Cd(II).

Table 1. Binding energies and relative contents of elements on biochar surface.

Sample	Binding Energy (eV)			Surface Atomic Composition (%)		
	C1s	Si2p	S2p	C1s	Si2p	S2p
Initial biochar	284.80	102.07	164.39	53.75	9.31	0.17
Actived biochar	284.79	102.69	164.41	65.29	13.92	0.29
No-elution biochar	284.79	103.72	163.01	67.75	12.63	2.84
Imprinted biochar	284.81	103.73	163.89	69.33	13.28	3.03

3.2. Adsorption Experiment Results

3.2.1. Adsorption Kinetics

Figure 2 shows that the uptake of Cd(II) by IB is rapid during the initial 120 min and the equilibrium is reached within 960 min (q_{960} was similar to q_{1440}). In Figure 2, the adsorption data match the pseudo-second-order kinetic model well (R^2 = 0.9964), and the parameters of pseudo-second-order kinetics are fitted. The calculated q_e (6.76 mg·g^{-1})from the pseudo-first-order kinetic model agrees very well with the experimental data. Thus, the adsorption process is a chemisorption process.

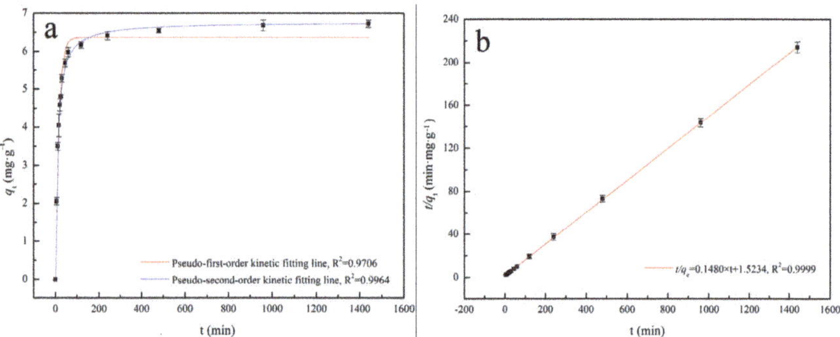

Figure 2. (**a**) Relation between Cd(II) adsorption amount and the contact time; and (**b**) the kinetic fitting line.

3.2.2. Adsorption Isotherms

The equilibrium adsorption isotherm is fundamental in describing the interactive behavior between solute and sorbent, and it is important for the design of adsorption system. Figure 3a shows the Cd(II) adsorption capacity on IB at different initial Cd(II) concentrations and temperature. It is evident that the initial Cd(II) concentrations affect the adsorption capacity of IB to Cd(II): the adsorption capacity of IB increase with the initial Cd(II) concentration, and adsorption favors higher temperatures.

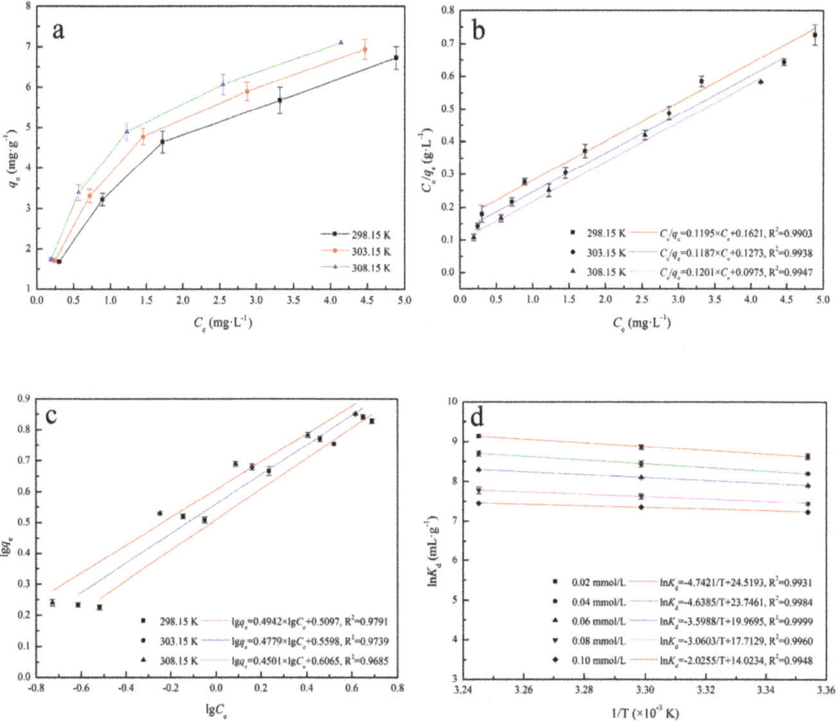

Figure 3. Adsorption isotherms parameters: (**a**) Adsorption isotherms of Cd(II) on IB; (**b**) The fitting line of Langmuir model; (**c**) The fitting line of Freundlich model; (**d**) Plots of $\ln K_d - 1/T$ of IB.

Figure 3b,c (Langmuir and Freundlich, respectively) shows the isotherm constants and the correlation coefficients (R^2) obtained by linear regression. It indicates that the adsorption process is well described by the Langmuir model ($R^2_{Langmuir} > R^2_{Freundlich}$). The fact that the Langmuir isotherm fitted the experimental data very well may be due to homogenous distribution of active sites on the IB surface. ΔS^0 and ΔH^0 were calculated from the slope and intercept of Van't Hoff plots of $\ln K_d$ versus $1/T$ (Figure 3d). In Table 2, negative ΔG^0 and positive ΔH^0 indicate that the adsorption process is spontaneous and endothermic. The positive ΔS^0 reflects an increase in randomness at the solid/solution interface during Cd(II) adsorption on IB.

Table 2. Thermodynamic parameters for adsorption of Cd(II) on the imprinted biochar.

C_0 (mmol·L^{-1})	ΔH^0 (kJ·mol^{-1})	ΔS^0 (J·mol^{-1}·K^{-1})	ΔG^0 (kJ·mol^{-1})		
			298.15 K	303.15 K	308.15 K
0.02	39.43	203.87	−21.35	−22.37	−23.39
0.04	38.57	197.44	−20.30	−21.29	−22.27
0.06	29.92	166.04	−19.58	−20.41	−21.24
0.08	25.44	147.27	−18.46	−19.20	−19.94
0.10	16.84	116.60	−17.92	−18.51	−19.09

3.2.3. Stoichiometric Displacement Theory of Adsorption

However, these gas–solid adsorption equations were originally derived only for gas–solid adsorption systems, and they have not been related to the strong interactions that exist among the solute, solvent and solid sorbent in a liquid–solid adsorption process. The adsorption mechanisms of liquid–solid systems are more complex than those in gas–solid systems. In addition, the volume of the adsorption layer on solid sorbent surface cannot be estimated accurately, so the values of ΔH, ΔS and ΔG cannot be calculated accurately. Recently, stoichiometric displacement theory of adsorption (SDT-A) [29–31] has been used to explain the adsorption mechanisms of solute in liquid–solid system.

The data in Table 3 and Figure 4 indicate that: (1) At the same initial Cd(II) concentrations, the negative ΔG_A indicates that Cd(II) adsorption on the IB was a spontaneous and heat release process, while increasing temperature benefits the adsorption. ΔG_T was negative and decreases, indicating that the increasing temperature was beneficial to the Cd(II) adsorption. (2) At the same adsorption temperature, the constant ΔG_A and ΔH_A indicate that the adsorption process was a spontaneous and exothermic process, which was unaffected by the Cd(II) concentrations. (3) Under different conditions, the function variable of ΔS_T indicated that the degree of disorder of the whole system increased. Compared to the thermodynamic parameters derived from Langmuir, the values of ΔG varied little, however, the values of ΔH and ΔS greatly differ.

Table 3. Thermodynamic parameters of SDT-A for Cd(II) adsorption on IB.

C_0 (mmol·L^{-1})	ΔH_T (kJ·mol^{-1})			ΔS_T (J·mol^{-1}·K^{-1})			ΔG_T (kJ·mol^{-1})		
	298.15	303.15	308.15	298.15	303.15	308.15	298.15	303.15	308.15
0.02	21.05	21.70	22.68	142.83	146.21	150.29	−21.53	−22.54	−23.64
0.04	17.42	18.15	18.95	126.11	129.48	133.16	−21.19	−22.22	−22.08
0.06	15.20	15.78	16.35	115.93	118.56	121.18	−19.36	−19.94	−20.76
0.08	13.00	13.48	13.89	105.78	108.00	109.90	−18.54	−19.26	−19.97
0.10	11.69	12.00	12.25	99.78	101.19	102.35	−18.06	−18.68	−19.29

ΔH_A = 17.03 kJ·mol^{-1}, ΔS_A = 124.33 J·mol^{-1}·K^{-1}, ΔG_A = −20.04 kJ·mol^{-1} (298.15 K), −20.66 kJ·mol^{-1} (303.15 K), −21.28 kJ·mol^{-1} (308.15 K).

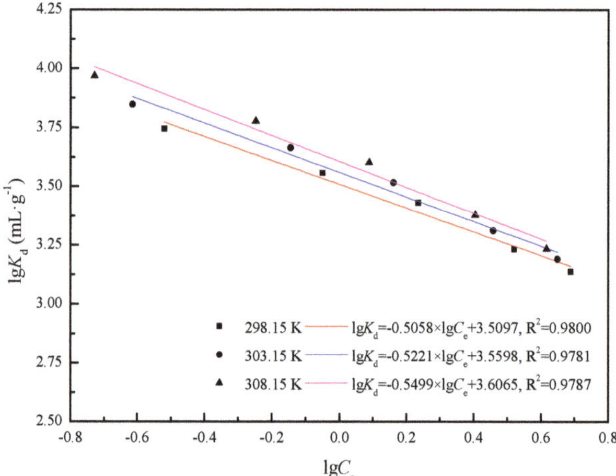

Figure 4. Plots of $\ln K_d - \lg C_e$ of IB.

3.2.4. Effect of pH

Figure 5a presents the effect of initial pH of Cd(II) solution on the adsorption capacity of IB. It indicates that the initial pH of Cd(II) solution affects the adsorption process, especially the adsorption capacity. At the initial pH range of about 3.0–8.0, the adsorption capacity of IB to Cd(II) remains approximately stable (average about 6.26 mg·g^{-1}), and there is a sharp decrease at pH > 8.0. At different pH environments, Cd(II) solution exhibited in different forms during the adsorption process, and the different forms affected the adsorption behaviors. The Cd(II) solution chemistry could be calculated to generate the concentration logarithmic diagram of each component at different pH. The relative species distribution of cadmium is calculated from the hydrolysis constants ($\log\beta_1$ = 4.17, $\log\beta_2$ = 8.33, $\log\beta_3$ = 9.02, $\log\beta_4$ = 8.62) [32]. The precipitation curve of cadmium is calculated from the precipitation constant of Cd(OH)$_{2(s)}$ (K_{sp} = 5.27 × 10^{-15}) and the initial Cd(II) concentration (1.0 × 10^{-4} mol·L^{-1}). Cd(II) is found to form precipitation at pH ≈ 7.75 (the equations are shown in the Supplementary Materials). Figure 5b shows the lgC-pH of the Cd(II) hydrolysis components as a function of pH when the Cd(II) concentration is 0.1 mmol·L^{-1}. It indicates that the main component at pH = 3.0–8.0 is in the form of Cd(II) ion, while the main component is Cd(OH)$_{2(s)}$ at pH > 8.0, when the reaction with IB is week. At lower pH (<3.0), the Cd(II) has to compete with hydrogen ion among the exchange sites.

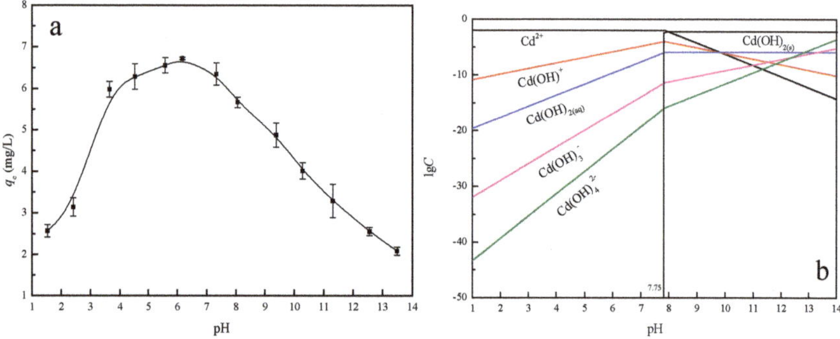

Figure 5. (**a**) The effect of initial pH of Cd(II) solution on the adsorption capacity of IB; and (**b**) Logarithmic diagram of Cd(II) hydrolysis components (c = 0.1 mmol·L^{-1}).

3.2.5. Effect of Sorbent Dosage

The effect of sorbent dosage on adsorption capacity of IB to Cd(II) is shown in Figure S4. It indicates that the sorbent dosage is an important factor to the adsorption capacity. Figure S4 shows that the removal efficiency increases from 26.05% to 78.66% when the sorbent dosage increases from 0.5 to 5.0 g·L^{-1} (Figure S5). With higher adsorbent dosage, more active sites compete for the same amount of Cd(II) ions, therefore only the higher affinity active sites will be occupied. resulting in a decrease in adsorption capacity. For a constant initial Cd(II) concentration, the increase sorbent dosage provides more functional groups and active sites, thus leading to the removal efficiency of Cd(II) increasing. When the dosage is 2.0 g·L^{-1}, the adsorption amount is 6.72 mg·g^{-1} and the removal efficiency is 73.34%; however, the removal efficiency does not change with the increasing dosage. Thus, 2.0 g·L^{-1} adsorbent is selected as the optimum dose.

3.2.6. Selective Adsorption

Competitive adsorption of Cd(II)/Cu(II), Cd(II)/Zn(II), Cd(II)/Co(II) and Cd(II)/Pb(II) were investigated in their double mixture systems. Relative selectivity coefficients ($\eta = k_{IB}{:}k_{NIB}$) for Cd(II)/Cu(II), Cd(II)/Zn(II), Cd(II)/Co(II) and Cd(II)/Pb(II) are 6.06, 5.98, 6.81 and 8.05, respectively, and the relative selectivity coefficients ($\eta = k_{IB}{:}k_{BS}$) are 5.37, 5.85, 6.50 and 6.10 (Table S1).

The results indicate that IB has a higher selectivity for Cd(II), even in the presence of Co(II), Pb(II), Zn(II) and Cu(II) interferences in the same medium due to the coordination geometry selectivity of IB, which could provide ligand groups arranged in a suitable way for coordination of Cd(II). Although some ions have similar size with Cd(II) and some ions have high affinity with the ligand, IB still exhibits high selectivity for extraction of Cd(II) in the presence of other metal ions. The results also indicate that the adsorption capacity and selectivity of biochar based (BS) is low.

3.2.7. Desorption and Repeated Use

Desorption of Cd(II) from IB is studied in a batch experimental set-up. The best desorption time is found to be 8 h. With a single washing, desorption ratio is up to 89.19% (0.5 h: 33.67%; 1.0 h: 48.56%; 2.0 h: 65.96%; 4.0 h: 79.33%; 8.0 h: 89.19%; 12 h: 89.67%; 16 h: 90.13%; and 24 h: 90.18%).

To verify the reusability of IB, the cycle of adsorption–desorption was repeated seven times using the same sample. The results reveal that IB could be used repeatedly without significantly losing its adsorption capacities. Adsorption capacity of IB decreased only 15.64% after five adsorption–desorption cycles (Figure S6).

4. Conclusions

A biochar prepared by agricultural waste was modified using 3-Mercaptopropyltrimethoxysilane Epoxy-chloropropane via Ionic-imprinted Technique. IB adsorbed Cd(II) at suitable pH of about 3.0–8.0 and dosage of 2.0 g·L^{-1}. During the initial 120 min, the removal of Cd(II) from IB was rapid, and the equilibrium time was about 960 min. The adsorption process matched well with pseudo-second-order kinetic model and Langmuir model. The results of adsorption isotherms and SDT-A informed that IB adsorption capacity increased with the initial Cd(II) concentration and the adsorption temperature, and the adsorption process was spontaneous and endothermic. IB had a higher selectivity for Cd(II) in the presence of Co(II), Pb(II), Zn(II) and Cu(II), which provided ligand groups arranged in a suitable way for coordination of Cd(II). Overall, 6.0 mol·L^{-1} hydrochloric acid could remove Cd(II) from IB effectively, and the desorption ratio was up to 89.19% with a single washing. Adsorption capacity of IB decreased only 15.64% after five adsorption–desorption cycles.

Supplementary Materials: The following are available online at www.mdpi.com/1996-1944/11/2/299/s1, Figure S1: XRD patterns of initial biochar and activated biochar. Figure S2: Zeta-potentials of activated biochar and imprinted biochar. Figure S3: FT-IR spectra of MPS, activated biochar, and imprinted biochar. Figure S4: Effect of sorbent dosage on adsorption capacity of SRFB to Cd(II). Figure S5: The relation between desorption ratios and desorption time. Figure S6: The relation between adsorption-desorption cycle and absorption capacity. Table S1: The selectivity parameters of Cd(II) adsorption on IB and NIB.

Acknowledgments: This study was supported by the National Natural Science Foundation of China (51708301); Natural Science Foundation of Tianjin, China (17JCZDJC39500); 2017 Science and Technology Demonstration Project of Industrial Integration and Development, Tianjin, China (17ZXYENC00100); 2017 Jinnan District Science and Technology Project of Tianjin, China-Research on Malodorous Gas Pollution Treatment with Microorganism deodorant; and the Scientific Research Fund of Education Department of Sichuan Province, China (17ZA0399). The authors appreciate the financial support and thank the editor and reviewers for their very useful suggestions and comments.

Author Contributions: Shiqiu Zhang, Le Liu and Meiting Ju conceived and designed the experiments; Shiqiu Zhang and Xue Yang performed the experiments; Shiqiu Zhang, Le Liu and Meiting Ju analyzed the data; Xue Yang and Kui Zheng contributed reagents/materials/analysis tools; Shiqiu Zhang wrote the paper.

Conflicts of Interest: The authors declare no conflict of interest.

References

1. Gozde, D.; Cagdas, O.; Suat, U.; Ralph, S.; Jale, Y. The slow and fast pyrolysis of cherry seed. *Bioresour. Technol.* **2011**, *102*, 1869–1878.
2. Chen, R.; Ye, C. Land management: Resolving soil pollution in China. *Nature* **2014**, *505*, 483. [CrossRef] [PubMed]
3. Li, Z.; Ma, Z.; Yuan, Z.; Huang, L. A review of soil heavy metal pollution from mines in China: pollution and health risk assessment. *Sci. Total Environ.* **2014**, *468*, 843–853. [CrossRef] [PubMed]
4. Zhang, Y.N.; Chu, C.L.; Li, T.; Xu, S.; Liu, L.; Ju, M. A water quality management strategy for regionally protected water through health risk assessment and spatial distribution of heavy metal pollution in 3 marine reserves. *Sci. Total Environ.* **2017**, *599–600*, 721–731. [CrossRef] [PubMed]
5. Almuktar, S.A.A.A.N.; Scholz, M.; Al-Isawi, R.H.K.; Sani, A. Recycling of domestic wastewater treated by vertical-flow wetlands for irrigating chillies and sweet peppers. *Agric. Water Manag.* **2015**, *149*, 1–122. [CrossRef]
6. Huang, Y.; Qiu, W.; Yu, Z.; Song, Z. Toxic effect of cadmium adsorbed by different sizes of nano-hydroxyapatite on the growth of rice seedlings. *Environ. Toxicol. Pharmacol.* **2017**, *52*, 1–7. [CrossRef] [PubMed]
7. Das, P.; Samantaray, S.; Rout, G.R. Studies on cadmium toxicity in plants: A review. *Environ. Pollut.* **1997**, *98*, 29–36. [CrossRef]
8. Asgher, M.; Khan, M.I.; Anjum, N.A.; Khan, N.A. Minimising toxicity of cadmium in plants-role of plant growth regulators. *Protoplasma* **2015**, *252*, 399–413. [CrossRef] [PubMed]
9. Benavides, M.P.; Gallego, S.M.; Tomaro, M.L. Cadmium toxicity in plants. *Braz. J. Plant Physiol.* **2005**, *17*, 27–34. [CrossRef]
10. Flanagan, P.R.; McLellan, J.S.; Haist, J.; Cherian, G.; Chamberlain, M.J.; Valberg, L.S. Increased dietary cadmium absorption in mice and human subjects with iron deficiency. *Gastroenterology* **1978**, *74*, 841–846. [PubMed]
11. Kaewsarn, P.; Yu, Q. Cadmium(II) removal from aqueous solutions by pre-treated biomass of marine alga Padina sp. *Environ. Pollut.* **2001**, *112*, 209–213. [CrossRef]
12. Singh, G.; Rana, D.; Matsuura, T.; Ramakrishna, S.; Narbaitz, R. Removal of disinfection byproducts from water by carbonized electrospun nanofibrous membranes. *Sep. Purif. Technol.* **2010**, *74*, 202–212. [CrossRef]
13. Demirbas, A. Heavy metal adsorption onto agro-based waste materials: A review. *J. Hazard. Mater.* **2008**, *157*, 220–229. [CrossRef] [PubMed]
14. Lehmann, J. A handful of carbon. *Nature* **2007**, *447*, 143–144. [CrossRef] [PubMed]
15. Yong, S.K.; Kobayashi, M.; Fumiaki, T.; Hideaki, S.; Takami, S.; Kentaro, T.; Ryusuke, H.; Takayoshi, K. Greenhouse gas emissions after a prescribed fire in white birch-dwarf bamboo stands in northern Japan, focusing on the role of charcoal. *Eur. J. For. Res.* **2011**, *130*, 1031–1044.
16. Laird, D.; Fleming, P.; Wang, B.; Horton, R.; Karlen, D. Biochar impact on nutrient leaching from a Midwestern agricultural soil. *Geoderma* **2010**, *158*, 436–442. [CrossRef]
17. Chen, X.C.; Chen, G.G.; Chen, L.G.; Chen, Y.X.; Lehmann, J.; McBride, M.B.; Hay, A.G. Adsorption of copper and zinc by biochars produced from pyrolysis of hardwood and corn straw in aqueous solution. *Bioresour. Technol.* **2011**, *102*, 8877–8884. [CrossRef] [PubMed]

18. Agrafioti, E.; Kalderis, D.; Diamadopoulos, E. Arsenic and chromium removal from water using biochars derived from rice husk, organic solid wastes and sewage sludge. *J. Environ. Manag.* **2014**, *133*, 309–314. [CrossRef] [PubMed]

19. Zaidi, S.A. Molecular imprinting polymers and their composites: A promising material for diverse applications. *Biomater. Sci.* **2017**, *5*, 388–402. [CrossRef] [PubMed]

20. Gao, B.J.; Wang, J.; An, F.Q.; Liu, Q. Molecular imprinted material prepared by novel surface imprinting technique for selective adsorption of pirimicarb. *Polymer* **2008**, *49*, 1230–1238. [CrossRef]

21. Fang, G.Z.; Tan, J.; Yan, X.P. An ion-imprinted functionalized silica gel sorbent prepared by a surface imprinting technique combined with a sol-gel process for selective solid-phase extraction of cadmium(II). *Anal. Chem.* **2005**, *77*, 1734–1739. [CrossRef] [PubMed]

22. Liu, Y.H.; Cao, X.H.; Hua, R.; Wang, Y.Q.; Liu, Y.T.; Pang, C.; Wang, Y. Selective adsorption of uranyl ion on ion-imprinted chitosan/PVA cross-linked hydrogel. *Hydrometallurgy* **2010**, *104*, 150–155. [CrossRef]

23. Kupai, J.; Razali, M.; Buyuktiryaki, S.; Kecili, R.; Szekely, G. Long-term stability and reusability of molecularly imprinted polymers. *Polym. Chem.* **2017**, *8*, 666–673. [CrossRef] [PubMed]

24. Zhou, Y.H.; Gu, Z.N. Study on hydroxyl in quartz glass and quartz raw materials. *J. Chin. Ceram. Soc.* **2002**, *30*, 357–361.

25. Rovetta, M.R.; Blacic, J.D.; Hervig, R.L.; Holloway, J.R. An experimental study of hydroxyl in quartz using infrared spectroscopy and ion microprobe techniques. *J. Geophys. Res. Solid Earth* **1989**, *94*, 5840–5850. [CrossRef]

26. Wang, W.Q.; Zhu, Y.G.; Zhang, S.Q.; Deng, J.; Huang, Y.; Yan, W. Flotation behaviors of perovskite, titanaugite, and magnesium aluminate spinel using octyl hydroxamic acid as the collector. *Minerals* **2017**, *7*, 134. [CrossRef]

27. Ren, Y.; Zhang, M.; Zhao, D. Synthesis and properties of magnetic Cu(II) ion imprinted composite adsorbent for selective removal of copper. *Desalination* **2008**, *228*, 135–149. [CrossRef]

28. Arzu, E.; Rıdvan, S.; Adil, D. Ni(II) ion-imprinted solid-phase extraction and preconcentration in aqueous solutions by packed-bed columns. *Anal. Chim. Acta* **2004**, *502*, 91–97.

29. Geng, X.D.; Zebolsky, D.M. The stoichiometric displacement model and Langmuir and Freundlich adsorption. *J. Chem. Educ.* **2002**, *79*, 385–388. [CrossRef]

30. Wang, Y.; Geng, X.D. A quantitative relationship between the affinity of component to adsorbent in liquid–solid system, β_a, and composition of bulk solution. *Thermochim. Acta* **2003**, *404*, 109–115. [CrossRef]

31. Song, Z.H.; Geng, X.D. A Study on the retention mechanism of solute of liquid-solid chromatography by stoichiometric displacement model. *Acta Chim. Sin.* **1990**, *48*, 237–241.

32. Xu, R.; Zhou, G.Y.; Tang, Y.H.; Chu, L.; Liu, C.B.; Zeng, Z.B.; Luo, S.L. New double network hydrogel adsorbent: Highly efficient removal of Cd(II) and Mn(II) ions in aqueous solution. *Chem. Eng. J.* **2015**, *275*, 179–188. [CrossRef]

Article

Influence of Surface Chemistry on the Electrochemical Performance of Biomass-Derived Carbon Electrodes for its Use as Supercapacitors

Abdelhakim Elmouwahidi [1], Esther Bailón-García [1], Luis A. Romero-Cano [2], Ana I. Zárate-Guzmán [3], Agustín F. Pérez-Cadenas [1,*] and Francisco Carrasco-Marín [1]

[1] Research Group in Carbon Materials, Inorganic Chemistry Department, Faculty of Sciences, University of Granada, Campus Fuente Nueva s/n. 18071 Granada, Spain

[2] Facultad de Ciencias Químicas, Universidad Autónoma de Guadalajara, Av. Patria 1201, Zapopan, Jalisco C. P. 45129, Mexico

[3] Centro de Investigación y Desarrollo Tecnológico en Electroquímica (CIDETEQ) S.C., Parque Tecnológico Sanfandila, Pedro Escobedo, Querétaro 760703, Mexico

*   Correspondence: afperez@ugr.es; Tel.: +34-958243316

Received: 28 June 2019; Accepted: 1 August 2019; Published: 2 August 2019

Abstract: Activated carbons prepared by chemical activation from three different types of waste woods were treated with four agents: melamine, ammonium carbamate, nitric acid, and ammonium persulfate, for the introduction of nitrogen and oxygen groups on the surface of materials. The results indicate that the presence of the heteroatoms enhances the capacitance, energy density, and power density of all samples. The samples treated with ammonium persulfate show the maximum of capacitance of 290 F g^{-1} while for the melamine, ammonium carbamate, and nitric acid treatments, the samples reached the maximum capacitances values of 283, 280, and 455 F g^{-1} respectively. This remarkable electro-chemical performance, as the high specific capacitances can be due to several reasons: i) The excellent and adequate textural characteristics makes possible a large adsorption interface for electrolyte to form the electrical double layer, leading to a great electrochemical double layer capacitance. ii) The doping with hetero-atoms enhances the surface interaction of these materials with the aqueous electrolyte, increasing the accessibility of electrolyte ions. iii) The hetero-atoms groups can also provide considerable pseudo-capacitance improving the overall capacitance.

Keywords: nitrogen and oxygen doped activated carbon; surface chemistry; supercapacitor capacitance; energy power density

---

1. Introduction

As a consequence of the change in the energy model to which we are involved, together with the challenge of mitigating climate change, one of the main technologies with serious possibilities of being implemented are high-performance energy storage and conversion devices as fuel cells, batteries or supercapacitors [1–3]. Supercapacitors are electrochemical devices capable of provide an unusually high energy amount, that is high power density. During the rechargeable electrochemical cycles the charge carriers migrate reciprocally between electrolytes and electrodes. Supercapacitors have also excellent reversibility together with long cycle life, therefore supercapacitors are specifically good candidates for large-scale applications of portable and automotive electronic systems. Nevertheless, one inconvenience of supercapacitors is the relatively low energy density, an in this line, many efforts are being made by designing and optimizing the materials for the corresponding electrodes.

Carbon based materials are probably the best candidates for this type of electrochemical applications taking in account the extended literature published during the last years [4–6]. On the other hand, a different type of precursors are used for the preparation of activated carbon from waste agriculture products such as olive stone [7–9], melia azedarach stones [10], argan seed shells [11], coconut [12,13], etc., by using different methods of chemical and physical activation [14,15]. The use of biomass wastes for carbon electrodes preparation and its use as supercapacitors is one of the best available options, not only for their very high electro-chemical performance but also for the economy of the synthesis process [3]. Indeed, for the electrochemical applications, the pore structure and an adequate surface chemistry are crucial, and carbon-based materials obtained from woods are easy tunable in both ways. For example, the capacitive behavior of carbon materials can be further improved by the presence of active species that contribute to the total specific capacitance by the pseudo-capacitive effect [16,17]. Others works have found that functional groups containing heteroatoms such as O and N are very favourable to improve the capacitance, and these groups can be introduced using different doping methods, as chemical oxidations, plasma treatment, or electrochemical treatments. [18–20]. The main aim of oxidation of a carbon surface is obtaining a more hydrophilic surface structure with groups such as carboxyl groups [21,22]. When oxygen containing groups are present on the surface of the activated carbon they affect the capacitance of this material mainly enhancing its wettability, and therefore increasing the capacitance and providing high energy and power densities; these type of groups can also produce pseudo capacitance effects. Several types of compounds have been used as oxidizers: ammonium per sulfate, sodium hypochlorite and permanganate, concentrated nitric or sulfuric acid or hydrogen peroxide. Furthermore, it has been reported that nitrogen groups modify the electron donor/acceptor characteristics of carbon materials depending on the type of interactions between the nitrogen and carbon atoms. The nitrogen-containing groups generally provide basic property, which could improve the interaction between carbon materials and acid molecules, such as dipole-dipole, H-bonding, covalent bonding, among others. The nitrogen groups can be formed by treatment with urea, melamine, nitric acid, and others types of containing nitrogen molecules [16,17,23].

In this work we present the treatment of three types of activated carbons, produced from chemical activation of waste woods, with four agents: nitric acid and ammonium persulfate for the introduction of oxygen groups; and melamine and ammonium carbamate for the introduction of nitrogen functionalities on the surface of the activated carbons and finally, the effects of these oxygen and nitrogen groups on the electrochemical performances of the corresponding electrodes have been comparatively studied and discussed, showing these materials as excellent candidates to form part of applicable supercapacitors.

2. Materials and Methods

2.1. Synthesis of Modified Activated Carbons Samples

The activated carbons (ACs) were prepared by chemical activation with KOH of three different woods: custard apple, fig tree, and olive tree following the method described previously [11]. The samples were prepared by impregnation with KOH in a weight ratio of 1:1. The mixtures were heated at 60 °C to dryness and after that at 110 °C until evaporation total of the water. The solids produced were carbonized under N_2 flow (300 cm^3 min^{-1}) and heating rate of 5 °C min^{-1}, at 300 °C for 1 h followed by activation at 800 °C for 2 h. The produced activated carbons were washed with HCl (1 M) and with distilled water until neutralization of the washing water and total elimination of chloride. The samples were designated as CK, FK, and OK, indicating that they were obtained from custard apple, fig tree, and olive tree, respectively.

The pre-prepared samples were treated with two oxidative agents for the introduction of oxygen surface groups: nitric acid and ammonium persulfate; and with ammonium carbamate and melamine for the introduction of nitrogen groups.

In details, the treatment with nitric acid was carried with 1 M diluted nitric acid at boiling temperature. Nitric acid (1 M) was slowly added through the funnel. The oxidation was carried out at the boiling temperature for 2 h. The oxidized ACs were washed with distilled water until the absence of nitrates, and then dried overnight at 50 °C. The prepared samples were denoted FKN, OKN, and CKN.

The treatment with ammonium persulfate $(NH_4)_2S_2O_8$ to introduce surface oxygen functionalities was performed as reported by Moreno-Castilla [24]. In detail, the treatment was carried out with a saturated solution of this salt in H_2SO_4 1 M (1 g of carbon/10 mL of solution) at 25 °C for 48 h. After the treatment, the samples were washed with distilled water until absence of sulfates was reached. The prepared samples were denoted FKS, OKS, and CKS.

The treatment with melamine was prepared by mixing 0.5 g of activated carbon with 33 mg of melamine dissolved in 20 mL of ethanol as described in our previous work [11]. After stirring this slurry, the solvent was slowly removed by evaporation and the remaining residue was heat-treated at 750 °C for 1 h under N_2 flow (60 cm^3 min^{-1}). The corresponding samples were denoted FKM, OKM, and CKM.

Finally, the introduction of nitrogen functionality by using ammonium carbamate was carried by mixing 2 g of ammonium carbamate and 2 g of activated carbon in 20 mL of distillated water. After stirring this slurry, the residue was heated at 550 °C for 1 h and 600 °C for 1 h, both under N_2 flow (60 cm^3 min^{-1}). The prepared samples denoted FKC, OKC, and CKC.

2.2. Characterization

Textural characterization was carried out by gas adsorption, using N_2 and CO_2 at −196 °C and 0 °C, respectively, in a Quantachrome Autosorb-1 equipment (Anton Paar QuantaTec, Boynton Beach, FL, USA). The apparent surface area (S_{BET}) together with: the micropore volume (W_0), the mean micropore width (L_0) and the microporous surface (S_{mic}), were obtained applying the BET and Dubinin–Radushkevich equations, respectively. The total pore volume, V_{total}, was considered as the volume of N_2 adsorbed at $P/P_0 = 0.95$ and the mesopore volume, V_{meso}, was obtained by the difference between V_{total} and the micropore volume obtained from nitrogen adsorption.

The surface chemistry of the activated carbons was studied by X-ray photoelectron spectroscopy (XPS) and temperature programmed desorption coupled with mass spectrometry (TPD). TPD and XPS analysis were carried out as described elsewhere [11]. For TPD a heating rate of 20 °C min^{-1} to 1000 °C was used. Total oxygen content, O_{TPD}, was measured counting the amount of CO and CO_2 evolved [11]. During the XPS analysis C_{1s}, O_{1s}, N_{1s}, and S_{2p} spectra were recorded and deconvoluted as described elsewhere [11].

2.3. Electrochemical Measurements

The electrochemical measurements were performed in a two electrodes system as described elsewhere [25,26]. The working temperature was 25 °C using H_2SO_4 (1 M) as electrolyte. Glass fibrous material was used as a separator. The preparation of the working electrodes was carried out as described elsewhere [26]. The voltage window was 0–0.9 V in H_2SO_4 (1 M). In the mentioned voltage interval, Cyclic voltammetry (CV) was performed at different scan rates (0.5, 2.5, 5, 10 and 20 mV s^{-1}). The gravimetric capacitance obtained from CV, C_{CV} (F g^{-1}), the gravimetric capacitance obtained from galvanostatic charge–discharge analyses, C_{GD} (F g^{-1}), the capacitance value, C_{max}, obtained from impedance spectroscopy measurements, and the electrical energies, E (Wh Kg^{-1}) and power densities, P (W Kg^{-1}) for two-electrode cell were calculated as described elsewhere [26]. Finally, the stability of supercapacitors was also monitored by charge–discharge cycles as described elsewhere [25].

3. Results and Discussion

3.1. Structural and Textural Characterization

Table 1 shows the characterization data obtained from N_2 adsorption–desorption and CO_2 adsorption isotherms; the results indicate a change after the modification with different treatments for the introduction of oxygen and nitrogen functionalities. Figure S1 collects the N_2 adsorption–desorption isotherms at 77 K for CK-series as example.

From the data it can be seen that the treatment with melamine and ammonium carbamate, for all samples, increase the surface area and also the pore volume increase, which can be due to the thermal treatment at higher temperature and also the modification can cause the degradation of non-carbon impurities, resulting in an increase in surface area (CKC = 1706 m^2 g^{-1}, FKC = 1669 m^2 g^{-1}, OKC = 1314 m^2 g^{-1}). The only difference between the three samples was that the FK series shows a decreasing of the pore diameter; however, both CK and OK series show an increasing of the pore diameter. On the other hand, the results indicate a decreasing of micropore volume after the treatment with ammonium carbamate but a slow increasing with the melamine treatment. Furthermore, all samples show an increasing of the micropore diameter indicating a destruction, or fusion of microporous structure.

Table 1. Textural characteristics of modified activated carbons.

Sample	N_2					CO_2	
	S_{BET} m^2/g	$W_0(N_2)$ cm^3/g	$L_0(N_2)$ nm	V_{total} cm^3/g	V_{mes} cm^3/g	$W_0(CO_2)$ cm^3/g	$L_0(CO_2)$ nm
CK	1504	0.59	1.20	—	—	0.35	0.70
CKM	1525	0.60	1.16	0.75	0.15	0.37	0.70
CKC	1706	0.68	1.27	0.85	0.17	0.06	0.95
CKN	46	0.01	4.84	0.10	0.09	0.13	0.46
CKS	1042	0.41	1.16	0.54	0.13	0.32	0.61
FK	1024	0.41	1.30	—	—	0.31	0.70
FKM	1575	0.63	1.26	0.80	0.17	0.35	0.67
FKC	1669	0.67	1.28	0.84	0.17	0.06	0.99
FKN	287	0.11	1.40	0.20	0.09	0.07	0.93
FKS	1114	0.44	1.36	0.59	0.15	0.32	0.62
OK	1273	0.49	1.30	—	—	0.34	0.70
OKM	1400	0.56	1.34	0.77	0.21	0.23	0.63
OKC	1314	0.53	1.32	0.69	0.16	0.30	0.71
OKN	153	0.06	2.81	0.15	0.09	0.14	0.47
OKS	1075	0.42	1.30	0.58	0.16	0.25	0.61

For all samples treated with melamine and ammonium carbamate, $W_0(N_2)$ is inferior to $W_0(CO_2)$ indicating the presence of constriction at micropores entrances and partial accessibility of N_2 molecule at −196 °C. So, the treatment with melamine and ammonium carbamate for the introduction of nitrogen functionalities has the more significant effects on the microporous structure of all the three samples treated.

The results of the oxidation with the treatment with ammonium persulfate indicate decreasing in the surface area, mesoporous and microporous volume, and mesoporous and microporous diameter by partially destroying micro and mesopore walls. The oxidation with nitric acid indicates destruction of the pore structure and of mesopore and microporous volume. This is due to the destruction of pore walls and micropore blocking by oxygen-containing groups introduced during the chemical modification. The only exception is that the sample FKN shows lower pore destruction (0.11 cm^3 g^{-1}) than the both two other samples CKN (0.01 cm^3 g^{-1}) and OKN (0.01 cm^3 g^{-1}). Furthermore, of the three series of samples, all the samples with the same treatment have a similar value of the mesopore volume (0.09 cm^3 g^{-1}).

The surface chemistry of all activated carbon samples was characterized by XPS and TPD experiments. Table 2 contains the O and N surface contents determined by XPS as well as the quantification of the amount of CO and CO_2 desorbed in these experiments. The results of the TPD indicate that samples have a wide range of surface oxygen groups being, in all the cases, the CO evolved groups larger than the CO_2. The treatment of the three activated carbons CK, FK, and OK showed similar higher oxygen contents, CO/CO_2 ratio. The samples treated with ammonium persulfate and nitric acid has a very higher amount of oxygen varying from 14.08 to 27.02%.

Table 2. Surface chemistry of the modified activated carbons.

Sample	O_{xps} (wt.%)	N_{xps} (wt.%)	O_{TPD} (wt.%)	CO (mmol g^{-1})	CO_2 (mmol g^{-1})	CO/CO_2
CKM	1.7	1.6	1.0	0.11	0.26	0.42
CKC	3.7	0.8	2.1	0.45	0.43	1.05
CKN	21.4	1.1	25.5	1.16	7.41	0.16
CKS	15.5	0.9	14.0	0.87	3.96	0.22
FKM	1.9	2.1	1.0	0.19	0.23	0.83
FKC	4.3	0.7	2.1	0.48	0.42	1.14
FKN	20.6	0.9	25.3	1.22	7.32	0.17
FKS	16.0	0.5	15.1	0.94	4.26	0.22
OKM	6.1	1.0	3.7	0.26	1.03	0.25
OKC	5.7	0.9	7.0	0.41	1.98	0.21
OKN	22.0	1.4	27.0	1.78	7.85	0.23
OKS	15.0	0.2	14.4	0.89	4.02	0.22

All the samples treated with nitric acid show lower O_{XPS} than O_{TPD} that is, there is a non-uniform oxygen surface groups distribution; on the contrary, ammonium persulfate treated samples present a uniform distribution of the oxygen content, O_{XPS} (CKS = 15.5 wt.%; FKS = 16.0 wt.%; OKS = 15.0 wt.%) are similar to O_{TPD} values (CKS = 14.0 wt.%; FKS = 15.1 wt.%; OKS = 14.4 wt.%). It is important to note that for oxidized samples, CO/CO_2 ratio was generally lower than in those treated with melamine or ammonium carbamate because the oxidation mainly increased the amount of CO_2-evolving groups such as carboxyl acid groups, which usually increases during oxidation treatments [24]. This increase in the oxygen surface functionalities is accompanied by a decrease in the hydrophobicity because oxygen functionalities with large polarity have been introduced, e.g., carboxyl groups. The different types of oxygen groups present on the surface of carbon materials decompose upon heating producing CO and CO_2 at different temperatures. In this line, CO_2 evolves at low temperatures as a consequence of the decomposition of the acidic groups, typically carboxylic groups and/or lactones [26]. However, the CO evolution takes place at higher temperatures and it is related to the decomposition of basic or neutral groups such as carbonyls, phenols, and ethers.

The treatment of the activated samples with melamine and carbamate ammonium increased the content on nitrogen, which was fixed forming part of pyridinic (N-6), pyrrolic, and/or pyridonic (N-5) and quaternary-N (N-Q) groups [23]. The treatment greatly increased the amount of N-6 functionalities and reduced the amount of N-Q functionalities. The results indicate that the melamine treatment fix more nitrogen contents than the ammonium carbamate and also that show higher oxygen contents in the surface than those treated with melamine.

In order to characterize the surface chemistry in the outermost layer of the materials, XPS analysis was performed. The deconvolution of C_{1s}, O_{1s}, N_{1s}, and S_{2p} signals, and the corresponding peaks fitting showed the presence of diverse contributions to Binding Energies (BEs) that are displayed in Supplementary Figures S2–S4 and in Table 3, together with their corresponding percentages.

Table 3. XPS data obtained after deconvolution of the high-resolution XP spectra.

Sample	C_{1s} (eV)	FWHM (eV)	Peak (%)	O_{1s} (eV)	Peak (%)	N_{1S} (eV)	Peak (%)	$S2p_{3/2}$ (eV)	Peak (%)
CKM	284.5	1.34	65	531.6	33	398.3	33		
	285.7		17	532.9	67	399.4	28		
	286.9		7			400.5	24		
	288.4		5			401.6	15		
	290.2		4						
	291.7		2						
CKC	284.5	1.48	65	531.3	49	398.4	20		
	285.8		18	533.4	51	399.4	31		
	287.2		7			400.5	31		
	288.5		5			401.6	18		
	290.3		4						
	291.6		1						
CKN	284.5	1.45	60	531.4	39	399.3	16		
	285.8		19	533.0	61	400.5	11		
	286.9		6			401.5	34		
	288.6		12			405.5	40		
	290.1		3						
	291.5		1						
CKS	284.6	1.40	61	531.5	36			168.2	62
	285.8		19	533.0	64			169.5	38
	287.0		6						
	288.5		11						
	290.3		3						
	291.7		1						
FKM	284.6	1.35	62	531.4	36	398.4	28		
	285.6		19	533.0	64	399.4	31		
	286.8		8			400.5	28		
	288.3		5			401.9	13		
	290.2		4						
	291.6		1						
FKC	284.6	1.35	65	531.1	36	398.4	28		
	285.8		18	533.3	64	399.4	31		
	287.1		7			400.5	28		
	288.5		5			401.9	13		
	290.3		4						
	291.5		1						
FKN	284.6	1.43	58	531.5	36	399.6	12		
	285.8		21	533.1	64	400.5	13		
	286.9		5			401.6	35		
	288.6		12			405.7	40		
	290.1		2						
	291.5		1						
FKS	284.5	1.43	62	531.6	40			168.2	67
	285.9		16	533.0	60			169.4	33
	287.0		7						
	288.5		11						
	290.3		4						
	291.9		1						

Table 3. *Cont.*

Sample	C_{1s} (eV)	FWHM (eV)	Peak (%)	O_{1s} (eV)	Peak (%)	N_{1S} (eV)	Peak (%)	$S2p_{3/2}$ (eV)	Peak (%)
OKM	284.6	1.37	66	531.4	53	398.3	27		
	285.7		17	533.0	47	399.4	31		
	286.9		7			400.5	27		
	288.5		4			401.6	16		
	290.1		4						
	291.5		2						
OKC	284.5	1.38	66	531.2	48	398.3	20		
	285.8		17	533.0	52	399.5	40		
	287.1		6			400.5	32		
	288.6		5			401.9	8		
	290.2		4						
	291.6		1						
OKN	284.6	1.45	58	531.5	39	399.5	16		
	285.7		21	533.1	61	400.5	14		
	287.0		6			401.6	26		
	288.6		12			405.6	45		
	290.1		2						
	291.3		1						
OKS	284.5	1.42	62	531.5	48			168.2	71
	285.9		18	533.1	52			169.2	29
	287.0		7						
	288.5		9						
	290.2		4						
	291.6		1						

The C_{1s} spectrum for the all the modified samples contains two main peaks centered at 284.5 ± 0.1 (~63% on average) and 285.7 ± 0.2 eV (~18% on average) corresponding to C=C and C-C bonds, respectively [26–28]. Only minor peaks were detected at higher BE, which is typical for an activated carbon with a low oxygen content. It is important to note that after treatment with nitric acid and ammonium persulfate, samples that have been oxidized show an increase in the presence of O-C=O groups (288.6 ± 0.1 eV) [26–28] going from ~ 5% to ~ 11% on average.

The deconvolution of the spectra O_{1s} is presented in the Table 3, which can be deconvoluted into two peaks. The first one centered at 531.6 ± 0.4 eV attributed to the presence of the oxygen double bonded C=O groups. The second peak at 532.9 ± 0.5 eV indicates the presence of the singly bonded oxygen (-O-) in C-O [26–28]. An important observation is made when comparing the precursors used for the preparation of the activated carbons. For the samples prepared from custard apple and fig tree, the same distribution of oxygenated species is observed (40% of C=O groups and 60% of C-O groups, on average), however, for samples prepared from wood of olive tree this proportion changes (50% C=O and 50% C-O, on average)

The chemical state of nitrogen present in the treated samples with melamine and ammonium carbamate is further discussed on the basis of the XPS results. Deconvolution of the N_{1s} spectra resulted in four peaks. The first, centered at 398.3 ± 0.1 eV, designated N-6, is attributed to pyridinic-N, with the nitrogen atom in a six-membered ring and contributing with one p-electron to the aromatic π−system (~26% on average). The second peak at 399.4 eV can be attributed to the presence of nitrile groups present in the samples (~32% on average). The third peak centered at 400.5 eV, which is ascribed to pyrrolic-N or pyridone-N and will be referred as N-5 (~28% on average). Both groups have similar chemical environment for the nitrogen atom, with two p-electrons contributing to the π−system. Finally, the fourth peak at 401.6 ± 0.3 eV is attributed to quaternary nitrogen (N-Q) (~14% on average), which compared to pyridinic-N is defined as relatively more positively charge nitrogen incorporated in a graphene layer [29–32]. All samples showed a uniform distribution of the nitrogen groups introduced.

As expected, the samples treated with oxidizing agents showed a different distribution of these groups. For the samples oxidized with nitric acid, two notable changes are observed in comparison with the aforementioned samples: a) the peak attributed to pyridinic-N groups close to 398.3 eV is not appreciable; and b) the presence of a broad peak centered at 405.6 eV is observed (~42% on average), which can be attributed to nitrogen of the NO_2 groups [33]. It is important to highlight that for the samples treated with ammonium persulfate, the presence of nitrogen groups in not detectable by means of XP spectroscopy analysis.

Finally, only for the samples treated with ammonium persulfate a small amount of sulphur was detected on the surface due to the epoxy groups formed during the synthesis of the materials. The S_{2p} peak could be deconvoluted with a doublet of sulfate ($S2p_{3/2}$ at 168.2 eV and $S2p_{1/2}$ at 169.4 eV) indicating the presence of groups SO_2 in the surface of these materials [34].

3.2. Electrochemical Characterization

The electrochemical characterization was evaluated in H_2SO_4 1 M with two electrode system. Figure 1 shows the CV curves of all the electrodes tested between 0 and 0.9 V at the scan rate of 0.5 mV s^{-1}. It can be seen that the samples treated with melamine, ammonium carbamate, and ammonium persulfate, show a quasi-rectangular shape, which is a feature of electrochemical double-layer capacitors. In contrast, the samples treated with nitric acid show non-rectangular shape due to the effect of the presence of the higher content of oxygen surface groups evolved as CO_2 which induces some diffusional restrictions due to the interaction between strong oxygen surface groups and the ions of the electrolyte, setting aside the CVs curves from the purely capacitive shape.

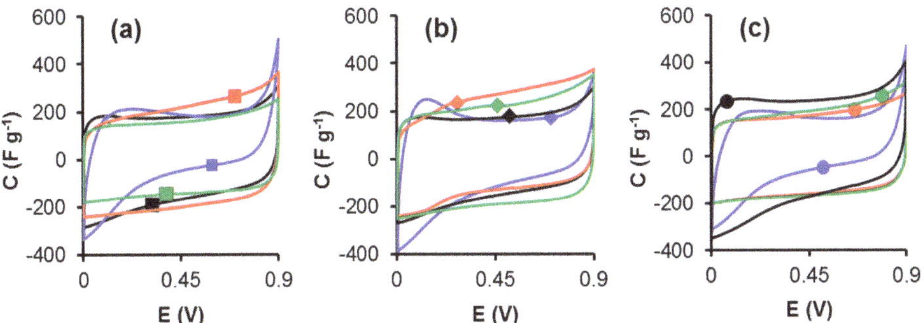

Figure 1. Cyclic voltammograms at 0.5 mVs^{-1} of all samples: (**a**) CK-series, (**b**) FK-series and (**c**) OK-series in H_2SO_4 1 M. Treatments: melamine (red), ammonium carbamate (green), nitric acid (blue), and ammonium persulfate (black).

Figure 2 shows the galvanostatic charge–discharge curves, it is clear that the curves of all oxidized samples exhibit a slightly distorted triangular shape, due to the pseudo-capacitive behavior of the oxygen functional groups. This distortion was more significant for the nitric acid treated samples. In contrary, the nitrogen doped samples show a symmetric triangular shape indicating a good diffusion inside the pore structure.

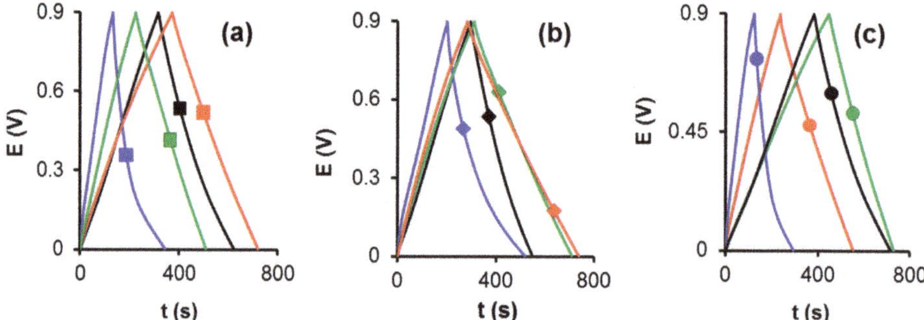

Figure 2. Galvanostatic charge-discharge curves at 125 mA g^{-1} for all samples: (**a**) CK-series, (**b**) FK-series and (**c**) OK-series. Treatments: melamine (red), ammonium carbamate (green), nitric acid (blue), and ammonium persulfate (black).

The specific capacitance versus current density for all samples was further collected to study the rate performance. The results are presented in Figure 3 and indicate that all the treatment process give samples with a good stability at higher current density of 10 A g^{-1} except for samples treated with nitric acid, which present a lower stability and a fast decreasing of the capacitance with the increasing of the current density. On the other hand, it is found that the samples oxidized by ammonium persulfate show the largest specific capacitances at current densities between 250 mA g^{-1} and 10 A g^{-1} for all modified samples except for FK series at high current densities. The maximum specific capacitance of 312 F g^{-1} was attained at a current density of 125 mA g^{-1} for OKS, which is much higher than those of FKS (237 F g^{-1}) and CKS (290 F g^{-1}).

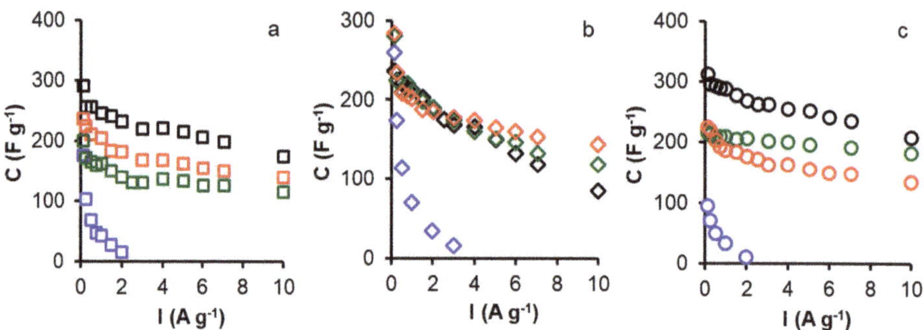

Figure 3. Variation of the specific capacitance with current density in H$_2$SO$_4$ 1 M for all samples: (**a**) CK-series, (**b**) FK-series, and (**c**) OK-series. Treatments: melamine (red), ammonium carbamate (green), nitric acid (blue), and ammonium persulfate (black).

Remarkably, the specific capacitance of ammonium persulfate treated samples cannot still be maintained at higher current density, which is due to the effect of oxygen functionalities. The retention ratio of those samples from 125 mA g^{-1} to 10 A g^{-1} was between 36% and 67%. These above results clearly show and confirm that the introduction of a large amount of oxygen-containing groups and the increased amount of mesopores are very effective enhancing the electrolyte accessibility, leading to fast ion response and higher capacitance and that the presence of surface quinone groups increases the capacitance of oxidized activated carbons by introducing pseudo-capacitance effects; nevertheless, the oxidation of the surface no always produced one way effects because also fixes carboxyl groups which thereby increasing its ohmic resistance [35] due to their high polarity, bind water molecules that hinder and retard electrolyte diffusion into the microporosity.

The nitric acid treated samples show a fast decreasing in the capacitance at higher current densities due to the destruction of the pore structure and also to the higher values of oxygen present in the surface functionalities of treated samples.

It is known that the increase in the population of the CO-desorbing complexes can has positive effect on the capacitance, while the CO_2-desorbing complexes show a negative effect in double-layer formation. Upon heat-treating the oxidized carbon, most of the CO_2-desorbing complexes were removed while the population of CO-desorbing complexes reached a maximum. This treatment has produced electrodes with the highest capacitance. Cyclic voltammetry showed that the presence of the CO desorbing complexes significantly enhanced the double-layer formation and thus the capacitance. This indicates that due to the local changes of electronic charge density a proton adsorbed by a carbonyl or quinone-type site facilitates an excess specific double-layer capacitance. The faradic current increased with the total number of oxygen atoms on the surface, indicating that both the CO- and CO_2-desorbing complexes enhanced the redox process.

It has been reported that surface N functionalities are electrochemically active because they are electron-rich [36,37]. In this way, protons can be attracted to the electrode surface, producing pseudo-capacitive interactions [38]. For the melamine nitrogen doped samples, the result indicate that the sample FKM presents the higher capacitance 283 F g^{-1} followed by CKM 236 F g^{-1} and OKM 225 F g^{-1}, this difference of the capacitance can be explained by the nitrogen content of all the samples (Table 3) which indicates that the samples FKM has more than 2% of nitrogen contrary to both the two others ones which have a nitrogen content of 1.6% for CKM and 1% for OKM. This can also be explained by the difference of surface area between all samples. The results (Table 2) indicate that those samples show good retention stability between 50 and 60% due to an adequate pore structure of all samples facilitating a good penetration of the electrolyte inside the porosity at high current densities.

For the ammonium carbamate modified samples, all the samples present the seam values of the nitrogen content 0.8%, so the difference of the capacitance and the electrochemical performance can be explained by the difference of the porous structure. The results (Table 4) indicate that the sample FKC has the higher capacitance of 280 F g^{-1} compared to both of the two other samples CKC 199 F g^{-1} and OKC 216 F g^{-1}. This difference can be explained by the difference in pore volume and especially the micropores volume. The sample FKC has a micropore volume of 0.67 cm^3 g^{-1} which is higher than both the other ones. Those samples present capacitance retention at current density of 10 A g^{-1} varying from 42% for FKC to 84% for OKC and 57% for CKC. This difference is due in one part to the pore diameter and also to the difference of oxygen amount. It is known that the presence of oxygen in the surface chemistry can affect the capacitance retention and especially at higher current densities. This remarkable electro-chemical performance, as the high specific capacitances can be due to the excellent and adequate textural characteristics which makes possible a large adsorption interface for electrolyte to form the electrical double layer, leading to a great electrochemical double layer capacitance; but also the doping with hetero-atoms enhances the surface interaction of these materials with the aqueous electrolyte, increasing the accessibility of electrolyte ions. The hetero-atoms groups can also provide considerable pseudo-capacitance improving the overall capacitance.

In order to get insights on the influence of surface chemistry on the electrochemical performance of the electrodes, electrochemical impedance spectroscopy (EIS) was used. The Nyquist plots and their corresponding equivalent circuits are shown in Figure 4 (the continuous line represents the adjustment of the experimental data to the equivalent circuit model). In order to obtain kinetic parameters, impedance data were fitting to the equivalent circuit proposed by Zhang et al. [39] using ZVIEW software, version 2.7, the results are shown in Table 5. The equivalent circuit proposed is based in a fractional-order model, which consists of a series resistor (R_s), a parallel resistor (R_{ct}), a Constant-Phase Element (CPE), and a Walburg-like element (W). The ionic resistance of the electrolyte, the intrinsic resistance of the active material, and the contact resistance at the electrode/current collector interface are contained in the R_s. Therefore an R_{ct} represents the faradic charge transfer resistance at the interface between the current collector and the active material. Finally, the W represents the diffusive resistance.

To evaluate the goodness of fit of the experimental data to the equivalent circuit, the statistic X_i^2 was used. We obtained values in the range of 1.0×10^{-3} for all of the fitted data (Table 5). Thus, we determined that the proposed equivalent circuit fits the experimental data reasonably well.

Table 4. Electrochemical capacitances ($F g^{-1}$) of modified samples in H_2SO_4 1 M. Retention capacitance at $10 A g^{-1}$ referred to $125 mA g^{-1}$.

Sample	Cv 0.5 mV s^{-1}	Ccp 125 mA g^{-1}	Ccp 2 A g^{-1}	Ccp 10 mA g^{-1}	Retention (%)
CKM	205	236	181	139	59
CKC	153	199	139	115	58
CKN	153	176	16	-	8*
CKS	180	290	231	174	60
FKM	200	283	186	144	51
FKC	210	280	190	119	43
FKN	169	259	35	-	14*
FKS	171	237	189	85	36
OKM	167	225	176	135	60
OKC	180	216	206	183	85
OKN	143	145	11	-	48*
OKS	218	312	268	208	67

* Retention capacitance at $2 A g^{-1}$.

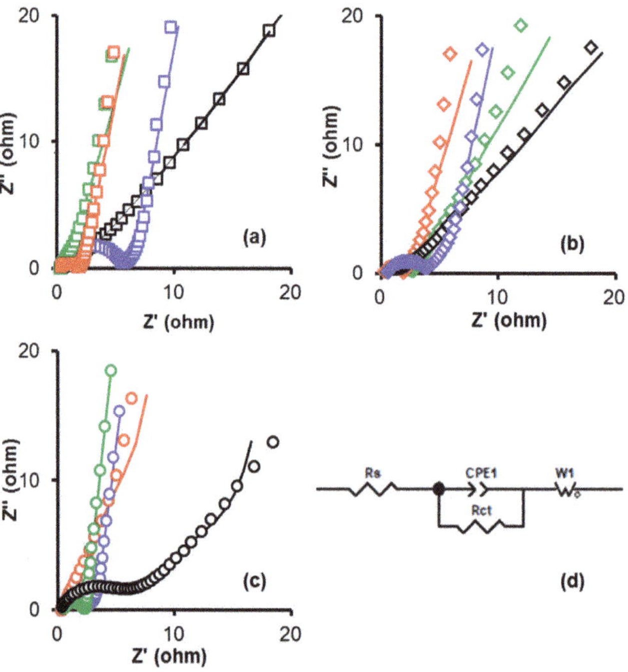

Figure 4. Nyquist plots obtained from EIS experiments on H_2SO_4 1 M for all the samples: (**a**) CK-series, (**b**) FK-series, (**c**) OK-series, and (**d**) equivalent circuit model. Treatments: melamine (red), ammonium carbamate (green), nitric acid (blue), and ammonium persulfate (black).

Once oxygen and nitrogen have been introduced to the material, R_s decreases dramatically, so that the formation is no longer perceptible in the figure, indicating a decrease in the formation of the solid-electrolyte interface layer. For all the samples treated with melamine the R_s value is the lowest (0.2 Ω), so that this effect can be attributed to a decrease in the formation of the solid-electrolyte interface layer due to the increase in the nitrogen functional groups which improve the hydrophobicity of the material, thus making the surface more wettable with electrolyte [40,41].

The charge transfer resistances (R_{ct}) of all nitrogen treated samples are varying between 0.05–2.18 Ω. A faster ion diffusion and lower impedance on the electrode/electrolyte are taking place, being this deduced from the small semicircles formed when the faradic charge transfer resistance (R_{ct}) of electrodes in H_2SO_4 electrolyte is represented. Those lower resistances due to the lower change in the pore structure after the melamine and ammonium carbamate treatments. In contrast, the oxygenated samples show a higher internal resistance compared to nitrogenated ones (for example: OKM = 0.05 Ω, OKC = 1.74 Ω vs. OKN = 5.04 Ω, OKS = 2.00 Ω). This higher resistance due to the lower pore diameter resultants from the treatment with nitric acid and ammonium persulfate. In addition, all curves show a Walburg-like element angle higher than 45°, indicating the suitability of the electrode materials for supercapacitors.

Table 5. Equivalent series resistance (R_s), charge transfer resistance (R_{CT}), and C_{max} at 1 mHz, from EIS.

Sample	R_s (Ω)	R_{ct} (Ω)	$X_i{}^2$	τ (s)	$C'max$ (F g^{-1})
CKM	0.20	1.45	2.1×10^{-3}	0.62	170
CKC	0.21	1.32	3.1×10^{-3}	0.62	134
CKN	0.30	4.45	5.0×10^{-3}	159.17	196
CKS	0.24	5.18	9.0×10^{-4}	1.98	118
FKM	0.25	1.48	3.7×10^{-3}	1.11	171
FKC	0.51	2.18	1.7×10^{-3}	3.55	177
FKN	0.42	1.40	1.5×10^{-3}	49.41	173
FKS	0.55	2.94	1.7×10^{-3}	1.48	166
OKM	0.29	0.05	1.4×10^{-3}	1.48	141
OKC	0.48	1.74	4.0×10^{-4}	0.46	157
OKN	0.38	5.04	3.9×10^{-3}	159.17	134
OKS	0.22	2.00	9.0×10^{-4}	0.62	141

The relaxation time constant (τ) is a quantitative measure of the speed with which the device can be discharged and this can be calculated using the equation $\tau = 1/(2\ f_0)$, being f_0 the transition frequency between a pure capacitive and a pure resistive behavior that can be obtained from the maximum within the variation of the imaginary part of the capacitance (C") against the frequency. Results collected in Table 5 show that nitrogen doped samples present the faster discharging time (1.3 s on average) compared to the oxygenated ones and that the acid nitric treated samples (61.97 s on average) due to the lower microporosity.

The Ragone plots of all the electrodes tested are displayed in Figure 5 and the maximum and minimum energies and power densities are shown in Table 6. When analyzing the results, it is clear that both the maximum energy density (E_{max}) and the maximum power density (P_{max}) have been considerably improved with the introduction of nitrogen groups on the surface, since the series of materials treated with melamine and sodium carbamate show better results (average E_{max} = 6.25 Wh kg^{-1}; average P_{max}, = 2363 W kg^{-1}) than the oxidized samples (average E_{max} = 1.89 Wh kg^{-1}; average P_{max}, = 854 W kg^{-1}). Additionally, these results are adequately related to the textural properties of the materials, since alkaline treatments increase the volume of mesopores. Since the power density is influenced by the transport of internal pore ions, this behavior can be explained to be due to the conditions of the porosity channels and the hydrophobicity of the surface.

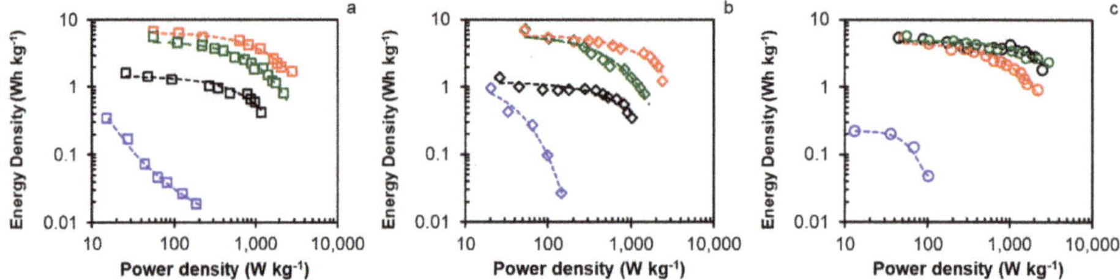

Figure 5. Ragone plots on H_2SO_4 1 M for all samples: (**a**) CK-series, (**b**) FK-series, and (**c**) OK-series. Treatments: melamine (red), ammonium carbamate (green), nitric acid (blue), and ammonium persulfate (black).

Table 6. Maximum and minimum energy densities (Wh Kg^{-1}) and power densities (W Kg^{-1}) of all samples from Ragone's plots.

Sample	P_{max} W kg^{-1}	E_{min} Wh kg^{-1}	E_{max} Wh kg^{-1}	P_{min} W kg^{-1}
CKM	2825	1.72	6.61	57
CKC	2245	0.81	5.57	57
CKN	184	0.02	0.35	15
CKS	1180	0.42	1.64	26
FKM	2396	1.23	6.90	53
FKC	1461	0.79	7.10	54
FKN	147	0.03	0.95	21
FKS	997	0.12	1.40	26
OKM	2214	0.92	5.44	53
OKC	3039	2.35	5.88	56
OKN	111	0.01	0.22	13
OKS	2506	1.82	5.41	45

The energy released decreased at higher power density; but the energy densities of nitrogen doped samples were acceptable at higher power density. The long-term stability of electrodes is a very important property that can limit the application of any materials as supercapacitors for practical applications. Figure 6 shows the variation in the gravimetric capacitance with the number of charge–discharge cycles at a constant current density of 1 A g^{-1} employing H_2SO_4 1 M as electrolyte. After 10,000 cycles the retention capacity for the all modified samples are between 97.0%, and 100%.

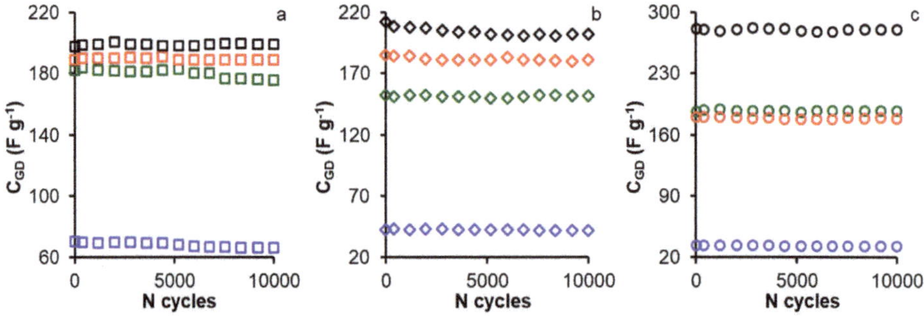

Figure 6. Variation of the gravimetric capacitance (C_{GD}) with the number of charge discharge cycles at 1 A g^{-1} in the potential window between 0 and 0.9 V in H_2SO_4 1 M for all samples: (**a**) CK-series, (**b**) FK-series, and (**c**) OK-series. Treatments: melamine (red), ammonium carbamate (green), nitric acid (blue), and ammonium persulfate (black).

4. Conclusions

In this study, three series of ACs were prepared by KOH activation of different woods: Custard apple tree (CK-series), Fig tree (FK-series) and Olive tree (OK-series). The ACs were treated with four agents: melamine, ammonium carbamate, nitric acid, and ammonium persulfate, for the introduction of nitrogen and oxygen groups on the surface of materials with the aim to study the influence of surface chemistry on the electrochemical performance of biomass-derived carbon electrodes for its use as supercapacitors. The results showed that the treatments introduce different nitrogen functionalities, such as pyridine quaternary-N and oxidized nitrogen, which improve the wettability and the ions transfers. The effectiveness of the activation and doping methods is very good, obtaining comparable materials in porosity and relative chemical properties, in spite of the different origins of the woods. The obtained electro-chemical results are also very remarkable, since after 10,000 cycles, the retention capacity for the all modified samples are between 97.0% and 100%.; with the advantage that very cheap waste materials can be used for the supercapacitor development. Nevertheless, it should be clarified that the treatment with nitric acid, although it is also very reproducible in its effects, is not an advisable doping treatment because it destroys the microporosity and, therefore, reduces the electrochemical performance. Finally, the high electro-chemical performance, such as the very remarkable specific capacitances, can be probably due to the excellent and adequate textural characteristics, as high surface areas, which makes possible a large adsorption interface for electrolyte to form the electrical double layer, leading to a great electrochemical double layer capacitance; as well as the doping with hetero-atoms which enhances the surface interaction of these materials with the aqueous electrolyte, increasing the accessibility of electrolyte ions.

Supplementary Materials: The following are available online at http://www.mdpi.com/1996-1944/12/15/2458/s1, Figure S1. N_2 adsorption and desorption isotherms at 77K of CK-series samples. Figure S2: High resolution XPS deconvoluted spectra in the corresponding regions: (a) C_{1s}, (b) O_{1s}, (c) N_{1s} and (d) $S2p_{3/2}$ for the activated carbons prepared from Custard apple tree wood (CK-Serie). Figure S3: High resolution XPS deconvoluted spectra in the corresponding regions: (a) C_{1s}, (b) O_{1s}, (c) N_{1s} and (d) $S2p_{3/2}$ for the activated carbons prepared from Fig tree wood (FK-Serie). Figure S4: High resolution XPS deconvoluted spectra in the corresponding regions: (a) C_{1s}, (b) O_{1s}, (c) N_{1s} and (d) $S2p_{3/2}$ for the activated carbons prepared from Olive tree wood (OK-Serie).

Author Contributions: Conceptualization: F.C.M. & A.F.P.C.; Methodology: A.E. and E.B.G.; Analysis and discussion of results: All the authors; EIS data analysis: F.C.M. & A.I.Z.G.; Writing–All the authors; Project Administration: A.F.P.C.; Funding Acquisition: F.C.M & A.F.P.C

Acknowledgments: This work was supported by FEDER and Spanish MINECO (grant number CTQ-2013-44789-R); and Junta de Andalucía (grant numbers P12-RNM-2892, RNM172).

References

1. Luo, X.; Wang, J.; Dooner, M.; Clarke, J. Overview of current development in electrical energy storage technologies and the application potential in power system operation. *Appl. Energy* **2015**, *137*, 511–536. [CrossRef]
2. Faraji, S.; Ani, F.N. The development supercapacitor from activated carbon by electroless plating—A review. *Renew. Sustain. Energy Rev.* **2015**, *42*, 823–834. [CrossRef]
3. Weinstein, L.; Dash, R. Supercapacitor carbons: Have exotic carbons failed? *Mater. Today* **2013**, *16*, 356–357. [CrossRef]
4. Lin, Z.; Goikolea, E.; Balducci, A.; Naoi, K.; Taberna, P.L.; Salanne, M.; Yushin, G.; Simon, P. Materials for supercapacitors: When Li-ion battery power is not enough. *Mater. Today* **2018**, *21*, 419–436. [CrossRef]
5. Kostoglou, N.; Koczwara, C.; Prehal, C.; Terziyska, V.; Babic, B.; Matovic, B.; Constantinides, G.; Tampaxis, C.; Charalambopoulou, G.; Steriotis, T.; et al. Nanoporous activated carbon cloth as a versatile material for hydrogen adsorption, selective gas separation and electrochemical energy storage. *Nano Energy* **2017**, *40*, 49–64. [CrossRef]

6. Lee, J.; Jäckel, N.; Kim, D.; Widmaier, M.; Sathyamoorthi, S.; Srimuk, P.; Kim, C.; Fleischmann, S.; Zeiger, M.; Presser, V. Porous carbon as a quasi-reference electrode in aqueous electrolytes. *Electrochim. Acta* **2016**, *222*, 1800–1805. [CrossRef]

7. Yakout, S.M.; El-Deen, G.S. Characterization of activated carbon prepared by phosphoric acid activation of olive stones. *Arab. J. Chem.* **2016**, *9*, S1155–S1162. [CrossRef]

8. Ubago-Pérez, R.; Carrasco-Marín, F.; Fairén-Jiménez, D.; Moreno-Castilla, C. Granular and monolithic activated carbons from KOH-activation of olive stones. *Microporous Mesoporous Mater.* **2006**, *92*, 64–70. [CrossRef]

9. Moreno-Castilla, C.; Carrasco-Marín, F.; López-Ramón, M.V.; Alvarez-Merino, M.A. Chemical and physical activation of olive-mill waste water to produce activated carbons. *Carbon* **2001**, *39*, 1415–1420. [CrossRef]

10. Moreno-Castilla, C.; García-Rosero, H.; Carrasco-Marín, F. Symmetric supercapacitor electrodes from KOH activation of pristine, carbonized, and hydrothermally treated Melia azedarach stones. *Materials* **2017**, *10*, 745. [CrossRef]

11. Elmouwahidi, A.; Zapata-Benabithe, Z.; Carrasco-Marín, F.; Moreno-Castilla, C. Activated carbons from KOH-activation of argan (Argania spinosa) seed shells as supercapacitor electrodes. *Bioresour. Technol.* **2012**, *111*, 185–190. [CrossRef]

12. Yin, L.; Chen, Y.; Li, D.; Zhao, X.; Hou, B.; Cao, B. 3-Dimensional hierarchical porous activated carbon derived from coconut fibers with high-rate performance for symmetric supercapacitors. *Mater. Des.* **2016**, *111*, 44–50. [CrossRef]

13. Arena, N.; Lee, J.; Clift, R. Life Cycle Assessment of activated carbon production from coconut shells. *J. Clean. Prod.* **2016**, *125*, 68–77. [CrossRef]

14. Contreras, M.S.; Páez, C.A.; Zubizarreta, L.; Léonard, A.; Blacher, S.; Olivera-Fuentes, C.G.; Arenillas, A.; Pirard, J.P.; Job, N. A comparison of physical activation of carbon xerogels with carbon dioxide with chemical activation using hydroxides. *Carbon* **2010**, *48*, 3157–3168. [CrossRef]

15. Demiral, H.; Demiral, I.; Karabacakoĝlu, B.; Tümsek, F. Production of activated carbon from olive bagasse by physical activation. *Chem. Eng. Res. Des.* **2011**, *89*, 206–213. [CrossRef]

16. Hou, S.; Wang, M.; Xu, X.; Li, Y.; Li, Y.; Lu, T.; Pan, L. Nitrogen-doped carbon spheres: A new high-energy-density and long-life pseudo-capacitive electrode material for electrochemical flow capacitor. *J. Colloid Interface Sci.* **2017**, *491*, 161–166. [CrossRef]

17. He, D.; Niu, J.; Dou, M.; Ji, J.; Huang, Y.; Wang, F. Nitrogen and oxygen co-doped carbon networks with a mesopore-dominant hierarchical porosity for high energy and power density supercapacitors. *Electrochim. Acta* **2017**, *238*, 310–318. [CrossRef]

18. Hsiao-Hsuan, S.; Chi-Chang, H. Capacitance Enhancement of Activated Carbon Modified in the Propylene Carbonate Electrolyte. *J. Electrochem. Soc.* **2014**, *161*, A1828–A1835. [CrossRef]

19. Kakhki, R.M.Z.; Heydari, S. A simple conductometric method for trace level determination of brilliant green in water based on β-cyclodextrin and silver nitrate and determination of their thermodynamic parameters. *Arab. J. Chem.* **2013**, *7*, 1086–1090. [CrossRef]

20. Zhang, Y.; Li, X.; Huang, J.; Xing, W.; Yan, Z. Functionalization of Petroleum Coke-Derived Carbon for Synergistically Enhanced Capacitive Performance. *Nanoscale Res. Lett.* **2016**, *11*, 163. [CrossRef]

21. Leng, C.; Sun, K.; Li, J.; Jiang, J. The reconstruction of char surface by oxidized quantum-size carbon dots under the ultrasonic energy to prepare modified activated carbon materials as electrodes for supercapacitors. *J. Alloys Compd.* **2017**, *714*, 443–452. [CrossRef]

22. Li, J.; Liu, W.; Xiao, D.; Wang, X. Oxygen-rich hierarchical porous carbon made from pomelo peel fiber as electrode material for supercapacitor. *Appl. Surf. Sci.* **2017**, *416*, 918–924. [CrossRef]

23. Pels, J.R.; Kapteijn, F.; Moulijn, J.A.; Zhu, Q.; Thomas, K.M. Evolution of nitrogen functionalities in carbonaceous materials during pyrolysis. *Carbon* **1995**, *33*, 1641–1653. [CrossRef]

24. Moreno-Castilla, C.; Ferro-García, M.A.; Joly, J.P.; Bautista-Toledo, I.; Carrasco-Marín, F.; Rivera-Utrilla, J. Activated Carbon Surface Modifications by Nitrcc Acid, Hydrogen Peroxide, and Ammonium Peroxydisulfate Treatmenss. *Langmuir* **1995**, *11*, 4386–4392. [CrossRef]

25. Conway, B.E. AC Impedance Behavior of Electrochemical Capacitors and Other Electrochemical Systems. In *Book Electrochemical Supercapacitors: Scientific Fundamentals and Technological Applications*, 1st ed.; Springer: Boston, MA, USA, 2013; Volume 1, pp. 479–524. [CrossRef]

26. Elmouwahidi, A.; Bailón-García, E.; Pérez-Cadenas, A.F.; Maldonado-Hódar, F.J.; Carrasco-Marín, F. Activated carbons from KOH and H_3PO_4-activation of olive residues and its application as supercapacitor electrodes. *Electrochimi. Acta* **2017**, *229*, 219–228. [CrossRef]

27. Figueiredo, J.; Pereira, M.F.; Freitas, M.M.; Órfão, J.J. Modification of the surface chemistry of activated carbons. *Carbon* **1999**, *37*, 1379–1389. [CrossRef]

28. Moreno-Castilla, C.; López-Ramón, M.V.; Carrasco-Marín, F. Changes in surface chemistry of activated carbons by wet oxidation. *Carbon* **2000**, *38*, 1995–2001. [CrossRef]

29. Gorgulho, H.F.; Gonçalves, F.; Pereira, M.F.R.; Figueiredo, J.L. Synthesis and characterization of nitrogen-doped carbon xerogels. *Carbon* **2009**, *47*, 2032–2039. [CrossRef]

30. Stańczyk, K.; Dziembaj, R.; Piwowarska, Z.; Witkowski, S. Transformation of nitrogen structures in carbonization of model compounds determined by XPS. *Carbon* **1995**, *33*, 1383–1392. [CrossRef]

31. Raymundo-Piñero, E.; Cazorla-Amorós, D.; Linares-Solano, A.; Find, J.; Wild, U.; Schlögl, R. Structural characterization of N-containing activated carbon fibers prepared from a low softening point petroleum pitch and a melamine resin. *Carbon* **2002**, *40*, 597–608. [CrossRef]

32. Jurewicz, K.; Babeł, K.; Pietrzak, R.; Delpeux, S.; Wachowska, H. Capacitance properties of multi-walled carbon nanotubes modified by activation and ammoxidation. *Carbon* **2006**, *44*, 2368–2375. [CrossRef]

33. Ando, R.A.; Do Nascimento, G.M.; Landers, R.; Santos, P.S. Spectroscopic investigation of conjugated polymers derived from nitroanilines. *Spectrochim. Acta Part A* **2008**, *69*, 319–326. [CrossRef]

34. Xie, W.; Weng, L.T.; Chan, C.K.; Yeung, K.L.; Chan, C.M. Reactions of SO2 and NH3 with epoxy groups on the surface of graphite oxide powder. *Phys. Chem. Chem. Phys.* **2018**, *20*, 6431–6439. [CrossRef]

35. Guo, Y.; Qi, J.; Jiang, Y.; Yang, S.; Wang, Z.; Xu, H. Performance of electrical double layer capacitors with porous carbons derived from rice husk. *Mater. Chem. Phys.* **2003**, *80*, 704–709. [CrossRef]

36. Frackowiak, E. Carbon materials for supercapacitor application. *Phys. Chem. Chem. Phys.* **2007**, *9*, 1774–1785. [CrossRef]

37. Rufford, T.E.; Hulicova-Jurcakova, D.; Zhu, Z.; Lu, G.Q. Nanoporous carbon electrode from waste coffee beans for high performance supercapacitors. *Electrochem. Commun.* **2008**, *10*, 1594–1597. [CrossRef]

38. Hulicova, D.; Yamashita, J.; Soneda, Y.; Hatori, H.; Kodama, M. Supercapacitors prepared from melamine-based carbon. *Chem. Mater.* **2005**, *17*, 1241–1247. [CrossRef]

39. Zhang, L.; Hu, X.; Wang, Z.; Sun, F.; Dorrell, D.G. A review of supercapacitor modeling, estimation, and applications: A control/management perspective. *Renew. Sustain. Energy Rev.* **2018**, *81*, 1868–1878. [CrossRef]

40. Liu, X.; Wang, Y.; Zhan, L.; Qiao, W.; Liang, X.; Ling, L. Effect of oxygen-containing functional groups on the impedance behavior of activated carbon-based electric double-layer capacitors. *J. Solid State Electrochem.* **2011**, *15*, 413–419. [CrossRef]

41. Zárate-Guzmán, A.I.; Manríquez-Rocha, J.; Antaño-López, R.; Rodríguez-Valadez, F.J.; Godínez, L.A. Study of the Electrical Properties of a Packed Carbon Bed for Its Potential Application as a 3D-Cathode in Electrochemical Processes. *J. Electrochem. Soc.* **2018**, *165*, E460–E465. [CrossRef]

 *materials* 

Communication

Novel Synthesis of Nitrogen-Containing Bio-Phenol Resin and Its Molten Salt Activation of Porous Carbon for Supercapacitor Electrode

Tao Ai [1,2,*], Zhe Wang [1,*], Haoran Zhang [1], Fenghua Hong [1], Xin Yan [1] and Xinhua Su [1]

[1]   School of Materials Science & Engineering, Chang'an University, Xi'an 710061, China;
      solarann@163.com (H.Z.); 2017131041@chd.edu.cn (F.H.); xinyan@chd.edu.cn (X.Y.);
      suxinghua@chd.edu.cn (X.S.)
[2]   Engineering Research Center of Pavement Materials, Ministry of Education of P.R. China,
      Chang'an University, Xi'an 710061, China
*   Correspondence: aitao@chd.edu.cn (T.A.); 18729608722@163.com (Z.W.)

Received: 6 May 2019; Accepted: 19 June 2019; Published: 20 June 2019

Abstract: Nitrogen hybridization is an attractive way to enhance the wettability and electric conductivity of porous carbon, which increases the capacitance of carbon-based supercapacitor, however, there is lack of low-cost methods to prepare the nitrogen-doped porous carbon materials. Herein, a novel facile nitrogen-containing bio-phenolic resin was synthesized by polymerization of the carbamate bio-oil, Phenol and paraformaldehyde. As a precursor of nitrogen-doped porous carbon, the nitrogen-containing bio-phenol resin was activated by the one-step molten-salt method. The resultant nitrogen-doped porous carbon showed a high specific surface area up to 1401 m^2·g^{-1}. As a supercapacitor electrode, the nitrogen-doped porous carbons showed specific capacitance of 159 F·g^{-1} at 0.5 A·g^{-1}. It also exhibited high cyclic stability with 94.8% retention of the initial specific capacitance over 1000 charge-discharge cycles at 1.0 A·g^{-1}. The results suggest that these nitrogen-containing bio-phenol resin provide a new source of nitrogen-doped porous carbon for high-performance supercapacitor electrodes.

Keywords: nitrogen-doped; bio-phenol resin; porous carbon; molten salt; supercapacitor; electrode material

---

1. Introduction

A supercapacitor is a new generation of energy storage device, and its core component is electrode material. Among the electrode materials, porous carbon materials were the first to be studied and the most mature for technical application [1,2]. The incorporation of nitrogen into the porous carbon structure can improve its wettability, conductivity, and increases its specific capacitance [3]. There are two ways to incorporate the nitrogen element into the porous carbons [4]. One is the treatment of porous carbon by nitrogenous compound at high temperature; The other is carbonization of nitrogen-containing carbon precursors, for example, nitrogen-containing phenolic resins. Because the latter have higher nitrogen content and more stable cycling stability in super capacitors than the former, there is a lot of literature on nitrogen-containing phenolic resins for preparing nitrogen-doped porous carbons [5–9]. However, the preparation of nitrogen-containing phenolic resin is not effective, which hinders the mass production of nitrogen-doped porous carbons.

In recent years, with the development of sustainable chemical technology, the cheap bio-oil, which is generally produced by fast pyrolysis of rich lignocellulosic biomass, can often be used as partial substitute of phenol to synthesize bio-based phenolic resin [10–15]. Bio-oil provides a promising renewable resource to substitute petroleum-based phenol; however, few studies have used phenol-rich bio-oil to synthesize nitrogen-containing phenolic resin.

In this study, we present a novel synthesis of nitrogen-containing phenol resin by amino esterification bio-oil. The nitrogen-containing phenolic resin was facilely synthesized by phenol, formaldehyde and the amino esterification bio-oil, then, the bio-phenol resin as a precursor of nitrogen-doped porous carbon was simultaneously carbonized and activated in molten salt with one-step method. The nitrogen-doped porous carbons show excellent supercapacitor performance.

2. Materials and Methods

2.1. Chemicals and Materials

Bio-oil was produced from rapid pyrolysis of Poplar Sawdust at 500 °C. Urea, Phenol and Paraformaldehyde were analytical grade and purchased from Shanghai Macklin Biochemical Co., Ltd. (Shanghai, China).

2.2. Synthesis of Nitrogen-Containing Bio-Phenolic Resin

The 50 g of bio-oil was added into a 3-neck boiling flask. The flask was heated by an oil bath to 130 °C until melted. An appropriate amount of urea (bio-oil to mass urea ratio of 100:50) was added and stirred evenly then heated to 150 °C and held until no ammonia escaped. After the reaction finished, the products were naturally cooled to room temperature and the black bitumen solid, called carbamate bio-oil, was obtained. A specific amount of carbamate bio-oil was added into the boiling flask, heated by oil bath to 90 °C. Phenol and poly formaldehyde (50% of the quality of carbamate bio-oil) was added. The mixture was held at 90 °C for 5 h. The product became a bio-phenolic resin after cooling.

2.3. Preparation of Nitrogen-Doped Porous Carbon

The bio-phenolic resin was cured at 100 °C for 1 h then at 180 °C for 2 h. The 3 g of heat cured bio-phenolic resin was weighed and mixed with a precise molar ratio of NaCl-KCl-KOH (NaCl:KCl:KOH = 4:4:1) salt. The mixture was placed in a tube furnace and heated from room temperature to 900 °C in a high purity N_2 atmosphere. The mixture was held at 900 °C for 4 h before being allowed to cool to room temperature. The product was washed with a 0.1 mL/L solution of HCL and Deionized water repeatedly until the filtrate neutralized. The product was then dried at 80 °C for 12 h to form the porous carbon.

2.4. Characterization of Resin and Porous Carbon

The molecular structure of the bio-oil after amino esterification was determined by an infrared (IR) spectrometer (Nicolet IS10, Thermo Scientific, Waltham, MA, USA). The thermogravimetric (TG) analysis of the phenolic resin was carried out by the Thermogravimetric Analyzer (Netzsch TG209F1Libra®, Selb, German). The nitrogen content of phenolic resins was characterized by the elemental analyzer (Vario EL cube, Elementar, Langenselbold, German). The morphology of the specimens was determined using emission scanning electron microscopy (SEM) (Hitachi S-4800, Tokyo, Japan). The XRD patterns of specimens were investigated with a powder diffractometer (Bruker D8 Davinci, Leipzig, Germany). Raman spectra were recorded on a Raman spectrometer (JY HR800, Horiba, Montpellier, France). The pore structure of the specimen was determined by Nitrogen adsorption-desorption isotherms at 77 K on an Automatic adsorption instrument (Mike ASAP2460, Micromeritics Instrument Corp., Norcross, GA, USA).

2.5. Preparation of Working Electrode and Electrochemical Analysis

The working electrode was prepared as follow: The porous carbon material (80 wt.%), acetylene black (10 wt.%) and polytetrafluoroethylene (10 wt.%) were used to prepare a uniform paste. The paste was then coated on a nickel foam current collector.

The electrochemical analysis of the specimen was carried out using a three-electrode configuration on the electrochemical workstation (CHI CHI660E, Shanghai, China) in a 6 mol/L KOH electrolyte with a range of −1 to 0 V.

To further evaluate the porous carbon as an electrode in symmetrical supercapacitor device, a CR2032 coin-type cell was assembled using the porous carbon as symmetrical electrodes, using a separator with the electrolyte of 6 M KOH solution. The performance of the device in terms of its energy density (E) and power density (P), which can be estimated using the following equations: $E = 1/2CU^2$, $P = E/\Delta t$, where C represents the specific capacitance based on the galvanostatic charge-discharge results of supercapacitor, while U refers to the potential change within the discharge time Δt.

3. Results and Discussion

3.1. IR and TG Analysis

In order to testify the chemical structure change, carbamate bio-oil was characterized by IR. Figure 1a is the result of IR spectra of carbamate bio-oil. Compared with bio-oil, the carbonyl absorption peak of 1700 cm^{-1} increased noticeably after amino esterification. The 3180–3360 cm^{-1} absorption peak is the symmetric and asymmetric vibration of amino N-H, showing two adjacent strong absorption peaks, while in the bio-oil there is only the absorption peak of hydroxyl in 3300–3500 cm^{-1}. Based on the above analysis, it can be seen that carbamate was introduced, there are reactive amides, and the N element was successfully introduced.

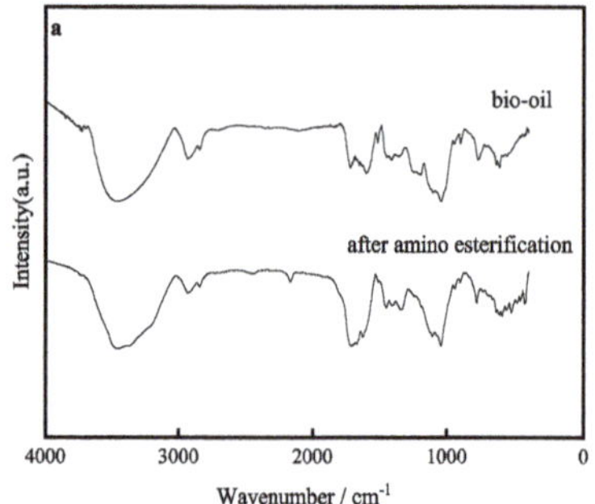

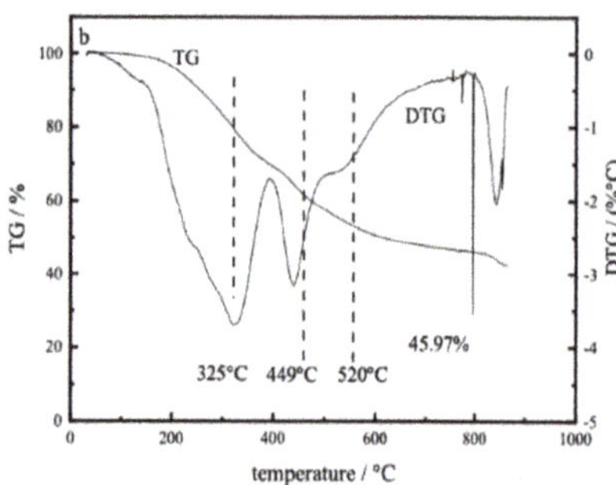

Figure 1. (**a**) IR spectra of bio-oil before and after carbamate; (**b**) TG-DTG curves of bio-phenolic resins.

Figure 1b is the result of thermogravimetry-derivative thermogravimetry (TG-DTG) curves of bio-phenolic resins. It can be seen that with the amination of bio-oil, the char yield of the bio-phenolic resin is nearly 46% at 800 °C. The higher char yield of the bio-phenolic resin means more resin carbon. These would good for the mass production of nitrogen-doped porous carbons.

3.2. SEM and Elemental Analysis

Molten salt one-step activation is a preparation method for activated porous carbon. Figure 2 is the SEM images of molten salt activated porous carbon specimen derived from bio-phenolic resins. It can be seen that the specimen has an abundant porous structure. Under a high-temperature molten salt environment, a resin precursor can be activated with a KOH activator. In the process of activation, the KOH in the molten salt system reacts with some of the carbon atoms, such as at the edge of the defect structure of the carbon atoms connected with the heteroatoms [16]. This activation makes the pore structure of the carbon material further developed. The microstructure of the activated carbon is changed by the complex carbonization and activation [17] so that it has a microporous and mesoporous structure.

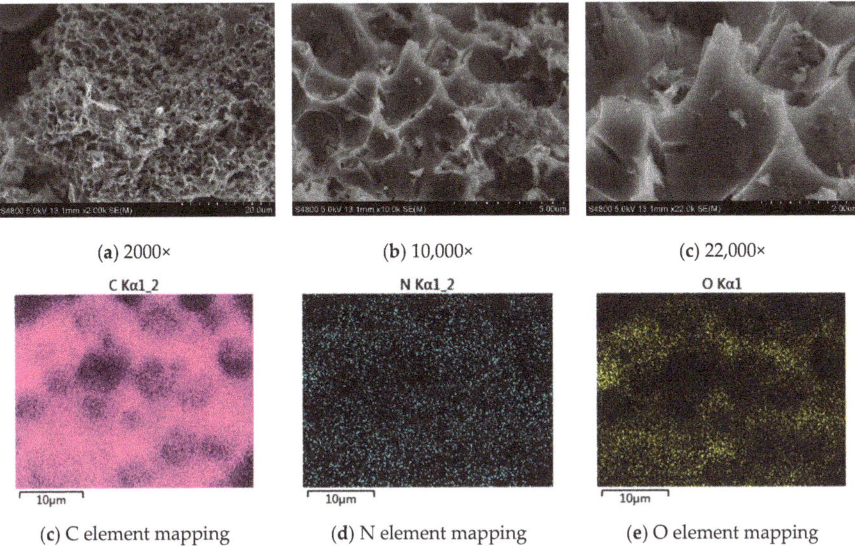

(**a**) 2000× (**b**) 10,000× (**c**) 22,000×

C Kα1_2 N Kα1_2 O Kα1

(**c**) C element mapping (**d**) N element mapping (**e**) O element mapping

Figure 2. Scanning electron microscopy (SEM) with different magnifications and element mapping images of porous carbon. (**a**) 2000×; (**b**) 10,000×; (**c**) 22,000×; (**d**) C element mapping; (**e**) N element mapping; (**f**) O element mapping.

The content of nitrogen in the bio-phenolic resins is important to a property of porous carbon. After testing, the nitrogen content in the bio-phenolic resins was as high as 9%. When these bio-phenolic resins were changed into activated carbon, the nitrogen content of the activated carbon still remains 5.6%. The mapping images of C, N and O elements of the activated carbon are shown in Figure 2. It is known that the elements C, N and O are distributed in large quantities evenly throughout the materials. The result indicates that the N and O elements in the bio-phenolic resin can be retained and the activated carbon is nitrogen-doped porous carbon.

3.3. XRD X-ray Diffraction and Raman Spectral Analysis

The XRD patterns of the nitrogen-doped porous carbon was shown in Figure 3a. It was showed that the carbon has a broad "steamed bun" diffraction peak at the 28° and 42°. The result represents the existence of the amorphous carbon and graphite structure [18]. Figure 3b is the Raman spectra of the nitrogen-doped porous carbon. It displayed apparent D and G peaks. The D peak is the characteristic absorption peak of amorphous carbon. The G peak is considered as the absorption peak of the graphite structure [19]; therefore, it is confirmed that the nitrogen-doped porous carbon belongs to the amorphous carbon and has a higher degree of graphitization.

3.4. Surface Area and Porosity Determination Using N_2 Adsorption

Figure 4a is the N_2 absorption-desorption isotherms of the porous carbon. The isotherms show the characteristics of the typical type I isotherms, indicating that the pore distribution is mainly microporous, and has a lower mesoporosity [20,21]. The pore size distribution of the porous carbon is shown in Figure 4b. It is further concluded that the pore size is mainly concentrated below 2 nm. Because of this fact, the activation with KOH produces a large number of micropores, which improves the specific capacitance of the carbon [22], we use this method to produce the porous carbon. The obtained carbon reached the specific surface area and pore volume of 1401 m^2/g and 0.61 cm^3/g (Table 1), respectively.

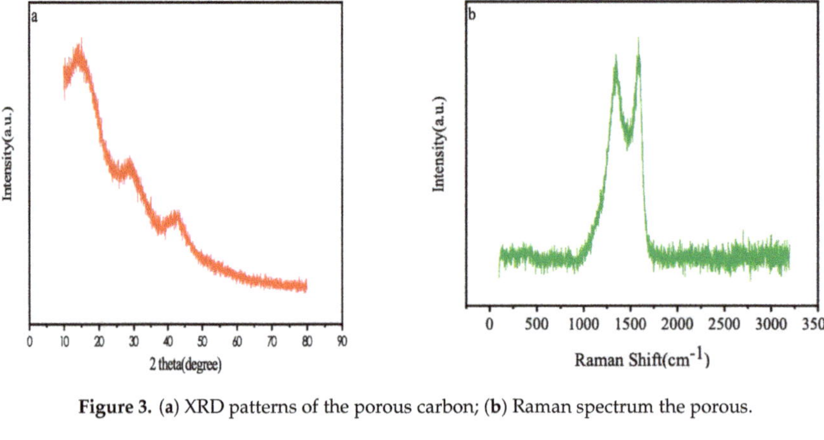

Figure 3. (**a**) XRD patterns of the porous carbon; (**b**) Raman spectrum the porous.

Figure 4. (**a**) Nitrogen adsorption-desorption isotherms; (**b**) pore size distributions.

Table 1. Textural properties of the carbon materials.

S_{BET} [a] (m^2/g)	S_{micro} [b] (m^2/g)	V_{total} [c] (cm^3/g)	V_{micro} [d] (cm^3/g)	D [e] (nm)
1401	1132	0.609	0.453	1.768

a = Brunauer-Emmett-Teller (BET) surface area. b = Micropore surface area, derived from the t-plot method. c = Total pore volume, measured at P/P0 = 0.98. d Micropore volume, derived from the Dubinin-Astakhov method e Micropore average diameter, calculated by the Barret-Joyner-Halenda (BJH) method.

3.5. Analysis of Electrochemical Energy Storage

Figure 5a is the cyclic voltammetry (CV) curve of the specimen at different scanning rates. It can be seen that the CV curve of the specimen presents a rectangular shape under cycling rate of 50 mV/s, indicating that the electric double-layer provides most of the capacitance. It also shows that the electrode has better conductivity and higher current response. With the sweep speed increasing to the 200 mv/s, the corresponding curve still has no obvious distortion and keeps the approximate rectangular shape. A gradual increase in the area of the curve shows the good electrochemical characteristics of the material [23].

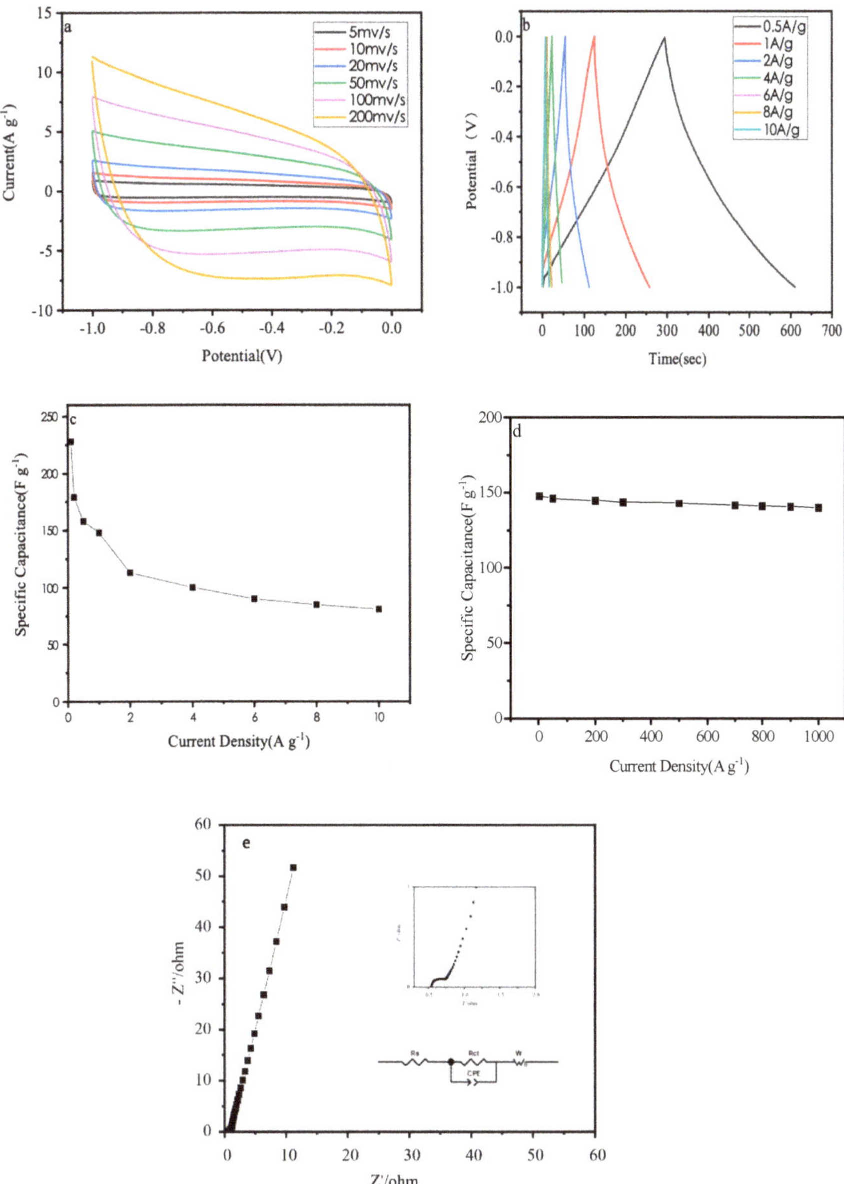

Figure 5. (**a**) cyclic voltammetry (CV) curves of porous carbon at various scan rates; (**b**) galvanostatic charge/discharge curves of porous carbon under various current densities; (**c**) specific capacitance versus current density of porous carbon; (**d**) cycling stability of porous carbon at 1A/g; (**e**) Nyquist plots.

Figure 5b is a galvanostatic charge/discharge curve with different current densities. The curve approximates an isosceles triangle. Because it shows excellent reversible and charge/discharge performance, this porous carbon can be used as electrode material for a supercapacitor. When the current density is 0.5, 1.0, 2.0, 6.0, 8.0 and 10.0 A/g, the specific capacitance is 159, 148, 113, 90, 85 and

81 F/g respectively. With the increase of current density, the capacitance began to show a certain degree of decline.

Figure 5c is the specific capacitance variation under different current densities. Because the porous structure of specimens is mainly microporous and less mesoporous, when the current density increases from 0.5 A/g to 10 A/g, the specific capacitance drops from 159 F/g to 81 F/g, and the capacitance retention rate is only 52.6%. Mesoporous carbon provides a channel for ion migration. At high current density, ions can migrate quickly in the channel, thus increasing the capacitance retention rate. The low ratio of mesoporous carbon in these specimens lead to unsatisfactory capacitance retention rate [24].

In order to test the cyclic stability of the specimen, 1.0 A/g current density is used to charge/discharge 1000 times. The capacitance retention curve is shown in Figure 5d. After 1000 cycles, the specific capacitance remained of 94.8%. This curve shows that the nitrogen-doped porous carbon electrode has excellent cycling stability [25].

The electrochemical impedance spectroscopy measurement is a useful method to test the conductivity of electrode materials. Measurement results can be shown by Nyquist plot. Figure 5e is the Nyquist plot of the carbon electrodes. It can be seen from the figure that the impedance curve in the high frequency region is a semicircle, reflecting the charge transfer process at the electrode/electrolyte interface. The equivalent impedance simulation analysis is performed on the measured impedance map. As shown in the figure, the RESR value (0.535 Ω) is indicated that the electrode material has a low internal resistance [26].

A two carbon electrode symmetrical supercapacitor was assembled to evaluate its electrochemical performance. Figure 6 shows the Ragone plot for the symmetrical supercapacitor with the calculated power density and energy density. Compared to other types of activated carbon reported in the literature [27–30], when the current density is 0.5 A/g, the device shows a good energy density of 6.11 Wh/kg at a power density of 258 W/kg.

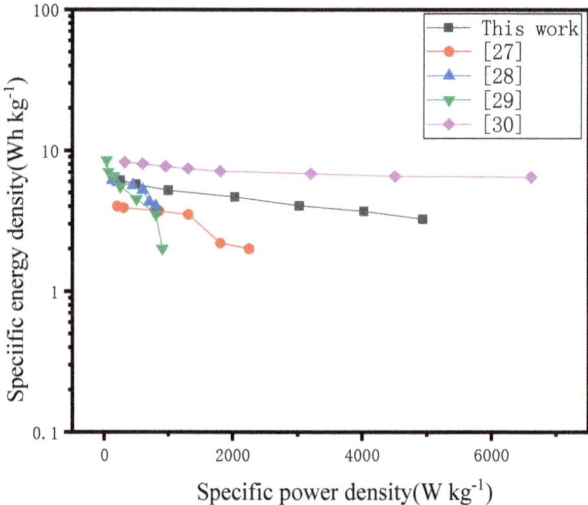

Figure 6. Ragone plot related to energy and power densities of carbon supercapacitor.

4. Conclusions

This study provides a novel facile synthesis method of nitrogen-containing bio-phenolic resin. The nitrogen-containing bio-phenolic resin is an ideal precursor for preparation of nitrogen-doped porous carbon. Future work should focus on enlarging the capacity of nitrogen-doped mesoporous carbon. The nitrogen-doped porous carbon electrode has excellent cycling stability and could be used for a wide range of applications in supercapacitor.

Author Contributions: Conceptualization, T.A.; methodology, F.H. and Z.W.; validation, X.S. and X.Y.; formal analysis, T.A.; investigation, Z.W.; resources, H.Z.; data curation, Z.W. and H.Z.; writing—original draft preparation, T.A.; writing—review and editing, T.A. and X.S.; supervision, T.A.; project administration, T.A.; funding acquisition, X.Y.

Funding: This research was funded by the International Project on Scientific and Technological Cooperation in Shaanxi Province of China (No. 2018KW-052).

Conflicts of Interest: The authors declare no conflict of interest.

References

1. Zhai, Y.; Dou, Y.; Zhao, D.; Fulvio, P.F.; Mayes, R.T.; Dai, S. Carbon materials for chemical capacitive energy storage. *Adv. Mater.* **2011**, *23*, 4828–4850. [CrossRef] [PubMed]
2. Molina-Sabio, M.; Rodrıguez-Reinoso, F. Role of chemical activation in the development of carbon porosity. *Colloid. Surf. A* **2004**, *241*, 15–25. [CrossRef]
3. Moussa, G.; Hajjar-Garreau, S.; Taberna, P.L.; Simon, P.; Matei Ghimbeu, C. Eco-Friendly Synthesis of Nitrogen-Doped Mesoporous Carbon for Supercapacitor Application. *C* **2018**, *4*, 20. [CrossRef]
4. Thirukumaran, P.; Atchudan, R.; Parveen, A.S.; Lee, Y.R.; Kim, S.C. Polybenzoxazine originated N-doped mesoporous carbon ropes as an electrode material for high-performance supercapacitors. *J. Alloys Compd.* **2018**, *750*, 384–391. [CrossRef]
5. Zhang, M.; Chen, M.; Reddeppa, N.; Xu, D.; Jing, Q.; Zha, R. Nitrogen self-doped carbon aerogels derived from trifunctional benzoxazine monomers as ultralight supercapacitor electrodes. *Nanoscale* **2018**, *10*, 6549–6557. [CrossRef] [PubMed]
6. Chen, H.; Zhou, M.; Wang, Z.; Zhao, S.; Guan, S. Rich nitrogen-doped ordered mesoporous phenolic resin-based carbon for supercapacitors. *Electrochim. Acta* **2014**, *148*, 187–194. [CrossRef]
7. Liu, S.; Zuo, P.; Wang, Y.; Li, X.; Zhang, W.; Xu, S.; Li, Z. Nitrogen-doped ordered mesoporous carbon microspheres made from m-aminophenol-formaldehyde resin as promising electrode materials for supercapacitors. *Micropor. Mesopor. Mater.* **2018**, *259*, 54–59. [CrossRef]
8. Xin, Z.; Fang, W.; Zhao, L. N-doped carbon foam constructed by liquid foam with hierarchical porous structure for supercapacitor. *J. Porous Mater.* **2018**, *25*, 1521–1529. [CrossRef]
9. Wei, J.; Zhou, D.; Sun, Z.; Deng, Y.; Xia, Y.; Zhao, D. A controllable synthesis of rich nitrogen-doped ordered mesoporous carbon for CO_2 capture and supercapacitors. *Adv. Funct. Mater.* **2013**, *23*, 2322–2328. [CrossRef]
10. Zhang, Y.; Yuan, Z.; Xu, C. Engineering biomass into formaldehyde-free phenolic resin for composite materials. *AIChE J.* **2015**, *61*, 1275–1283. [CrossRef]
11. Demirbaş, A. Mechanisms of liquefaction and pyrolysis reactions of biomass. *Energ. Convers. Manag.* **2000**, *41*, 633–646. [CrossRef]
12. Effendi, A.; Gerhauser, H.; Bridgwater, A.V. Production of renewable phenolic resins by thermochemical conversion of biomass: a review. *Renew. Sust. Energ. Rev.* **2008**, *12*, 2092–2116. [CrossRef]
13. Mohan, D.; Pittman, C.U.; Steele, P.H. Pyrolysis of wood/biomass for bio-oil: a critical review. *Energy Fuel* **2006**, *20*, 848–889. [CrossRef]
14. Cheng, S.; D'Cruz, I.; Yuan, Z.; Wang, M.; Anderson, M.; Leitch, M.; Xu, C. Use of biocrude derived from woody biomass to substitute phenol at a high-substitution level for the production of biobased phenolic resol resins. *J. Appl. Polym. Sci.* **2011**, *121*, 2743–2751. [CrossRef]
15. Li, B.; Yuan, Z.; Schmidt, J.; Xu, C.C. New foaming formulations for production of bio-phenol formaldehyde foams using raw kraft lignin. *Eur. Polym. J.* **2019**, *111*, 1–10. [CrossRef]
16. Wang, H.; Gao, Q.; Hu, J. High hydrogen storage capacity of porous carbons prepared by using activated carbon. *J. Am. Chem. Soc.* **2009**, *131*, 7016–7022. [CrossRef] [PubMed]
17. Tian, W.; Xueyan, W.U.; Wei, X. Preparation of Porous Carbon Material from Coffee Grounds and Its Application to Lithium Ion Batteries. *J. Jilin Univ.* **2014**, *52*, 802–806.
18. Kudin, K.N.; Ozbas, B.; Schniepp, H.C. Raman spectra of graphite oxide and functionalized graphene sheets. *Nano Lett.* **2008**, *8*, 36–41. [CrossRef]
19. Kruk, M.; Kohlhaas, K.M.; Dufour, B. Partially graphitic, high-surface-area mesoporous carbons from polyacrylonitrile templated by ordered and disordered mesoporous silicas. *Micropor. Mesopor. Mater.* **2007**, *102*, 178–187. [CrossRef]

20. Liu, X.; Zhang, W.; Zhang, Z. Preparation and characteristics of activated carbon from waste fiberboard and its use for adsorption of Cu(II). *Mater. Lett.* **2014**, *116*, 304–306. [CrossRef]
21. Li, W.; Liu, J.; Zhao, D. Mesoporous materials for energy conversion and storage devices. *Nat. Rev. Mater.* **2016**, *1*, 4465–4482. [CrossRef]
22. Wang, L.; Wang, R.; Zhao, H. High rate performance porous carbon prepared from coal for supercapacitors. *Mater. Lett.* **2015**, *149*, 85–88. [CrossRef]
23. Yang, W.; Yang, W.; Song, A.; Gao, L.; Su, L.; Shao, G. Supercapacitance of nitrogen-sulfur-oxygen co-doped 3D hierarchical porous carbon in aqueous and organic electrolyte. *J. Power Sources* **2017**, *359*, 556–567. [CrossRef]
24. Mori, T.; Iwamura, S.; Ogino, I.; Mukai, S.R. Cost-effective synthesis of activated carbons with high surface areas for electrodes of non-aqueous electric double layer capacitors. *Sep. Purif. Technol.* **2019**, *214*, 174–180. [CrossRef]
25. Chen, L.F.; Zhang, X.D.; Liang, H.W. Synthesis of nitrogen-doped porous carbon nanofibers as an efficient electrode material for supercapacitors. *ACS Nano* **2012**, *6*, 7092–7102. [CrossRef]
26. Hou, J.; Jiang, K.; Wei, R. Popcorn-Derived Porous Carbon Flakes with an Ultrahigh Specific Surface Area for Superior Performance Supercapacitors. *ACS Appl. Mater. Interf.* **2017**, *9*, 30626–30634. [CrossRef] [PubMed]
27. Zhou, L.; Cao, H.; Zhu, S.; Hou, L.; Yuan, C. Hierarchical micro-/mesoporous N-and O-enriched carbon derived from disposable cashmere: a competitive cost-effective material for high-performance electrochemical capacitors. *Green Chem.* **2015**, *17*, 2373–2382. [CrossRef]
28. Xing, W.; Qiao, S.Z.; Ding, R.G.; Li, F.; Lu, G.Q.; Yan, Z.F.; Cheng, H.M. Superior electric double layer capacitors using ordered mesoporous carbons. *Carbon* **2006**, *44*, 216–224. [CrossRef]
29. Zheng, Z.; Gao, Q. Hierarchical porous carbons prepared by an easy one-step carbonization and activation of phenol-formaldehyde resins with high performance for supercapacitors. *J. Power Sources* **2011**, *196*, 1615–1619. [CrossRef]
30. Zhang, J.; Zhang, W.; Han, M.; Pang, J. One pot synthesis of nitrogen-doped hierarchical porous carbon derived from phenolic formaldehyde resin with sodium citrate as activation agent for supercapacitors. *J. Mater. Sci.-Mater. Electron.* **2018**, *29*, 4639–4648. [CrossRef]

 *materials* 

Article

Carbon Xerogels Hydrothermally Doped with Bimetal Oxides for Oxygen Reduction Reaction

Abdalla Abdelwahab [1],*, Francisco Carrasco-Marín [2] and Agustín F. Pérez-Cadenas [2],*

[1] Materials Science and Nanotechnology Department, Faculty of Postgraduate Studies for Advanced Sciences, Beni-Suef University, Beni-Suef 62511, Egypt
[2] Carbon Materials Research Group, Department of Inorganic Chemistry, Faculty of Sciences, University of Granada, Campus Fuentenueva s/n, ES18071 Granada, Spain
* Correspondence: aabdelwahab@psas.bsu.edu.eg (A.A.); afperez@ugr.es (A.F.P.-C.)

Received: 27 June 2019; Accepted: 27 July 2019; Published: 31 July 2019

Abstract: A total of two carbon xerogels doped with cobalt and nickel were prepared by the sol–gel method. The obtained carbon xerogels underwent further surface modification with three binary metal oxides namely: nickel cobaltite, nickel ferrite, and cobalt ferrite through the hydrothermal method. The mesopore volumes of these materials ranged between 0.24 and 0.40 cm^3/g. Moreover, there was a morphology transformation for the carbon xerogels doped with nickel cobaltite, which is in the form of nano-needles after the hydrothermal process. Whereas the carbon xerogels doped with nickel ferrite and cobalt ferrite maintained the normal carbon xerogel structure after the hydrothermal process. The prepared materials were tested as electrocatalysts for oxygen reduction reaction using 0.1 M KOH. Among the prepared carbon xerogels cobalt-doped carbon xerogel had better electrocatalytic performance than the nickel-doped ones. Moreover, the carbon xerogels doped with nickel cobaltite showed excellent activity for oxygen reduction reaction due to mesoporosity development. $NiCo_2O_4$/Co-CX showed to be the best electrocatalyst of all the prepared electrocatalysts for oxygen reduction reaction application, exhibiting the highest electrocatalytic activity, lowest onset potential E_{onset} of -0.06 V, and the lowest equivalent series resistance (ESR) of 2.74 Ω.

Keywords: carbon gels; mesoporosity; electrocatalysis; oxygen reduction reaction

---

1. Introduction

The energy problem is one of the most important challenges the world is facing right now. Finding new sources of energy production and how to store this energy has become a major challenge. Lithium ion batteries are a kind of batteries that are used in portable electronics and electric vehicles [1]. Although lithium ion battery produces electricity with high energy density and low self-discharge, it presents some hazards as it contains a flammable electrolyte.

Fuel cells are electrochemical devices that are able to convert chemical energy into electrical energy when fuel and oxidant are supplied [2]. According to their working mechanism, fuel cell bears similarities in both batteries and engines, however, it has superior advantages as it does not need recharging and generates drinking water when the used fuel is hydrogen, so, it is considered as a "zero emission engine." Because it is environmentally friendly, fuel cells find commercial application in transportation, stationary power generation, and in low power portable devices. Fuel cells are facing some difficulties that delay its entry into the market. These difficulties can be attributed to economic factors, materials designing problem, and inadequacies in electrochemical devices operation [2,3].

The fuel cell is a galvanic cell that consists of two electrodes, anode and cathode. The anodic fuel cell reaction is either the direct oxidation of hydrogen or the oxidation of methanol. The cathodic fuel cell reaction is usually oxygen reduction reaction (ORR) and in most cases the source of oxygen is air. The major factor that limits the fuel cell performance is the cathodic oxygen reduction reaction

(ORR), as it consists of several steps in which molecular oxygen dissociates at the catalyst surface and combines with hydrogen ions [4,5]. Different factors can influence the reaction kinetics at the electrode surface, but still the electrocatalyst itself has the major effect. There are two pathways for the oxygen reduction reaction in aqueous electrolyte: four-electron and two-electron pathways. The direct four-electron pathway is preferable because the Faradaic efficiency of the reaction is greater; also it does not involve peroxide species in the solution.

To date, Pt and its alloys are the best known electrocatalysts for the ORR [6–9]. In order to reduce the cost of using pure Pt metal as a catalyst, alloying Pt with another metal reduces the cost; however the metal leaches away gradually [9,10], resulting in the loss of performance that reduces the total fuel cell efficiency and limits their market use. The current fuel cell technology is based on the development of non-precious metals and Pt-free electrocatalysts [10]. Transition metal oxides are new materials that exhibit excellent activity in many applications because of their rich redox reactions, higher conductivity than simple oxides, and availability of active sites. Wang et al. [11] fabricated graphene-nickel cobaltite nanocomposite (GNCC) that was used as a positive electrode in supercapacitors application. Higher capacitance was obtained for GNCC (618 $F \cdot g^{-1}$) compared with graphene-Co_3O_4 (340 $F \cdot g^{-1}$) or graphene-NiO (375 $F \cdot g^{-1}$) due to the rich numbers of Faradaic reactions on the nickel cobaltite. Moreover, Genqiang Zhang et al. [12] synthesized $NiCo_2O_4$-rGO hybrid nanosheets electrocatalysts for the oxygen reduction reaction. He found a comparable current density and onset potential with those of commercial Pt/C catalysts. In another study [13], cobalt ferrite thin films were prepared and tested as anode for lithium-ion batteries.

Carbon nanomaterials such as carbon nanotubes and graphene were applied as electrocatalysts for ORR application [14,15] and exhibited good performance due to their surface active sites that are necessary for reactants adsorption, bond-breaking and new bond-formation, and products desorption. A new emerging class of carbon nanomaterials is carbon gel [16,17], which has good electrochemical properties and high surface areas. Recently, carbon gels were tested as electrodes in methanol oxidation [18], supercapacitors [19], environmental applications [20,21], and oxygen reduction reaction (ORR) [22,23]. Carbon gels, in both of its forms xerogels and aerogels, offer the opportunity to be used as it is or doped with metals which enhances its electrochemical activity as the metal doping influences the obtained surface area and morphology [24–26].

To the best of our knowledge, bimetal oxides-doped carbon xerogels have not been tested before as electrocatalysts for oxygen reduction reaction in basic medium. So, in this work carbon xerogels were prepared by sol–gel process using two metal salts as polymerization catalysts namely cobalt acetate and nickel acetate to investigate the role of the metal catalyst. Moreover, the resultant two carbon xerogels were further doped with different binary metal oxides of nickel cobaltite, nickel ferrite, and cobalt ferrite through the hydrothermal method. The prepared samples were employed as electrocatalysts for oxygen reduction reaction, as this is one of our main goals to study the effect of bimetal oxides on the morphology of carbon gels and its activity toward ORR.

2. Materials and Methods

2.1. Preparation of Carbon Xerogel

The used monomers for preparation of carbon xerogels were resorcinol (R) and formaldehyde (F) with a molar ratio of R/F = 1/2. These monomers were dissolved in water (W) in the presence of cobalt acetate and nickel acetate as the polymerization catalysts. The amount of cobalt and nickel in the final carbon structure was calculated to be 6 wt.% and the used molar ratio between resorcinol and water R/W was 1/17. After stirring, the clear solution was filled in glass molds and placed in the oven for one day at 40 °C then five days at 80 °C. The obtained organic gel was placed in acetone for 3 days to allow solvent exchange to save the porosity during the drying method. The organic gels were dried using the microwave drying method (domestic Samsung microwave F600G, Samsung, MWF600G, Beijing, China) under Ar-gas flow at power of 10% for 10 min) to get their corresponding organic xerogels, followed by the carbonization process in tube furnace (Carbolite Gero single zone EVA, Carbolite Gero Neuhausen, Germany) at 900 °C for 2 h with a heating rate of 1 °C/min to get the carbon xerogels.

2.2. Binary Metal Oxides Surface Modification (XY$_2$O$_4$/CX)

The obtained cobalt and nickel-doped carbon xerogels were further doped with three different bimetal oxides through the hydrothermal process. Nickel cobaltite ($NiCo_2O_4$), nickel ferrite ($NiFe_2O_4$), cobalt ferrite ($CoFe_2O_4$) were chosen to be used as the bimetal oxides for doping.

In the synthesis of $NiCo_2O_4$/Co-CX and $NiCo_2O_4$/Ni-CX, typically 120 mg of carbon xerogel was dispersed into 40 mL *N,N*-dimethylformamide (DMF). Then, $Ni(Ac)_2 \cdot 6H_2O$ (0.125 gm; 0.5 mmol), $Co(Ac)_2 \cdot 4H_2O$ (0.250 gm; 1 mmol), and urea (0.360 gm; 6 mmol) were dissolved in 30 mL solution of H_2O and ethylene glycol with a volumetric ratio of 1:2. The two solutions were sonicated for 15 min and were placed in a polytetrafluoroethylene lined stainless steel autoclave at 180 °C for 12 h. The black precipitate obtained after the hydrothermal reaction was collected by centrifugation (4000 rpm for 5 min), washed several times with water and ethanol, dried at 60 °C for 12 h, and finally calcined at 360 °C for 3 h with a heating rate of 5 °C/min.

The same procedure was followed for the preparation of nickel ferrite ($NiFe_2O_4$) and cobalt ferrite ($CoFe_2O_4$) doped carbon xerogel except with replacing the metal salts with the appropriate ones and maintaining the molar ratios between the two salts at 1:2.

2.3. Characterization

The prepared samples were fully characterized with different characterization techniques. Sample surface area and porosity were characterized using surface area analysis with N_2 adsorption at −196 °C using the Quantachrome instrument (Quadrasorb, Boynton Beach, FL, USA) and then by applying the Brunauer-Emmett-Teller (BET) equation the isotherms were obtained. Before porosity analysis, the samples were outgassed for 12 h at 110 °C under high vacuum of 10^{-6} mbar. The mesoporosity of the prepared materials were calculated by applying the density functional theory (DFT) equation to the adsorption part of N_2-isotherms. Moreover, the samples morphology and particle size distribution were analyzed using scanning electron microscopy (SEM) and high resolution transmission electron microscopy (HRTEM), respectively. SEM analysis was performed using Zeiss SUPRA40VP instrument (Carl Zeiss AG, Oberkochen, Germany), equipped with both SE and BSE detectors and X-Max 50 mm energy dispersive X-ray microanalysis system.

HRTEM was carried out with FEI Titan G2 60–300 microscope (FEI, Eindhoven, Netherlands) with a high brightness electron gun (X-FEG) operated at 300 KV and equipped with a Cs image corrector (CEOS) and analytical electron microscopy (AEM) with a SUPER-X silicon-drift window-less EDX detector.

XRD analysis was carried out with a BRUKER D8 DISCOVER diffractometer (BRUKER, Rivas-Vaciamadrid, Spain) equipped with a IµS Cu microsource, operating at 50 KV, 1 mA, and 50 W at 25 °C, using a Cu Kα (λ = 15,406 Å) radiation, a Multilayer Optics Monochromator (Quazar Optics: Montel type 2-dim beam shaping) (Incoatec, Geesthacht, Germany), and a PILATUS3R 100K-A detector (Dectris Ltd, Baden, Switzerland). Diffraction patterns were recorded between 10° and 80° (2θ) with a step of 0.02° and a time per step of 40 s. The average crystal size was determined using the Scherrer equation.

X-ray photoelectron spectroscopy (XPS) measurements were carried out with a physical electronics ESCA 5701 (PHI, Chanhassen, MN, USA) operating at 12 KV and 10 mA and equipped with a MgK X-ray source (*hv* = 1253.6 eV). The obtained binding energy values are referred to C1s, O1s, and N1s peaks at 284.6, 529.3, and 399.3, respectively.

2.4. Electrode Preparation for ORR

A total of 5 mg of the prepared carbon material was dispersed into 400 µL of isopropanol and 30 µL of nafion solution (5 wt.%), and then sonicated (sonication bath, Samarth electronics, for 15 min).

Ten microliter of the suspended solution was deposited into a glassy carbon electrode with a diameter of 3 mm and dried under an infrared lamp for 5 min (100 W, R95, Philips, Madrid, Spain).

2.5. Electrochemical Measurements

The electrochemical measurements were carried out using a biologic multichannel VMP3 potentiostat (BioLogic, Seyssinet-Pariset, France). A three-electrode electrochemical cell was used during the analysis of electrodes performance in which Ag/AgCl and Pt electrodes were used as reference and counter electrodes, respectively. The used electrolyte was 0.1 M potassium hydroxide (KOH), which is first saturated with nitrogen then saturated with O_2 to evaluate the electrocatalyst performance in the absence and presence of oxygen. Different electrochemical techniques were used in the electrode evaluation: (i) cyclic voltammetry (CV), (ii) linear sweep voltammetry (LSV), and (iii) electrochemical impedance spectroscopy (EIS).

The cyclic voltammetry and linear sweep voltammetry were carried out in a potential range between 0.4 to −0.8 V. Two scan rates of 5 mV·s^{-1} and 50 mV·s^{-1} were used with CV and different rotation speeds from 500 to 4000 rpm were employed for LSV at 5 mV·s^{-1} in order to be able to apply the Koutecky–Levich model for evaluating the electrocatalyst performance and calculating the number of transferred electrons.

$$\frac{1}{j} = \frac{1}{j_k} + \frac{1}{B\omega^{0.5}} \tag{1}$$

$$B = 0.2nF(D_{O_2})^{2/3}v^{-1/6}C_{O_2} \tag{2}$$

where j, current density; j_k, kinetic current density; ω, rotation speed; F, Faraday constant; D_{O2}, oxygen diffusion coefficient ($1.9 \cdot 10^{-5}$ cm^2·s^{-1}); v, viscosity (0.01 cm^2·s^{-1}); C_{O2}, oxygen concentration ($1.2 \cdot 10^{-6}$ mol·cm^{-3}).

3. Results

Table 1 is constructed by applying the BET equation for the obtained isotherms of N_2 adsorption at 77 K. The BET surface areas (S_{BET}) ranged between 50 to 156 m^2.g^{-1}, as can be seen in Table 1. The highest surface areas were obtained for the samples doped with nickel ferrite $NiFe_2O_4$, which means these samples have high microporosity as appears from their pore size (Lo) data.

Table 1. Surface area analysis.

Sample	S_{BET}	$W_0(N_2)$	$L_0(N_2)$	$V_{0.95}(N_2)$	$V_{meso}(N_2)$	S_{DFT}	V_{DFT}	$L_0(DFT)$
	m^2/g	cm^3/g	nm	cm^3/g	cm^3/g	m^2/g	cm^3/g	nm
$NiFe_2O_4$/Ni-CX	156	0.06	1.91	0.46	0.40	185	0.47	2.18
$NiCo_2O_4$/Ni-CX	57	0.02	2.20	0.27	0.24	103	0.31	2.60
$CoFe_2O_4$/Ni-CX	84	0.03	1.98	0.35	0.31	127	0.36	2.84
$NiFe_2O_4$/Co-CX	112	0.04	1.77	0.33	0.29	139	0.37	2.84
$NiCo_2O_4$/Co-CX	64	0.03	2.98	0.30	0.27	128	0.32	2.84
$CoFe_2O_4$/Co-CX	50	0.02	1.48	0.27	0.25	85	0.29	2.18

W_0 is the micropore volume, L_0 is the pore size, $V_{0.95}$ is the pore volume at relative pressure of 0.95, V_{meso} is the mesopore volume. S_{DFT}, V_{DFT}, and $L_0(DFT)$ are the surface area, pore volume, and pore size obtained from DFT calculations, respectively.

Doping of carbon xerogel with bimetal oxides promote mesoporosity of carbon structure and this mesoporosity is confirmed by applying the DFT equation to the adsorption part for the obtained N_2- isotherms (Figure 1). The DFT pores diameters ($L_0(DFT)$) ranged from 2.18–2.84 nm and the samples doped with nickel cobaltite, $NiCo_2O_4$, showed homogenous particle size distribution. The mesoporous character of these samples is confirmed by the type-IV shape of their corresponding adsorption–desorption isotherms (Figure S1 in Supplementary Materials), also showing all of their significant hysteresis cycles.

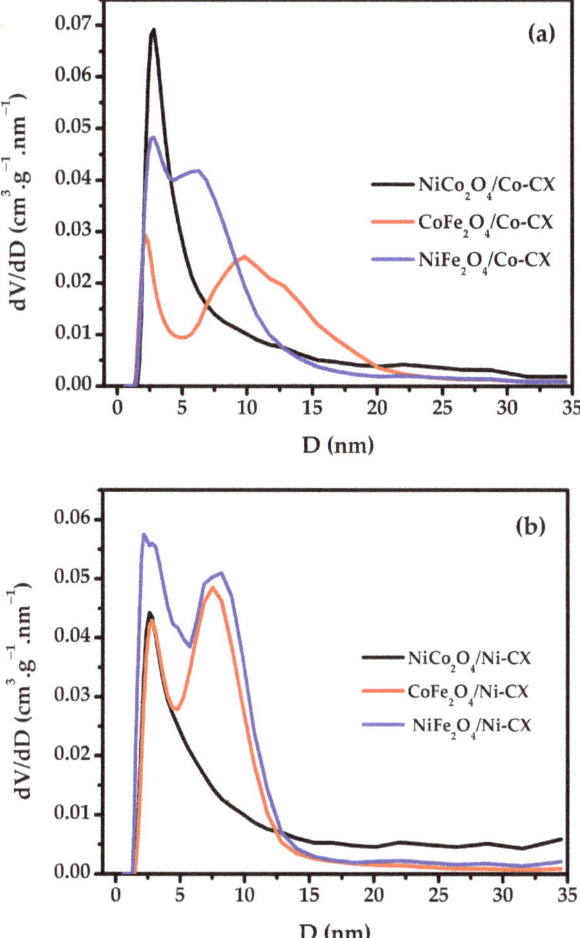

Figure 1. DFT analysis for adsorption part of N_2-adsorption/desorption isotherms, (**a**) bimetal oxides doped cobalt xerogels, (**b**) bimetal oxides doped nickel xerogels.

The existence of the binary metal oxides inside the carbon matrix is confirmed by XRD analysis (Figure 2) in which the crystallinity of the bimetal oxides is confirmed by the XRD pattern with the absence of any contaminated peaks. The XRD pattern of nickel cobaltite doped carbon xerogels and their corresponding planes is presented in Figure 2c.

The morphology of the prepared samples is revealed from the SEM images (Figure 3). The carbon xerogel undergoes morphology transformation from continued connected spherical particles to nano-needle like structure when doped with nickel cobaltite (Figure 3a,b), while it maintains its original morphology when doped with nickel ferrite (Figure 3c,d) or cobalt ferrite (Figure 3e,f).

The homogeneity and dispersity of the doped bimetal oxides nanoparticles inside the carbon matrix is studied from the TEM images (Figure 4). The metal nanoparticles are well dispersed inside the carbon structure in case of the doped samples with nickel ferrite (Figure 4c,d) and cobalt ferrite (Figure 4e,f). In addition, the nano-needle structure for nickel cobaltite with different lengths and widths was formed in both cobalt-doped carbon xerogels (Co-CX, Figure 4a) and nickel-doped carbon xerogels (Ni-CX, Figure 4b) [18].

Figure 5 shows the XP spectra of the prepared electrocatalysts in the presence of the metal cations in divalent and trivalent oxidation states. The prepared electrodes were tested as electrocatalysts for ORR application and their cyclic voltammograms for nickel cobaltite doped ones are presented in Figure 6.

The linear sweep voltammetry (LSV) technique is used with rotating desk electrode (Figure 7), and the electrocatalytic activity at different rotation speed is shown in Figure 8. The data obtained from LSV were used in order to apply the Koutecky–Levich model for the determination of the number of electrons transferred (Figure 9).

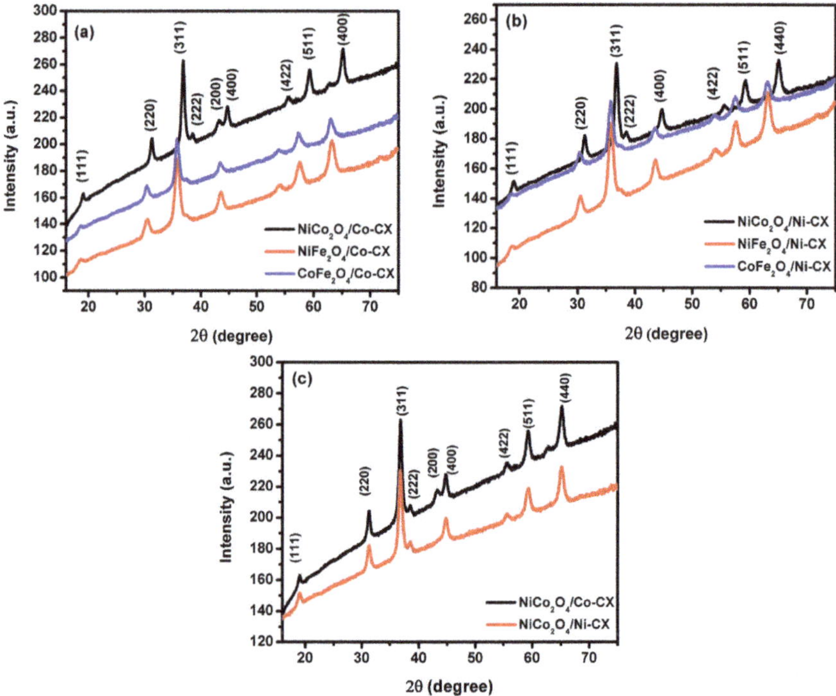

Figure 2. The XRD patterns for (**a**) bimetal oxides doped cobalt xerogels, (**b**) bimetal oxides doped nickel xerogels and (**c**) nickel cobaltite doped cobalt and nickel carbon xerogels.

Figure 3. Scanning electron microscopy (SEM) images for (**a**) $NiCo_2O_4$/Co-CX, (**b**) $NiCo_2O_4$/Ni-CX, (**c**) $NiFe_2O_4$/Co-CX, (**d**) $NiFe_2O_4$/Ni-CX, (**e**) $CoFe_2O_4$/Co-CX, and (**f**) $CoFe_2O_4$/Ni-CX.

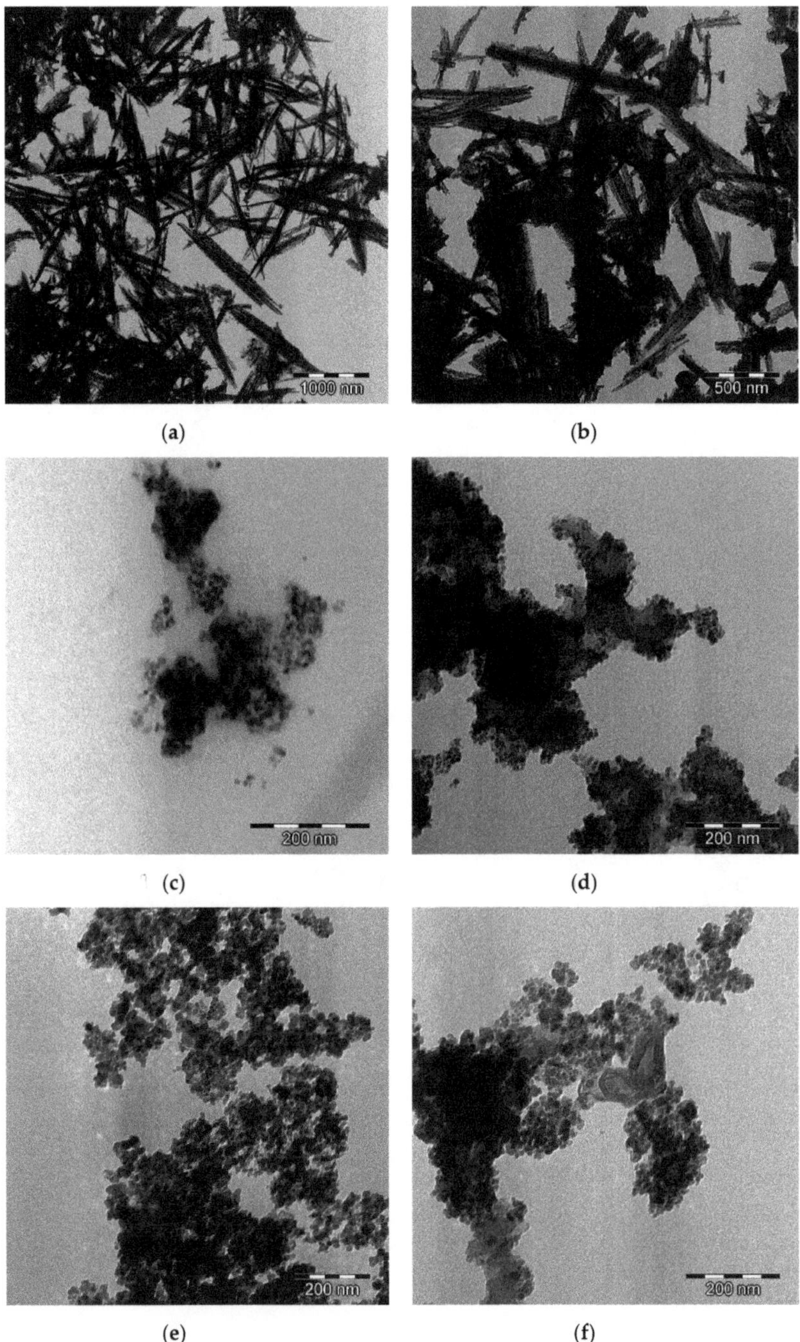

Figure 4. High resolution transmission electron microscopy (HRTEM) images for (**a**) NiCo$_2$O$_4$/Co-CX, (**b**) NiCo$_2$O$_4$/Ni-CX, (**c**) NiFe$_2$O$_4$/Co-CX, (**d**) NiFe$_2$O$_4$/Ni-CX, (**e**) CoFe$_2$O$_4$/Co-CX, and (**f**) CoFe$_2$O$_4$/Ni-CX.

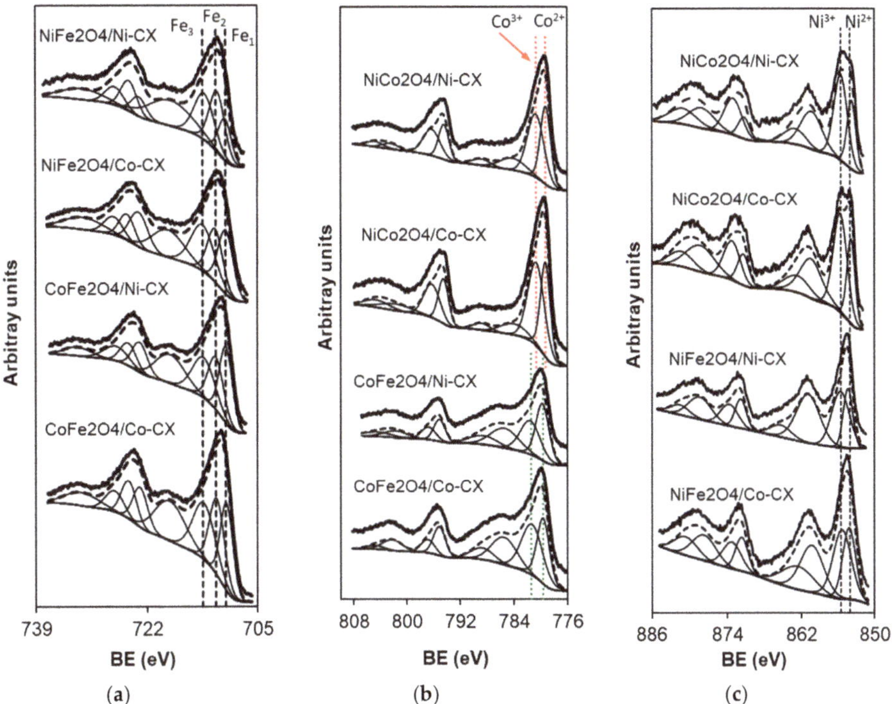

Figure 5. Deconvolution of the XP spectra for the prepared materials. (**a**) $NiFe_2O_4$ and $CoFe_2O_4$ doped carbon xerogels, (**b**) $NiCo_2O_4$ and $CoFe_2O_4$ doped carbon xerogels and (**c**) $NiCo_2O_4$ and $NiFe_2O_4$ doped carbon xerogels.

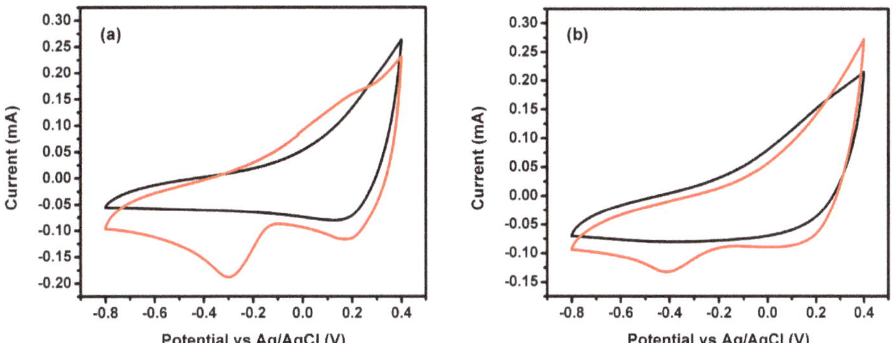

Figure 6. Cyclic voltammograms (CV) of (**a**) $NiCo_2O_4$/Co-CX in both nitrogen (black) and oxygen (red) saturated electrolyte and (**b**) $NiCo_2O_4$/Ni-CX in both nitrogen (black) and oxygen (red) saturated electrolyte.

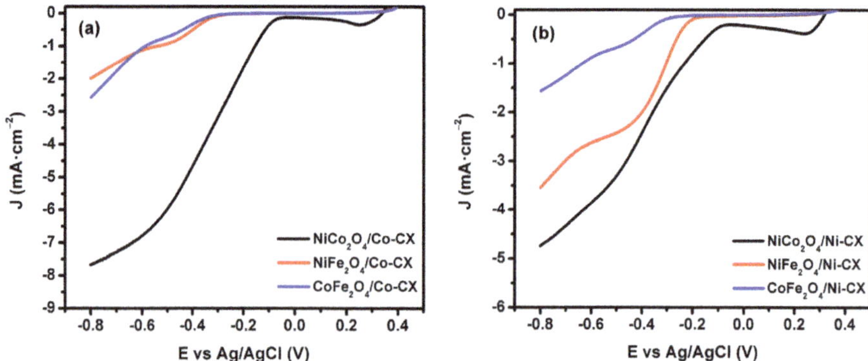

Figure 7. Linear sweep voltammograms (LSV) for (**a**) Co-CX and (**b**) Ni-CX doped with different bimetal oxides.

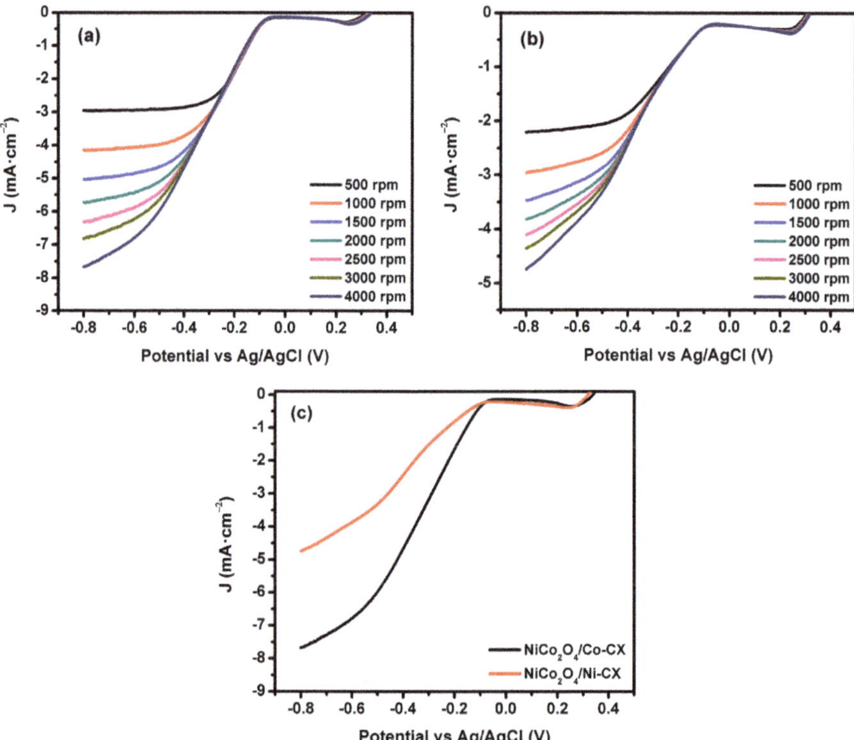

Figure 8. Linear sweep voltammograms (LSV) for (**a**) $NiCo_2O_4$/Co-CX, (**b**) $NiCo_2O_4$/Ni-CX at 5 mV.s^{-1} with different speeds from 500 rpm to 4000 rpm and (**c**) comparing the LSV for $NiCo_2O_4$/Co-CX and $NiCo_2O_4$/Ni-CX at 4000 rpm.

Electrochemical impedance spectroscopy (EIS) is an important technique for the evaluation of the performance of an electrode in certain applications by calculating the electrode resistance and equivalent series resistance (ESR). The EIS was performed by applying a frequency range from 100 KHz to 1 mHz with a sinusoidal signal amplitude of 10 mV, and the data obtained from EIS is shown in Figure 10.

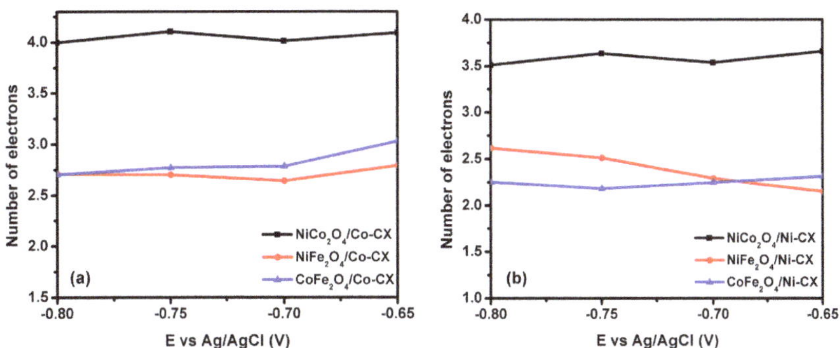

Figure 9. Variation of number of electron transferred with E vs. Ag/AgCl for bimetal oxides doped (**a**) Co-CX and (**b**) Ni-CX.

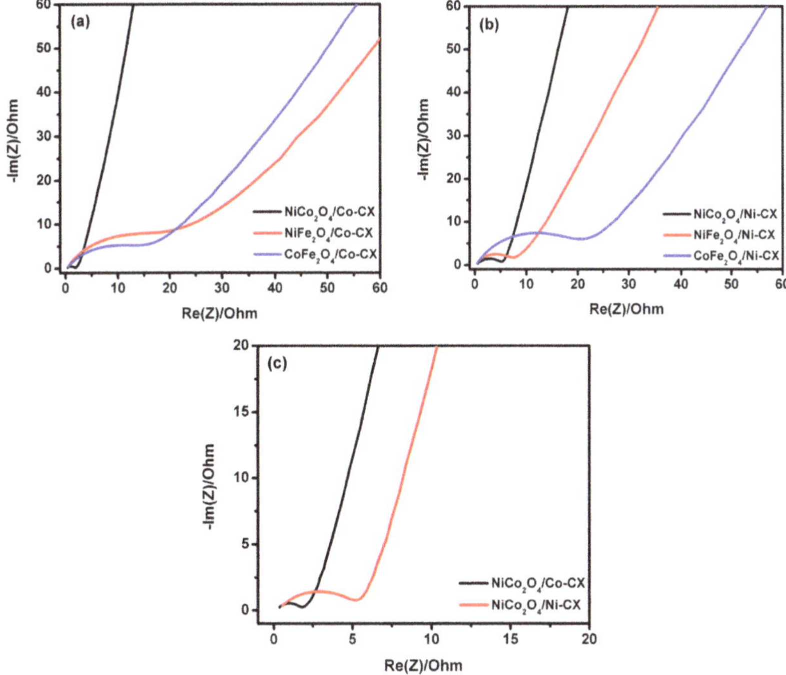

Figure 10. Nyquist plots obtained from EIS for bimetal oxides doped (**a**) Co-CX, (**b**) Ni-CX and (**c**) nickel cobaltite doped cobalt and nickel carbon xerogels.

4. Discussion

Samples doped with nickel cobaltite have well developed mesoporosity and in case of $NiCo_2O_4$/Co-CX the mean pore size is 2.98 nm, while for $NiCo_2O_4$/Ni-CX is 2.20 nm. The mesoporosity development is an indication for better accessibility of electrolyte ions inside the carbon structure, which in turn make these materials good electrocatalysts in catalysis application [27–29].

Energy-dispersive X-ray spectroscopy (EDXS) analysis also confirmed the presence of the different metals in the samples. Figure S2 contains the analysis carried out on the sample $NiCo_2O_4$-CoCX, as an example.

Determination of mean particle sizes for the prepared samples was carried out by applying Scherrer equation for the obtained XRD patterns (Table 2). Higher particle sizes were obtained for the nickel cobaltite doped carbon xerogels. The higher mean particle sizes for nickel cobaltite doped samples indicates higher active sites in these samples which promote the electrocatalytic reduction of oxygen. For example, in $NiCo_2O_4$/Co-CX the mean particle size is about 25.5 nm while for $NiFe_2O_4$/Co-CX and $CoFe_2O_4$/Co-CX is 19.8 and 21.8 nm, respectively. Likewise for Ni-CX series the $NiCo_2O_4$/Ni-CX has the highest particle size of 24.1 nm.

Table 2. Mean particle size obtained from Scherrer equation.

Sample	d_{XRD} (nm)
$NiCo_2O_4$/Co-CX	25.5
$NiFe_2O_4$/Co-CX	19.8
$CoFe_2O_4$/Co-CX	21.8
$NiCo_2O_4$/Ni-CX	24.1
$NiFe_2O_4$/Ni-CX	21.1
$CoFe_2O_4$/Ni-CX	20.2

Table 3 collects the binding energies (B.E.) and chemical composition corresponding to carbon, oxygen, and nitrogen with respect to the chemical composition analysed by XPS. The oxygen peak at lowest B.E. 529.8 ± 0.3 eV corresponds to the oxygen atoms bond to transition metal cations with oxidation states +2 and +3.

On the other hand, the XPS results corresponding to the region Fe2p are collected in Table 4 and Figure 5. Fe2p$_{3/2}$ peaks centred at 709.9 ± 0.2, 711.2 ± 0.3, and 713.2 ± 0.2 eV have been assigned to Fe^{2+} situated in octahedral holes (Fe_1), Fe^{3+} situated in octahedral holes (Fe_2), and Fe^{3+} situated in tetrahedral holes (Fe_3), respectively. This means that iron is forming part of compounds type $M_x^{2+}M_{1-x}^{3+}[Fe_y^{2+}Fe_{1-y}^{3+}]O_4$ [30] being M = Co and/or Ni.

Table 3. Binding energies and chemical composition of C1s, O1s, and N1s.

Sample	C1s			O1s				N1s		
	eV	FWHM eV	Peak %	eV	Peak %	% (Mass)	% (Atomic)	eV	% (Mass)	% (Atomic)
$NiCo_2O_4$/Co-CX	284.6	1.4	70.9	529.3	36.7	28.8	32.7	399.3	0.2	0.3
	285.7		8.9	530.7	24.7			400.7		
	286.3		9.7	531.8	23.3					
	288.5		10.4	533.2	15.3					
$NiCo_2O_4$/Ni-CX	284.6	1.4	67.5	529.1	34.9	29.6	32.3	398.9	0.2	0.2
	285.6		11.0	530.7	27.8			400.4		
	286.4		10.4	531.8	22.4					
	288.5		11.1	533.2	14.9					
$NiFe_2O_4$/Co-CX	284.6	1.4	71.8	528.7	15.7	25.3	28.0	399.3	0.2	0.3
	285.7		8.8	530.1	50.4			400.7		
	286.3		9.4	531.6	24.4					
	288.5		10.1	533.1	9.5					
$NiFe_2O_4$/Ni-CX	284.6	1.4	62.5	530.1	40.0	21.9	21.6	398.7	0.6	0.7
	285.7		20.0	531.9	43.9			400.3		
	286.8		10.5	533.6	16.2					
	288.9		7.0							
$CoFe_2O_4$/Co-CX	284.6	1.6	66.9	529.7	63.9	28.6	38.5	399.4	0.4	0.7
	285.6		17.3	531.3	25.9			400.3		
	286.6		5.6	533.0	10.2					
	288.5		10.2							
$CoFe_2O_4$/Ni-CX	284.5	1.5	67.5	529.6	52.4	26.9	30.9	399.3	0.4	0.5
	285.6		12.5	531.3	35.2			400.5		
	286.4		8.3	533.1	12.5					
	288.4		11.7							

Table 4. XPS results collected after deconvolution of Peaks.

Sample	Fe2p$_{3/2}$	Fe	Co2p$_{3/2}$	Peak	Co	Ni2p$_{3/2}$	Peak	Ni	%Fe(II)	%Fe(III)	%Fe(III)	%Fe(III)
	eV	%(Mass)	eV	%	%(Mass)	eV	%	%(Mass)	Oct	Total	Oct	Teth
NiCo$_2$O$_4$/Co-CX			779.2	43.0	21.6	854.3	34.4	12.1				
			780.6	57.0		855.9	65.6					
			783.9			860.9						
			788.8			863.2						
			794.4			871.9						
			796.2			873.7						
			801.6			878.7						
			804.5			881.7						
NiCo$_2$O$_4$/Ni-CX			779.2	43.7	18.7	854.2	32.3	11.2				
			780.6	56.3		856.0	67.7					
			784.1			860.8						
			788.8			863.5						
			794.4			871.7						
			796.2			873.5						
			802.4			878.5						
			804.8			881.5						
NiFe$_2$O$_4$/Co-CX	709.9	22.6				854.2	43.7	10.2	32.9	67.1	43.2	56.8
	711.5					855.6	56.3					
	713.4					860.3						
	718.7					862.5						
	723.2					871.7						
	725.1					873.2						
	727.1					877.7						
	731.8					880.6						
NiFe$_2$O$_4$/Ni-CX	710.0	15.6				854.3	39.3	7.8	21.9	78.1	50.2	49.8
	711.2					855.6	60.7					
	713.1					860.9						
	718.4					865.3						
	723.2					871.8						
	724.7					873.6						
	726.9					878.3						
	731.9					881.5						
CoFe$_2$O$_4$/Co-CX	709.8	32.7	779.6	48.7	14.2				30.7	69.3	52.7	47.3
	711.3		781.3	51.3								
	713.3		785.4									
	718.5		788.7									
	723.1		795.1									
	724.8		796.8									
	727.0		802.2									
	732.1		804.8									
CoFe$_2$O$_4$/Ni-CX	709.9	24.8	779.7	53.1	10.7			0.0	35.9	64.1	47.1	52.9
	711.4		781.5	46.9								
	713.3		785.4									
	718.4		788.5									
	723.1		795.1									
	724.9		796.7									
	726.7		801.7									
	732.6		804.5									

Different Co2p$_{3/2}$ peaks with respect to Co2p spectra peaks have been deconvoluted and assigned as the following: In the case of CoFe$_2$O$_4$ phases the peaks centred at 779.6 ± 0.1 correspond to Co^{2+} situated in tetrahedral holes whereas those centred at 781.5 ± 0.2 correspond to Co^{2+} situated in octahedral holes [30], therefore these metals deposited on the surface of the samples are forming part of the compounds type $(Co_x^{2+}Fe_y^{3+})[Fe_z^{2+}Fe_{1-y}^{3+}Co_{1-x}^{2+}]O_4$ where cations in parenthesis are situated in tetrahedral positions while cations in brackets are situated in octahedral positions. However, in the case of NiCo$_2$O$_4$ phases the peaks centred at 779.2 ± 0.1 correspond to Co^{2+} situated in octahedral holes whereas those centred at 780.6 ± 0.2 correspond to Co^{3+} situated in tetrahedral holes [4,31].

Finally, the XPS spectra of Ni2p region shows peaks at 854.2 ± 0.1 y 871.7 ± 0.2 eV which correspond to Ni^{2+} as well as peaks at 855.9 ± 0.3 y 873.5 ± 0.2 eV corresponding to Ni^{3+} cations [31]. Therefore, XPS results show the metals as divalent or trivalent species in all the cases. Taking into account all this XPS analysis, we can conclude that the different phases that have been synthetized and supported on the different samples of this work correspond with the stoichiometries collected in Table 5.

Table 5. Samples stoichiometries obtained from XPS.

Sample	Stoichiometry
$NiCo_2O_4/Co\text{-}CX$	$(Co_{0.86}Ni_{0.14})[Co_{1.14}Ni_{0.86}]$
$NiCo_2O_4/Ni\text{-}CX$	$(Co_{0.87}Ni_{0.13})[Co_{1.13}Ni_{0.87}]$
$NiFe_2O_4/Co\text{-}CX$	$(Fe_{0.66}Ni_{0.34})Fe_{1.24}Ni_{0.76}]$
$NiFe_2O_4/Ni\text{-}CX$	$(Fe_{0.78}Ni_{0.22})[Fe_{1.22}Ni_{0.78}]$
$CoFe_2O_4/Co\text{-}CX$	$(Co_{0.34}Fe_{0.66})[Fe_{1.34}Co_{0.66}]$
$CoFe_2O_4/Ni\text{-}CX$	$(Co_{0.32}Fe_{0.68})[Fe_{1.32}Co_{0.68}]$

For ORR application, there is a reduction peak for oxygen saturated electrolyte for both $NiCo_2O_4/Co\text{-}CX$ (Figure 6a) and $NiCo_2O_4/Ni\text{-}CX$ (Figure 6b). This reduction peak is absent when the electrolyte is saturated with nitrogen (Black line), which means that these electrodes have electroactivity toward oxygen reduction reaction. Also, as it can be seen from the linear sweep voltammograms (LSV) (Figure 7) that the electrocatalytic activity for the nickel cobaltite doped carbon xerogels is higher than that for samples doped with nickel ferrite or cobalt ferrite and the onset potential for that sample is lower because of the increase in mesoporosity that allows higher accessibility for the electrolytic ions to access the pores (Table 1). The onset potentials E_{onset} for all samples are compiled in Table 6, in which the lowest onset potentials E_{onset} of -0.06 V is obtained for $NiCo_2O_4/Co\text{-}CX$. Similar data was obtained for $NiCo_2O_4/Ni\text{-}CX$ with onset potential E_{onset} of -0.07 V. As it can be seen, samples doped with nickel ferrite and cobalt ferrite have comparable onset potentials.

Table 6. Parameters obtained from LSV at 4000 rpm (values of n refer to K-L fitting for data at -0.8 V) and equivalent series resistance (ESR) calculated from the Nyquist plot.

Sample	E_{onset}	n	ESR
	V		Ω
$NiCo_2O_4/Co\text{-}CX$	-0.06	4.0	2.74
$NiFe_2O_4/Co\text{-}CX$	-0.31	2.7	23.26
$CoFe_2O_4/Co\text{-}CX$	-0.32	2.7	15.90
$NiCo_2O_4/Ni\text{-}CX$	-0.07	3.5	6.18
$NiFe_2O_4/Ni\text{-}CX$	-0.19	2.6	10.57
$CoFe_2O_4/Ni\text{-}CX$	-0.28	2.2	22.03

In order to evaluate the number of transferred electrons during the reaction for each electrocatalyst, linear sweep voltammetry was carried out at 5 mV·s^{-1} at different rotating speeds from 500 rpm to 4000 rpm, in order to apply the Koutecky–Levich model (Figure 8). The LSV for $NiCo_2O_4/Co\text{-}CX$ is presented in Figure 8a, in which the activity is promoted by increasing the rotating speed due to better diffusion of the electrolyte ions inside the pores. LSV data for $NiCo_2O_4/Ni\text{-}CX$ are shown in Figure 8b. At the same rotating speed of 4000 rpm (Figure 8c), the activity of $NiCo_2O_4/Co\text{-}CX$ to oxygen reduction is higher than that of $NiCo_2O_4/Ni\text{-}CX$, indicating their current densities.

The data obtained from fitting the linear sweep voltammograms to the Koutecky–Levich model (Figure 9) confirms that there is a promotion in the number of electrons transferred during the oxygen reduction reaction by doping the carbon xerogels with $NiCo_2O_4$. For example, in case of $NiCo_2O_4/Co\text{-}CX$ (Figure 9a) the reaction takes place by the four electron pathway which is the favored one for oxygen reduction reaction (Table 6). While in case of $NiCo_2O_4/Ni\text{-}CX$ and the rest of the electrocatalysts, the reaction occurs by both two and four electrons transfer pathways.

The electrochemical impedance spectroscopy (EIS), were performed for all prepared samples using the two electrode configuration in which 6 M KOH was used as the electrolyte in order to evaluate the electrode resistance and equivalent series resistance ESR (Figure 10). The activity of nickel cobaltite doped carbon xerogels toward ORR can also be attributed to the good electrical conductivity of nickel cobaltite relative to nickel ferrite or cobalt ferrite [32] as it can be seen from the Nyquist plots with lower electrode resistance for these samples (Figure 10). The equivalent series resistance ESR for the prepared samples was calculated from the Nyquist plot and is compiled in Table 6. Figure 10a shows the Nyquist plots for cobalt doped carbon xerogels (Co-CX) with the three bimetal oxides, the lowest ESR was obtained for $NiCo_2O_4$/Co-CX which equals to 2.74 Ω that reveals higher electrical conductivity and higher electrochemical performance to oxygen reduction. Likewise in case of nickel doped carbon xerogels (Ni-CX) (Figure 10b), the lowest ESR was achieved for $NiCo_2O_4$/Ni-CX with 6.18 Ω. Moreover, comparing Co-CX and Ni-CX doped $NiCo_2O_4$ (Figure 10c), the activity of carbon xerogels doped with cobalt is higher than that of nickel doped one because of the development of mesoporosity and lower electrode resistance, this tendency is in good agreement with our previous published work [23]. On the other hand, by comparing our electrocatalyst $NiCo_2O_4$/Co-CX with the ones in previously published materials such as $NiCo_2O_4$-rGO hybrid nanosheets in the same conditions [12], lower onset potential of -0.06 V was found compared to -0.073 V vs. Ag/AgCl indicating higher electrocatalytic activity to ORR. In addition, the ORR current density at a rotating speed of 2500 rpm and at -0.8 V vs. Ag/AgCl for $NiCo_2O_4$/Co-CX is about -6.3 $mA.cm^{-2}$ while for $NiCo_2O_4$-rGO is about -2.0 $mA \cdot cm^{-2}$. The higher activity for ORR is also confirmed with the calculated number of electrons transferred which in case of $NiCo_2O_4$/Co-CX nanocomposite is n = 4.0 while for $NiCo_2O_4$-rGO hybrid nanosheets n = 3.8.

5. Conclusions

Binary metal oxides doped carbon xerogels were successfully prepared by the sol–gel process followed by a designed hydrothermal method. For all the prepared materials, the metal cations exist as divalent and trivalent species that occupy both the corresponding tetrahedral and octahedral positions in the crystal structure. The presence of metal cations inside the carbon xerogel structure develops the mesoporosity that makes these materials promising electrocatalysts for ORR. Nickel cobaltite doped carbon xerogels developed a new nano-needle like structure morphology and showed the highest electrocatalytic performance and lowest onset potential for oxygen reduction reaction. In general, development of mesoporosity of carbon xerogel together with increasing its electrical conductivity by bimetal oxides doping, especially nickel cobaltite based phase, improved the electrocatalytic performance in oxygen reduction reaction.

Supplementary Materials: The following are available online at http://www.mdpi.com/1996-1944/12/15/2446/s1, Figure S1: Figure S1: Nitrogen isotherms at $-196\,°C$ for samples: (**a**) $NiFe_2O_4$/Co-CX, □; $NiCo_2O_4$/Co-CX, ◇; $CoFe_2O_4$/Co-CX, ○ and (**b**) $NiFe_2O_4$/Ni-CX, □; $NiCo_2O_4$/Ni-CX, ◇; $CoFe_2O_4$/Ni-CX, ○. Adsorption curve—open symbols; desorption curve—closed symbols. Figure S2: EDXS analysis carried out on sample $NiCo_2O_4$-CoCX.

Author Contributions: Conceptualization: all the authors; methodology: A.A.; analysis and discussion of results: all the authors; XPS data analysis: F.C.-M.; writing—all the authors; project administration: A.A. and A.F.P.-C.; funding acquisition: A.A. and A.F.P.-C.

Funding: This research was funded by science and technology development fund [STDF] grant number [STF-25402] and from the project P12-RNM-2892 (Junta de Andalucía).

Acknowledgments: The authors greatly acknowledge the financial support from STDF and from Junta de Andalucía.

Conflicts of Interest: The authors declare no conflict of interest.

References

1. Winter, M.; Brodd, R.J. What are batteries, fuel cells and supercapacitors. *Chem. Rev.* **2004**, *104*, 4245–4270. [CrossRef] [PubMed]
2. Hoogers, G. *Fuel Cell Technology Handbook*; CRC Press: Boca Raton, FL, USA, 2003; pp. 4–23.
3. Carrette, L.; Friedrich, K.A.; Stimming, U. Fuel cells: principles, types fuels and applications. *ChemPhysChem* **2000**, *1*, 162–193. [CrossRef]
4. Mahala, C.; Basu, M. Nanosheets of $NiCo_2O_4$/NiO as efficient and stable electrocatalyst for oxygen evolution reaction. *ACS Omega* **2017**, *2*, 7559–7567. [CrossRef]

5. Shin, D.; An, X.; Choun, M.; Lee, J. Effect of transition metal induced pore structure on oxygen reduction reaction of electrospun fibrous carbon. *Catal. Today* **2016**, *260*, 82–88. [CrossRef]

6. Wang, Y.J.; Zhao, N.; Fang, B.; Li, H.; Bi, X.T.; Wang, H. Carbon-supported Pt-based alloy electrocatalysts for the oxygen reduction reaction in polymer electrolyte membrane fuel cells: particle size, shape, and composition manipulation and their impact to activity. *Chem. Rev.* **2015**, *115*, 3433–3467. [CrossRef]

7. Huang, X.; Zhao, Z.; Cao, L.; Chen, Y.; Zhu, E.; Lin, Z.; Li, M.; Yan, A.; Zettl, A.; Wang, Y.M. High-performance transition metal–doped Pt_3Ni octahedra for oxygen reduction reaction. *Science* **2015**, *348*, 1230–1234. [CrossRef] [PubMed]

8. Jayasayee, K.; Van Veen, J.R.; Manivasagam, T.G.; Celebi, S.; Hensen, E.J.; De Bruijn, F.A. Oxygen reduction reaction (ORR) activity and durability of carbon supported PtM (Co, Ni, Cu) alloys: Influence of particle size and non-noble metals. *Appl. Catal. B Environ.* **2012**, *111*, 515–526. [CrossRef]

9. Paulus, U.; Wokaun, A.; Scherer, G.; Schmidt, T.; Stamenkovic, V.; Radmilovic, V.; Markovic, N.; Ross, P. Oxygen reduction on carbon-supported Pt-Ni and Pt-Co alloy catalysts. *J. Phys. Chem. B* **2002**, *106*, 4181–4191. [CrossRef]

10. Wang, B. Recent development of non-platinum catalysts for oxygen reduction reaction. *J. Power Sources* **2005**, *152*, 1–15. [CrossRef]

11. Wang, H.; Holt, C.M.; Li, Z.; Tan, X.; Amirkhiz, B.S.; Xu, Z.; Olsen, B.C.; Stephenson, T.; Mitlin, D. Graphene-nickel cobaltite nanocomposite asymmetrical supercapacitor with commercial level mass loading. *Nano Res.* **2012**, *5*, 605–617. [CrossRef]

12. Zhang, G.; Xia, B.Y.; Wang, X.; Lou, X.W. Strongly coupled $NiCo_2O_4$-rGO hybrid nanosheets as a methanol-tolerant electrocatalyst for the oxygen reduction reaction. *Adv. Mater.* **2014**, *26*, 2408–2412. [CrossRef] [PubMed]

13. Chu, Y.-Q.; Fu, Z.-W.; Qin, Q.-Z. Cobalt ferrite thin films as anode material for lithium ion batteries. *Electrochim. Acta* **2004**, *49*, 4915–4921. [CrossRef]

14. Yasmin, S.; Cho, S.; Jeon, S. Electrochemically reduced graphene-oxide supported bimetallic nanoparticles highly efficient for oxygen reduction reaction with excellent methanol tolerance. *Appl. Surf. Sci.* **2018**, *434*, 905–912. [CrossRef]

15. Gong, K.; Du, F.; Xia, Z.; Durstock, M.; Dai, L. Nitrogen-doped carbon nanotube arrays with high electrocatalytic activity for oxygen reduction. *Science* **2009**, *323*, 760–764. [CrossRef]

16. Pekala, R.; Alviso, C.; Kong, F.; Hulsey, S. Aerogels derived from multifunctional organic monomers. *J. Non-Cryst. Solids* **1992**, *145*, 90–98. [CrossRef]

17. Pekala, R. Organic aerogels from the polycondensation of resorcinol with formaldehyde. *J. Mater. Sci.* **1989**, *24*, 3221–3227. [CrossRef]

18. EL-Deeb, M.M.; El Rouby, W.M.; Abdelwahab, A.; Farghali, A.A. Effect of pore geometry on the electrocatalytic performance of nickel cobaltite/carbon xerogel nanocomposite for methanol oxidation. *Electrochim. Acta* **2018**, *259*, 77–85. [CrossRef]

19. Zapata-Benabithe, Z.; Carrasco-Marín, F.; de Vicente, J.; Moreno-Castilla, C. Carbon xerogel microspheres and monoliths from resorcinol–formaldehyde mixtures with varying dilution ratios: Preparation, surface characteristics, and electrochemical double-layer capacitances. *Langmuir* **2013**, *29*, 6166–6173. [CrossRef] [PubMed]

20. Pérez-Cadenas, A.F.; Ros, C.H.; Morales-Torres, S.; Pérez-Cadenas, M.; Kooyman, P.J.; Moreno-Castilla, C.; Kapteijn, F. Metal-doped carbon xerogels for the electro-catalytic conversion of CO_2 to hydrocarbons. *Carbon* **2013**, *56*, 324–331. [CrossRef]

21. Abdelwahab, A.; Castelo-Quibén, J.; Pérez-Cadenas, M.; Elmouwahidi, A.; Maldonado-Hódar, F.J.; Carrasco-Marín, F.; Pérez-Cadenas, A.F. Cobalt-Doped Carbon Gels as Electro-Catalysts for the Reduction of CO_2 to Hydrocarbons. *Catalysts* **2017**, *7*, 25. [CrossRef]

22. Moreno-Castilla, C.; Maldonado-Hódar, F. Carbon aerogels for catalysis applications: An overview. *Carbon* **2005**, *43*, 455–465. [CrossRef]

23. Abdelwahab, A.; Castelo-Quibén, J.; Vivo-Vilches, J.F.; Pérez-Cadenas, M.; Maldonado-Hódar, F.J.; Carrasco-Marín, F.; Pérez-Cadenas, A.F. Electrodes Based on Carbon Aerogels Partially Graphitized by Doping with Transition Metals for Oxygen Reduction Reaction. *Nanomaterials* **2018**, *8*, 266. [CrossRef] [PubMed]

24. Maldonado-Hódar, F.; Moreno-Castilla, C.; Pérez-Cadenas, A. Surface morphology, metal dispersion, and pore texture of transition metal-doped monolithic carbon aerogels and steam-activated derivatives. *Microporous Mesoporous Mater.* **2004**, *69*, 119–125. [CrossRef]

25. Job, N.; Pirard, R.; Marien, J.; Pirard, J.-P. Synthesis of transition metal-doped carbon xerogels by solubilization of metal salts in resorcinol–formaldehyde aqueous solution. *Carbon* **2004**, *42*, 3217–3227. [CrossRef]

26. Moreno-Castilla, C.; Maldonado-Hódar, F.; Pérez-Cadenas, A. Physicochemical surface properties of Fe, Co, Ni, and Cu-doped monolithic organic aerogels. *Langmuir* **2003**, *19*, 5650–5655. [CrossRef]

27. Abdelwahab, A.; Castelo-Quibén, J.; Pérez-Cadenas, M.; Maldonado-Hódar, F.J.; Carrasco-Marín, F.; Pérez-Cadenas, A.F. Insight of the effect of graphitic cluster in the performance of carbon aerogels doped with nickel as electrodes for supercapacitors. *Carbon* **2018**, *139*, 888–895. [CrossRef]

28. Chmiola, J.; Yushin, G.; Dash, R.; Gogotsi, Y. Effect of pore size and surface area of carbide derived carbons on specific capacitance. *J. Power Sources* **2006**, *158*, 765–772. [CrossRef]

29. Gryglewicz, G.; Machnikowski, J.; Lorenc-Grabowska, E.; Lota, G.; Frackowiak, E. Effect of pore size distribution of coal-based activated carbons on double layer capacitance. *Electrochim. Acta* **2005**, *50*, 1197–1206. [CrossRef]

30. de Lima Alves, T.M.; Amorim, B.F.; Torres, M.A.M.; Bezerra, C.G.; de Medeiros, S.N.; Gastelois, P.L.; Outon, L.E.F.; de Almeida Macedo, W.A. Wasp-waisted behavior in magnetic hysteresis curves of $CoFe_2O_4$ nanopowder at a low temperature: experimental evidence and theoretical approach. *RSC Adv.* **2017**, *7*, 22187–22196. [CrossRef]

31. Yang, Y.; Zeng, D.; Yang, S.; Gu, L.; Liu, B.; Hao, S. Nickel cobaltite nanosheets coated on metal-organic framework-derived mesoporous carbon nanofibers for high-performance pseudocapacitors. *J. Colloid Interface Sci.* **2019**, *534*, 312–321. [CrossRef]

32. Jokar, E.; Shahrokhian, S. Synthesis and characterization of $NiCo_2O_4$ nanorods for preparation of supercapacitor electrodes. *J. Solid State Electrochem.* **2015**, *19*, 269–274. [CrossRef]

 *materials* 

Article

Green Synthesis of Three-Dimensional Hybrid N-Doped ORR Electro-Catalysts Derived from Apricot Sap

Ramesh Karunagaran [1], Campbell Coghlan [2], Cameron Shearer [3], Diana Tran [1], Karan Gulati [1], Tran Thanh Tung [1], Christian Doonan [2] and Dusan Losic [1,*]

[1]   School of Chemical Engineering, University of Adelaide, Adelaide, SA 5005, Australia;
      ramesh.karunagaran@adelaide.edu.au (R.K.); diana.tran@adelaide.edu.au (D.T.);
      k.gulati@griffith.edu.au (K.G.); tran.tung@adelaide.edu.au (T.T.T.)
[2]   School of Chemistry, University of Adelaide, Adelaide, SA 5005, Australia;
      cam.coghlan@adelaide.edu.au (C.C.); christian.doonan@adelaide.edu.au (C.D.)
[3]   School of Chemical and Physical Sciences, Flinders University, Adelaide, SA 5042, Australia;
      cameron.shearer@flinders.edu.au
*   Correspondence: dusan.losic@adelaide.edu.au; Tel.: +61-8-8013-4648

Received: 10 January 2018; Accepted: 26 January 2018; Published: 28 January 2018

Abstract: Rapid depletion of fossil fuel and increased energy demand has initiated a need for an alternative energy source to cater for the growing energy demand. Fuel cells are an enabling technology for the conversion of sustainable energy carriers (e.g., renewable hydrogen or bio-gas) into electrical power and heat. However, the hazardous raw materials and complicated experimental procedures used to produce electro-catalysts for the oxygen reduction reaction (ORR) in fuel cells has been a concern for the effective implementation of these catalysts. Therefore, environmentally friendly and low-cost oxygen reduction electro-catalysts synthesised from natural products are considered as an attractive alternative to currently used synthetic materials involving hazardous chemicals and waste. Herein, we describe a unique integrated oxygen reduction three-dimensional composite catalyst containing both nitrogen-doped carbon fibers (N-CF) and carbon microspheres (N-CMS) synthesised from apricot sap from an apricot tree. The synthesis was carried out via three-step process, including apricot sap resin preparation, hydrothermal treatment, and pyrolysis with a nitrogen precursor. The nitrogen-doped electro-catalysts synthesised were characterised by SEM, TEM, XRD, Raman, and BET techniques followed by electro-chemical testing for ORR catalysis activity. The obtained catalyst material shows high catalytic activity for ORR in the basic medium by facilitating the reaction via a four-electron transfer mechanism.

Keywords: oxygen reduction reaction (ORR); catalysis; carbon nanotubes; carbo microsphere; N–doped carbon

1. Introduction

The continued rise in global energy demand and the depletion of the world's non-renewable resources has initiated a global push towards renewable energy sources. Among the most promising methods for producing renewable energy are fuel cells, which have emerged as a promising avenue of research due to their ability to generate high power density [1]. Fuel cells are devices that electrochemically combine gaseous fuel (e.g., hydrogen) and an oxidant gas (e.g., oxygen) to produce electricity and heat by an oxygen reduction reaction (ORR) [1]. The efficiency of fuel cells and their practical applicability is dependent on the ORR catalyst present in the cell [2]. However, slow kinetics have hindered fuel cells from being utilised outside of a laboratory environment [3]. Currently, platinum (Pt) catalysts have outperformed all other catalysts in areas such as activity, stability, and selectivity [4] and have dominated the fuel cell industry as the preferred ORR catalysts [5–7]. However, these catalysts have been overlooked for industrial scale-up due to their high cost and low availability [8]. To overcome this problem, non-precious transition metals (Fe, Co, Ni) in addition to hetero atoms have been trialed to enhance the ORR activity [9–13]. The transition metals have the ability to significantly increase ORR catalytic activity by facilitating the incorporation of hetero atoms, such as nitrogen, in the carbon matrix during pyrolysis [14–17]. The incorporation of electro-negative nitrogen into a graphitic carbon framework has shown to induce high positive charge density on adjacent carbon atoms [18]. The electron donor properties of nitrogen-doped adjacent carbon atoms trigger a favourable diatomic O_2 adsorption and ultimately weaken the O_2 bond strength to facilitate ORR activity [15,18–21].

The high electrical conductivity of the mesoporous carbon materials [22] and their metal oxide hybrids has been utilised in applications such as lithium-ion batteries [23,24], super capacitors [25,26], and catalysts [27–29] in recent years. Various chemical approaches have been developed to synthesise nitrogen-doped carbon materials for ORR catalysis utilising materials such as graphene [30] and carbon nanotubes (CNTs) [31]. Carbon materials doped with nitrogen precursors, such as melamine ($C_3H_6N_6$) [32,33], ethylene diamine ($C_2H_4(NH_2)_2$) [34], o-phenylenediamine ($C_6H_8N_2$) [11], and ammonia (NH_3) [35], have shown high ORR activity. Mesoporous N-doped carbon spheres synthesised using multiple different methods, such as the one-pot soft template method [22], spray pyrolysis [36,37], and self-polymerisation [38], each show outstanding catalytic potential for ORR catalysis. However, high cost, hazardous chemical usage and waste has limited their translation into scale-up industrial applications [22,36–38].

To address this problem, green chemistry approaches using low-cost natural materials to synthesise mesoporous carbon (e.g., plant *Typha orientalis* [39], catkin [40], lignin [41], and soya chunks [42]) have been successfully demonstrated as efficient ORR catalysts. Apricot trees (*Prunus armeniaca* L.) are widely cultivated in areas where a scarcity of water remains the main obstacle for cultivation [43]. In many regions, apricot trees suffer from gummosis, a bark disease [44] resulting in the formation of sap which oozes out from the wounds caused from factors including weather, infection, insects, or mechanical damage. The sap commonly appears as an amber-coloured material, which contains various sugar components, such as xylose, arabinose, rhamnose, glucose, mannose, and galactose [45]. Analysis performed by Lluveras et al. [45] revealed that apricot sap consists of polysaccharides, primarily arabinose and galactose. Polysaccharides, such as galactose, glucose, sucrose, and starch, have been shown to undergo dehydration and subsequent aromatisation when hydrothermally treated at 160–200 °C, resulting in their conversion to char material with nano- or micrometer-size smooth carbon spheres [27,46].

Carbon microspheres (CMS) have recently attracted attention due to their unique applications, high density, and high strength in carbon product fabrication [47]. Carbon-spheres synthesised using polysaccharides have successfully been implemented in the application of catalysis for synthetic fuel [27], Li-ion batteries [48], and electrochemical capacitors [49]. Previously, hybrid CMS containing transition metals have been synthesised using polysaccharides [27,46]. The ability of the iron oxides to bind with the oxygen functional groups in the sugar molecules through coulombic and/or electrostatic interactions has resulted in the formation of iron oxide encapsulated carbon spheres [27]. Similar hybrid materials can be synthesised using the polysaccharides present in apricot sap, which have not been used for any catalytic application in the past.

This work explores the use of apricot sap containing sugar molecules as a natural source and method for the generation of a new type of three-dimensional (3D) hybrid N-doped ORR electro-catalysts composed of microspherical and nanotubular structures. These catalysts were synthesised through a three-step process as shown in Scheme 1. Firstly, an apricot sap resin suspension containing polysaccharides of arabinose and galactose was prepared. Secondly, the apricot resin solution was hydrothermally treated with iron oxide nanoparticle or cobalt precursors to obtain a char material with carbon microspheres embedded with magnetic nanoparticles. Finally, the char material was pyrolysed (950 °C) with a nitrogen precursor of melamine to dope the graphitic carbons with nitrogen. The pyrolysed composite material forms an integrated composite material with both carbon fibers (CFs) and CMS. A similar integrated structure was reported in our previous paper, where we hypothesised that the decomposition of melamine during pyrolysis causes disruption to the iron oxide magnetic nanoparticle clusters' (FeMNPC) surface that is embedded in the carbon sphere to diffuse FeMNPC particles out of the sphere to catalyse the formation of N-doped carbon fibers (N-CFs) [50]. This hybrid carbon catalyst contains N-CFs and N-doped carbon microspheres (N-CMS) with magnetic nanoparticles, forming a unique 3D intergrated morphology.

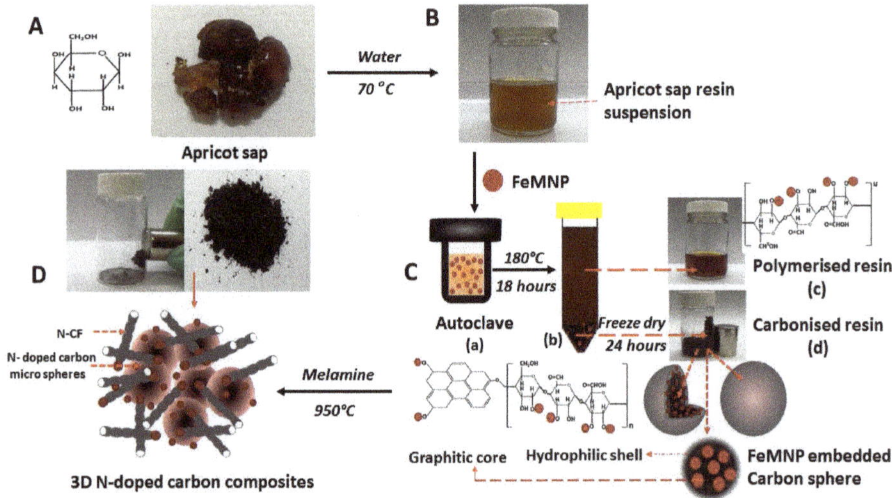

Scheme 1. Schematic procedure of three-dimensional (3D)-integrated N-doped carbon microspheres (CMS) and N-doped carbon fibers N-CFs catalysts from apricot sap. (**A**) Apricot sap collected from the apricot tree; (**B**) apricot sap dissolved in water (apricot resin suspension); (**C**) apricot resin suspension containing FeMNP, hydrothermally treated to produce magnetic insoluble char material (carbonised resin) with FeMNP embedded CMS structures (HT-APG-Fe); and (**D**) HT-APG-Fe pyrolised with melamine to form N-doped integrated structures containing N-CFs and N-CMS (N-APG-Fe). FeMNP: iron oxide magnetic nanoparticle.

The structural and chemical composition of the prepared N-doped 3D integrated catalyst with FeMNPs (N-APG-Fe) and cobalt nanoparticle clusters (CoMNPs) (N-APG-Co) were characterized with SEM, TEM, XRD, Raman, and BET followed by testing their electrochemical catalytic properties and ORR activity. The conversion process of the naturally occurring waste and apricot sap material into an effective electro-catalyst for ORR reaction is also described.

2. Results and Discussion

2.1. Formation of Integrated Morphology of N-CFs and N-CMS

The steps involved in the synthesis of N-APG-Fe are shown in Scheme 1. In the first step, apricot sap (Scheme 1A) was dissolved in water (70 °C) to make a translucent light-orange colour resin suspension (Scheme 1B). In the second step, the resin suspension was hydrothermally treated in the presence of FeMNPs. During this process, the oxygen functional groups (i.e., OH and C=O) bind to the iron oxide particles through Coulombic interactions to form a hybrid material [27]. During the hydrothermal process (Scheme 1(Ca)), sugar molecules polymerise (Scheme 1(Cb)) to form intermolecular crosslinks between linear or branched oligosaccharides due to dehydration [46]. As a result of dehydration and polymerisation, oxygen functional groups associated with the sugar molecules were reduced along with their negative charges to make the polymerised material more water-insoluble. Subsequently, the insoluble material (char) settles as FeMNPs-embedded spheres with a hydrophobic core and hydrophilic shell [46,51,52] (Scheme 1(Cc)). In the final process, the hydrothermally obtained char is pyrolysed in the presence of melamine at 950 °C to introduce N atoms into the carbon framework. The pyrolysed composite material forms an integrated composite material with both N-CF and N-microspheres (Scheme 1D). During pyrolysis, the decomposition of melamine caused disruption to the spheres' surface and caused the FeMNPCs embedded within the sphere to diffuse out [50], which catalysed the formation of N-CF [53]. The synthesised hybrid material, which consists of both N-CF and N-CMS, formed a unique 3D integrated morphology.

2.2. Structual and Chemical Characterisation of Prepared 3D N-Doped Carbon Composites

The morphology of the hydrothermally induced char material (HT-APG-Fe) was determined using SEM (Figure 1), which shows the formation of carbon microspheres (1–6 μm). The image of a broken sphere (Figure 1A, inset) shows FeMNPCs embedded within the microspheres. EDX analysis conducted on the particles (Figure S1 in supporting information (SI)) indicated that an average of 57% (wt %) of the material consisted of Fe, confirming the presence of FeMNPCs in the carbon sphere. The magnetic property of the material was confirmed by applying an external magnet to the sample.

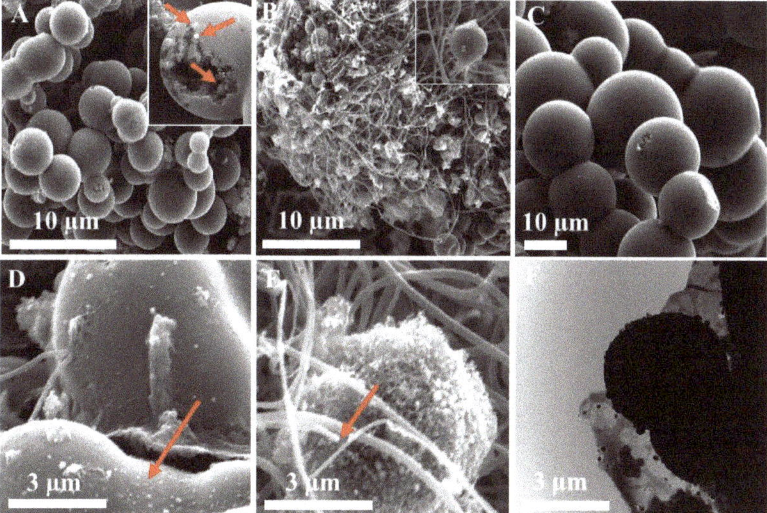

Figure 1. SEM images of (**A**) carbon microspheres formed from char material of hydrothermally treated apricot resin (HT-APG-Fe), (**B**) integrated structure composed of CMS and CFs of HT-APG-Fe pyrolysed with melamine at 950 °C (N-APG-Fe) (inset shows the presence of micro spheres and CFs), (**C**) carbon micro spheres of HT-APG-Fe pyrolysed without melamine at 950 °C (APG-Fe), (**D**) formation of CFs from FeMNPC from the sphere interior of N-APG-Fe (red arrow shows FeMNPC diffused out of the sphere after pyrolysis with melamine), (**E**) formation of CF from FeMNPC diffused out of the sphere in N-APG-Fe (red arrow shows the CF forming from the tip of FeMNPC), and (**F**) TEM image of FeMNPC diffused out of the sphere in N-APG-Fe.

In order to make HT-APG-Fe catalytically active, it was pyrolysed with a N precursor (melamine) (N-APG-Fe) at 950 °C to introduce N atoms into the carbon framework and improve catalytic properties (Figure 1B). The nitrogen doping eliminates the electro-neutrality of the carbon framework and generates favourable charged sites for oxygen adsorption [19,54]. Similarly, to compare the catalytic activity of the N-doped and non N-doped catalysts, HT-APG-Fe was pyrolysed without melamine (APG-Fe) at 950 °C and the SEM image is presented in Figure 1C. The SEM revealed that APG-Fe formed interconnected smooth microspheres in the range of 1–6 μm. Interestingly, in contrast to the smooth CMS formed in APG-Fe (Figure 1C), the catalysts pyrolysed with melamine (N-APG-Fe) (Figure 1B) formed an integrated composite material with both CF and CMSs. EDX analysis was performed to determine the N-doping on CMS and CF and revealed an average of 2.55 and 2.04 (At %) of N presented in the CMS and the CF, respectively. During pyrolysis, the decomposition of melamine causes disruption to the FeMNPC's surface that is embedded in the carbon sphere for the FeMNPC particles to diffuse out of the sphere and form N-CFs (Figure 1D,F).

To demonstrate if this synthetic procedure is generic for the formation of 3D integrated N-CMS and N-CF structures, we repeated the procedure using cobalt precursors, which is commonly used as an alternate transition metal to fabricate ORR catalysts. The SEM images of the hydrothermally produced structures (Figure S2A), pyrolysed with and without melamine (Figure S2B,C), respectively, revealed that integrated structures with carbon spheres and CFs, similar to N-APG-Fe, were produced when the hydrothermally reduced char materials containing cobalt oxide nanoparticles were pyrolyzed with melamine. The hydrothermally produced char material without any nanoparticles (HT-APG, Figure S3A), when pyrolysed with melamine (Figure S3B), did not produce any integrated products of CMS and CFs. As the integrated structures were only seen on the catalysts with transition metals (Figures 1B and S2B in SI), we deduce that both a transition metal oxide and a nitrogen precursor are needed for the synthesis of the integrated structure comprised of both N-CMS and N-CFs. Previously, we reported a 3D integrated structure of N-CMS and N-CFs using sugar galactose (N-GAL-Fe) [50]. A similar morphology was observed in N-APG-Fe, which demonstrates that the presence of galactose in the apricot sap also contributes significantly to the formation of the integrated structure.

The morphologies of N-CF in N-APG-Fe and N-APG-Co (Figure 2A,B) were further investigated with TEM. The images clearly illustrate that the N-CFs originate from the tip of the MNPC. An EDX analysis was conducted on the particles at the tip of the CF (Figure 2A,B) and revealed 59.20% and 18.50% (wt.%) of Fe and Co, respectively, suggesting that the CFs are formed by a metal-induced mechanism [53]. The TEM images of CF from N-APG-Fe (Figure 2C) and N-APG-Co (Figure 2D) show that the CFs possess an irregular corrugated morphology with a width of approximately 150–500 nm, similar to those reported by M. Terrones et al. [55]. To investigate the presence of N-doping on these CFs (which facilitate ORR) [31], an EDX elemental analysis was performed on CFs grown from Fe (Fe-CF) and Co (Co-CF). The N-content was found to be 2.04 and 6.32 At. % for Fe-CF and Co-CF, respectively, compared to 0% in the non-doped sample, confirming nitrogen doping on CF. This reveals that C and N precursors from pyrolysed melamine had diffused into the metal clusters to form the CF [53,56].

The XRD analysis conducted on N-APG, APGFe-N, and APGCo-N is shown in Figure 3A. The diffraction peaks seen at 25.78°, 42.66°, and 44.83° for APGFe-N and APGCo-N correspond to diffraction facets (002), (110), and (101), respectively, assigned to the presence of graphitic carbon [57–59]. Similarly, N-APG showed peaks at 23.96°, 42.53°, and 44.67° for diffraction facets (002), (110), and (101), respectively. The positive shift of the (002) peak of N-APG from 23.96° to 25.78° in N-APG-Fe and N-APG-Co can be assigned to the formation of the graphitic crystalline structure induced by the reduction of the oxygen functional group containing sugar molecules with a metal MNPC [60].

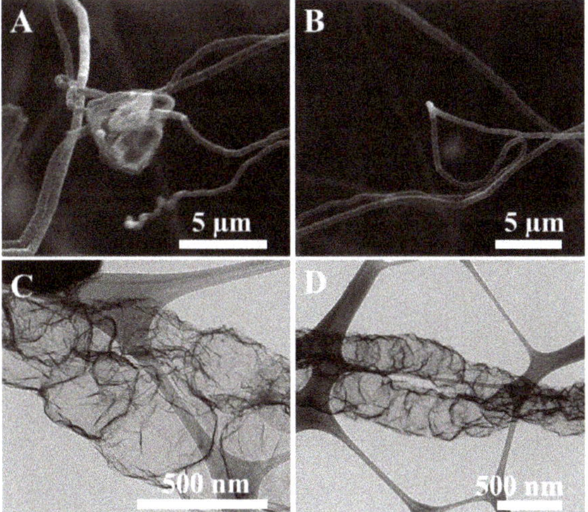

Figure 2. SEM images of N-CF obtained from (**A**) N-APG-Fe and (**B**) N-APG-Co. TEM image of (**C**) N-APG-Fe and (**D**) N-APG-Co.

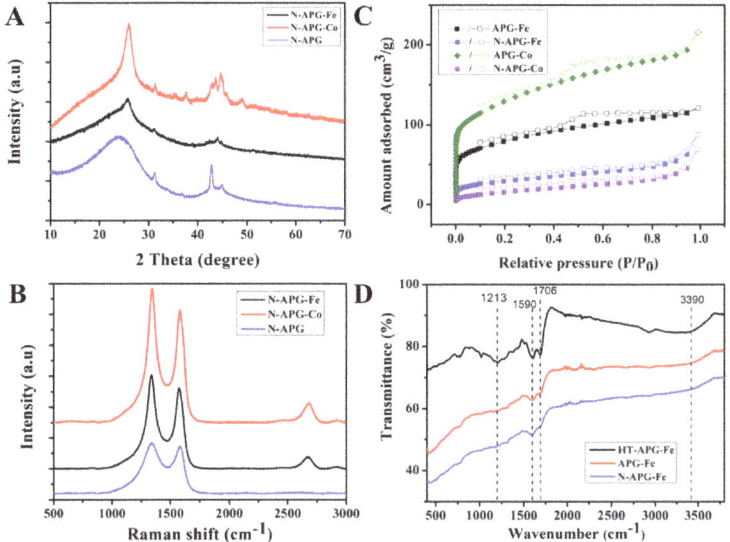

Figure 3. (**A**) XRD spectrum of N-APG, N-APG-Fe, and N-APG-Co, (**B**) Raman spectrum of N-APG, N-APG-Fe, and N-APG-Co, (**C**) N_2 adsorption/desorption isotherm of APG-Fe, N-APG-Fe, N-APG-Co, and N-APG-Co, (**D**) FTIR spectrum of HT-APG-Fe, APG-Fe, and N-APG-Fe.

The Raman spectrum performed on N-APG, N-APG-Fe, and N-APG-Co is shown in Figure 3B. N-APG-Fe and N-APG-Co show three characteristic peaks at 1342, 1581, and 2684 cm^{-1} for the D, G, and 2D bands, respectively, while N-APG showed only the D and the G band at 1338 and 1583 cm^{-1}, respectively. The additional 2D band indicates the presence of crystalline graphitic carbon material formed during the annealing process. This was facilitated by the reduction of oxygen groups in the sugar molecules by the addition of FeMNPC [61,62]. The I_D/I_G of N-APG (1.07), N-APG-Fe (1.16), and N-APG-Co (1.13) shows the disruption of sp^2 bonds and the formation of sp^3 defect sites [63,64], which are associated with the N-doping on the carbon framework. The higher I_D/I_G for N-APG-Fe and N-APG-Co compared to N-APG revealed that the transition metal particles have facilitated the incorporation of N atoms to the carbon framework to distort the graphitic framework [15,65]. This shows that the addition of transition metals formed greater positive sites on the adjacent carbon atoms to adsorb oxygen, thus enhancing the ORR activity.

The N$_2$ sorption isotherms of non-doped APG-Fe and APG-Co differ from the doped N-APG-Fe and N-APG-Co (Figure 3C). The characteristic type IV isotherm and H4 hysteresis loop for APG-Fe and APG-Co shows the presence of mesoporous slip-like pores [62,66,67] with mean pore size distributions of 4.64, 5.62, 7.68, and 13.84 Å (Figure S4). The surface area of the prepared catalysts was measured using Brunauer-Emmett-Teller (BET) and is shown in Table 1. The surface area measured for the doped catalysts was much lower than that of the non-doped catalysts. We hypothesise that the mesopores on the surface of the carbon microspheres on the non-doped catalysts contributed to the higher surface area. The significant reduction in the surface area of the doped sample can be assigned to the disruption of these mesopores or blocked pores due to the decomposition of melamine during pyrolysis.

Table 1. Surface area of doped and non-doped apricot catalysts with Fe and Co.

Catalyst	Surface Area (m^2/g)
APG-Fe	235.38
N-APG-Fe	73.15
APG-Co	375.62
N-APG-Co	39.86

The presence of any carbonyl groups was analysed by FTIR, which can form condensation products with melamine. The FTIR spectra of hydrothermally treated (HT-APG-Fe), pyrolysed without melamine (APG-Fe), and N-doped (N-APG-Fe) materials are presented in Figure 3D. HT-APG-Fe showed characteristic peaks at 1213, 1590, 1706, and 3390 cm^{-1} [68–70], which can be attributed to C-O and C-H stretching, the stretching vibration of C=O, carbonyl vibrations, and the stretching vibration of O-H, respectively. The presence of oxygen functional groups suggests that the carbon spheres were formed with a hydrophilic shell containing oxygen groups as suggested by Sun et al. [46] and Mer et al. [54]. When the peaks corresponding to the carbonyl groups of doped N-APG-Fe and non-doped APG-Fe were compared, a reduction of the intensity of the N-APG-Fe was observed. Since carbonyl groups can interact with the amine group of melamine [71], we hypothesised that the melamine was attached to the carbon spheres before undergoing complete decomposition. It is likely that these condensation products caused surface disruption of the microspheres and studies need to be undertaken to confirm this hypothesis.

XPS measurements were performed to determine the nitrogen species present in the N-APG-Fe catalyst. The high-resolution XPS C 1s spectrum (Figure S5) showed a variety of carbon bonds, including C-C (285.04 eV), C-N (286.03 eV), and O-C=O (290.03 eV) [72–74]. The high-resolution N 1s XPS spectra displayed in Figure S5B showed three distinct peaks, including pyridine-N (398.38 eV), graphitic-N (401.35 eV), and nitrogen oxide-N (403.12 eV) [75,76]. The high percentage (45.57 At. %) of pyridinic N (in the N1s analysis) along with 63.03 At. % of C-N (in the C1s analysis) in N-APG-Fe clearly demonstrates efficient N-doping on the carbon framework to facilitate O$_2$ adsorption. Similar peaks for C 1s and N 1s were observed for N-APG-Co (Figure S6). The N-doping on the graphitic carbon framework altered the electro-neutrality of the nano-carbon material. The pyridinic nitrogen with its strong electronic affinity induced high positive charge density on the adjacent carbon atoms. Thus, the electron donor properties of the N-doped adjacent carbon atoms are favourable for weakening the strength of the O-O bond to facilitate ORR activity [2,19,77]. However, the XPS analysis did not show any presence of Fe or Co. This showed that Fe-N-C sites were not formed and only N-C carbon has been formed.

2.3. Electrochemical Characterisation of Catalytic Performance

The electrochemical activity of N-doped (N-APG, N-APG-Fe, and N-APG-Co) catalysts were examined by cyclic voltammetry (CV) (Figure S7). The voltammograms between 0 and 1.2 V show well-defined cathodic peaks centered at 0.55, 0.67, and 0.74 V, respectively. The voltammograms of N-APG-Fe and N-APG-Co showed a higher positive overpotential shift, 120 and 190 mV, respectively, compared N-APG. These results indicate that the hybrid structures of N-CF and N-CMS formed by the introduction of Fe and Co have aided to increase ORR activity and O_2 uptake. Since the XPS did not detect any Fe-N-C active sites, these enhancements of ORR activity can be attributed to N-C catalytic sites merely in both the N-CF and N-CMS in the hybrid structure in N-APG-Fe and N-APG-Co.

To understand the reaction kinetics of N-APG, N-APGFe and N-APGCo, Rotating Ring Disc Electrode(RRDE) was employed to quantify the overall electron transfer number (n) and percentage of hydrogen peroxide (% HO_2^-). To explore the dependence on galactose sugar in the electron transfer kinetics, N-GAL-Fe was contrasted against these catalysts in the potential range between 0.10–1.15 V and the ring (Figure 4A) and disc (Figure 4B) currents. The onset potential measured for these catalysts (Table 2) showed a positive shift for N-APG-Fe (0.88 V) and N-APG-Co (0.86 V) compared to N-APG (0.84 V), revealing that the hybrid structures have initiated the ORR faster. However, the half-wave potential ($E_{1/2}$) of all of these catalysts shifted negatively compared to the standard Pt/C. The negative shift in the $E_{1/2}$ is due to the absence of any M-N-C catalytic active sites present in the catalysts. Liu et al. [78] described that M-N-C active sites perform the ORR reaction with a more positive $E_{1/2}$ compared to N-C active sites in the catalysts.

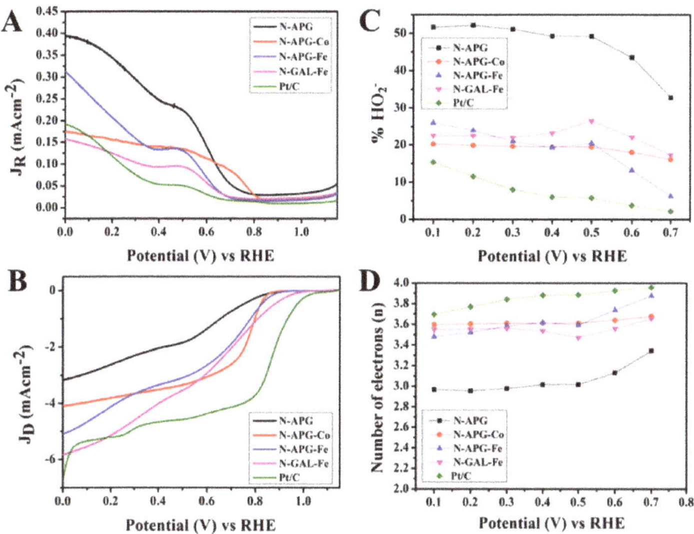

Figure 4. Rotating ring disc voltammograms of (**A**) ring current, (**B**) disc current of N-APG, N-APG-Co, N-APG-Fe, N-GAL-Fe, and Pt/C electrodes in oxygen-saturated 0.10 M KOH at 2000 rpm at a scan rate of 10 mV/s. (**C**) Percentage HO2- and (**D**) number of electrons of N-APG, N-APG-Fe, N-APG-Co, and Pt/C electrodes at various potential calculated according to RRDE data. RHE: reversible hydrogen electrode

Table 2. Surface area of doped and non-doped apricot catalysts with Fe and Co.

Product	Current density (mA/cm^2) at 0 V (RHE)	Onset Potential (RHE) (V)	Number of Electrons (n) (0.10–0.70 V) (RHE)	% HO$_2^-$ (0.10–0.70 V) (RHE)
N-APG	3.05	0.84	2.96–3.34	51.64–32.77
N-APG-Co	4.03	0.86	3.59–3.67	20.16–16.08
N-APG-Fe	4.91	0.88	3.48–3.87	25.99–6.19
N-GAL-Fe	5.81	0.96	3.54–3.65	22.54–17.25
Pt/C	6.70	1.04	3.69–3.95	15.20–2.15

The number of electrons transferred using N-APG, N-APG-Co, N-APG-Fe, and N-GAL-Fe catalyst electrodes within the potential region 0.10–0.70 V is shown in Table 2. The electron transfer number towards four of these catalysts reveals that the ORR reaction is carried out predominantly via a four-electron transfer mechanism. The catalytically analysed values and comparison chart of n and % HO$_2^-$ at 0.40 V (Figure S8) shows that both of the N-doped apricot and galactose catalysts follow a similar trend, showing the significance of galactose in the electron transfer mechanism. Unlike the doped catalysts, the non-doped catalysts did not perform effectively (Figure S9). The electro-chemical properties summarised in Table S1 in the SI of these catalysts showed a negative onset potential and lower electron transfer numbers than the doped catalysts. This shows that the doping of nitrogen has created more catalytically active sites for ORR. The stability of the N-APG-Co and N-APG-Fe was determined by cycling the catalysts between 0.00 V and 1.15 V at 100 mVS^{-1} in an oxygen-saturated 0.1 M KOH solution (Figure S10 in the SI). The results show that after 6000 cycles the onset overpotential had increased by 30 mV and 40 mV for N-APG-Co and N-APG-Fe, respectively, indicating only a slight deterioration of the catalysts.

The kinetics of electron transfer using the details obtained from RRDE and the scheme suggested by Damjanovic et al. [79] are described in the SI. The rate constants were calculated based on these equations for the N-APG, N-APG-Co, N-APG-Fe, and N-GAL-Fe in the potential region of 0.10–0.65 V (Figure S11 in SI). The calculated rate constants showed that N-APG-Co, N-APG-Fe, and N-GAL-Fe were predominantly driven by a four-electron k_1 kinetics, while in N-APG, the ORR was carried out via both the k_1 and k_2 pathways. The calculated value of the ratio of k_1/k_2 presented in Table S2 in the SI and Figure 5 showed $k_1/k_2 > 1$ for all catalysts. The higher values of k_1/k_2 for N-APG-Co, N-APG-Fe, and N-GAL-Fe compared to N-APG showed a dominant four-electron k_1 electron transfer pathway for these catalysts that demonstrates that the presence of the hybrid structure of N-CF and N-CMS significantly contributes to the generation of the active sites for oxygen adsorption.

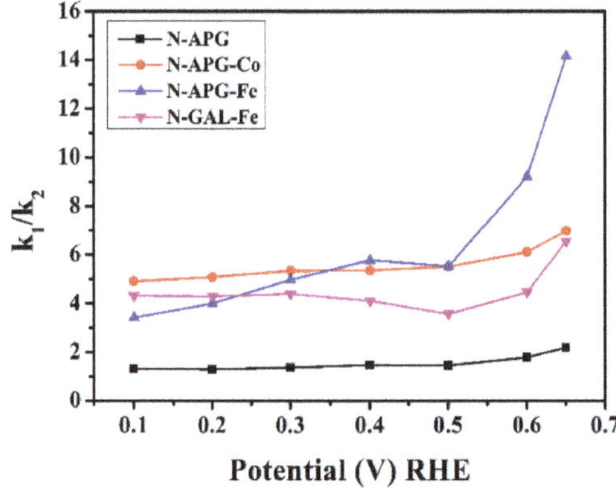

Figure 5. The ratio of rate constant k1/k2 for, N-APG, N-APG-Fe, N-APG-CO and N-GAL-Fe in the potential range of 0.10–0.65V.

In order to compare the efficiency of the N-doped apricot catalysts and galactose catalysts, the materials were contrasted against similar catalysts, and the comparison is presented in Table S3 in the SI. The comparison revealed that these catalysts had similar levels of activity compared with other synthetic material presented in the literature. While the performance of these materials is lower than the highest-performing Pt catalysts, the advantages of this approach are the scalable, stable, low-cost, and natural non-hazardous starting materials and the ease of their synthesis. However, the use of apricot sap in industrial large-scale production may be limited by its low yield. Our previous paper demonstrated the synthesis of a similar hydride structure comprising N-CF and N-CMS to fabricate C-N electrodes for ORR using galactose as the source. This approach has the potential to be implemented to synthesise C-N electrodes with similar integrated hybrid structures using natural and synthetic feedstocks containing polysaccharides. Furthermore, the production of efficient ORR catalysts at a lower cost to current catalysts, using natural resources and environment-friendly processes, may provide a step forward for natural products.

3. Materials and Methods

3.1. Materials

Apricot sap from an Apricot Moorpark tree (*Prunus armeniaca*) (South Australia) was collected from a local garden. Iron (II) chloride tetra hydrate ($FeCl_2 \cdot 4H_2O$) (Sigma Aldrich, St Louis, MO, USA), iron (III) chloride hexa hydrate ($FeCl_3 \cdot 6H_2O$) (Chem Supply, Gillman, Australia), hydrochloric acid (HCl) (Chem Supply, Gillman, Australia), ammonia (Chem Supply, Gillman, Australia), cobalt (II) acetate (Sigma Aldrich, St Louis, MO, USA), melamine (Sigma Aldrich, St Louis, MO, USA), and platinum standard catalyst (20 w% Vulcan XC-72) were used as purchased.

3.2. Methods

3.2.1. Synthesis of Carbonaceous Spheres from Apricot Sap (HT-APG)

The apricot sap (cca 100 g) was cut by a knife from a tree. The sap was washed with fresh water and dried in open air for 12 h. The sap (25 g) was dissolved in water (100 mL) and heated to 70 °C with manual stirring. The resin suspension was sealed and left for 24 h in an open environment. The obtained transparent light orange suspension was then filtered to obtain a contaminant-free resin suspension. The resin suspension (50 mL) was transferred in to an autoclave and heated at 180 °C for 18 h. The char was centrifuged and washed with distilled water (6 × 35 mL). The washed char was then freeze dried for 24 h. The final product weighed 1.62 g.

3.2.2. Synthesis of Cobalt Embedded Carbonaceous Spheres (HT-APG-Co)

Cobalt (II) acetate (150 mg) was dissolved with 50 mL of filtered resin suspension and stirred for 30 min. The product was then transferred to an autoclave and heated for 18 h at 180 °C. The product was cooled to room temperature and transferred in to a centrifuge tube and centrifuged with repeated washing with distilled water for six times (6 × 35 mL) and four times with 0.5 M H_2SO_4. The washed char was then freeze-dried for 24 h.

3.2.3. Synthesis of Maghemite Nanoparticles

Maghemite nanoparticles were synthesised according to the previously established method [80]. Briefly, $FeCl_2 \cdot 4H_2O$ (39.76 g) and $FeCl_3 \cdot 6H_2O$ (16.29 g) were dissolved in 1 M HCl (100 mL). The solution was stirred for 2 h and the pH adjusted to 9.8 using 2 M ammonia solution. Finally, the product was centrifuged and washed three times with distilled water (35 mL) and once with ethanol (35 mL) and dried for 6 h at 60–70 °C.

3.2.4. Synthesis of Fe-Embedded Carbonaceous Spheres (HT-APG-Fe)

Maghemite nanoparticles (200 mg) were suspended in the filtered resin suspension (50 mL) and stirred for 30 min, transferred to an autoclave, and heated for 18 h at 180 °C. The product was collected and centrifuged by washing with distilled water (6 × 35 mL) and four times with 0.5 M H_2SO_4. The product was then freeze-dried for 24 h and denoted as HT-APG-Fe.

3.2.5. Synthesis of Fe-Embedded Carbonaceous Spheres with Galactose (HT-GAL-Fe)

Maghemite nanoparticles (200 mg) were added into a suspension of 0.02 mole galactose in 40 mL water and mixed with stirring for 30 min. The mixture was transferred in to a Teflon autoclave and heated up to 180 °C for 18 h. Then, the product was collected, centrifuged, and repeatedly washed, six times with deionised water and four times with 0.5 M H_2SO_4. The product was collected and freeze-dried for 24 h (referred to as HT-GAL-Fe).

3.2.6. Pyrolysis of Carbonaceous Spheres with N-Precursor (N-Doped Carbon Spheres)

Each of the hydrothermally treated samples HT-APG, HT-APG-Fe, HT-APG-Co, and HT-GAL-Fe were ground together with melamine (1:10 w/w) using a mortar and a pestle. The mixture was placed in a tubular furnace and pyrolysed at 950 °C for 3 h under Ar at the rate of 10 °C/min. The N-doped products are referred to as N-APG, N-APG-Fe, N-APG-Co, and N-GAL-Fe, respectively.

3.2.7. Pyrolysis of Carbonaceous Spheres without N-Precursor

Hydrothermally synthesised HT-APG, HT-APG-Fe, HT-APG-Co, and HT-GAL-Fe were individually pyrolysed at 950 °C for 3 h under Ar at the rate of 10 °C/min in the tubular furnace. The pyrolysed products are referred to as APG, APG-Fe, APG-Co, and GAL-Fe, respectively.

3.2.8. Preparation of Catalytic Inks

Catalytic ink was prepared by ultra-sonication of each catalyst (2 mg) and suspended in Nafion suspension (1 mL of 1%). The prepared ink (10 μL) was carefully deposited on a glassy carbon rotating disc electrode (3 mm) and a rotating ring disc electrode (4 mm). The sample was then allowed to dry in air for 12 h.

3.3. Characterization

Scanning electron microscopy (SEM) images and energy-dispersive X-ray spectroscopy (EDX) were obtained using a Quanta 450 (FEI, Hillsboro, OR, USA) at an accelerating voltage of 10 kV. For EDX, three readings were obtained and the average was recorded. Transition electron microscopy (TEM) investigation was carried out using a Tecnai G2 Spirit (FEI, Hillsboro, OR, USA) operated at 120 kV. X-ray diffraction (XRD) was performed at 40 kV and 15 mA in the range of $2\theta = 10$–$70°$ at a speed of $10°$/min using a Miniflex 600 (Rigaku, Akishima, Tokyo, Japan). Gas adsorption isotherms were conducted using a Micromeritics 3-Flex or ASAP2020 analyser (Micro metrics Instruments Corporation, Norcross, GA, USA). The Brunauer–Emment–Teller (BET) surface area and pore size distribution were calculated using software on the Micromeritics 3-Flex or ASAP 2020 analyser (Beckman Coulter, Indianapolis, IN, USA). Fourier transform infrared (FTIR) spectroscopy was conducted using Spectrum 100 (Perkin Elmer, Waltham, MA, USA). Raman analysis was conducted using a LabRAM Evolution (Horiba Yvon, Kyoto, Japan) using a 532 nm wavelength. XPS was conducted on a custom-built SPECS instrument (Berlin, Germany). All XPS (X-ray photo electron spectroscopy) measurements were performed on sample prepared by drop-casting onto Si using a non-monochromatic Mg source operating at 120 kV and 200 W. High resolution XPS spectra were collected using a pass energy of 10 eV with an energy step of 0.1 eV.

Electrochemical Characterization

The ORR reactions were conducted utilising a Rotating Ring Disc Electrode (RRDE) apparatus connected to a bi potentiostat (CH 1760 C, CH Instruments Inc., Austin, TX, USA) in a standard three-electrode cell with an oxygen-saturated KOH (0.1 mol/L) solution. The glassy carbon electrode, platinum, and reversible hydrogen electrode (RHE) were used as the working, counter, and reference electrodes, respectively. The scan rate of the reaction was 0.01 Vs^{-1} in the range of 0 and 1.1 V. The cycle was repeated until stable voltammograms were obtained before the RRDE readings were derived at different speeds from 400 to 2400 rpm.

The reaction kinetics of the catalysts were examined by employing RRDE to quantify the overall electron transfer number (n) and percentage of hydrogen peroxide (% HO_2^-) at rotation speeds from 400 to 2400 rpm in an oxygen-saturated 0.1 M KOH solution. To elucidate the overall number of electrons (n) and % HO_2^- produced in the ring against the applied potential, Equations (1) and (2) were employed [10,81].

$$n = \frac{4I_D}{I_D + \frac{I_R}{N}} \tag{1}$$

$$\%H_2O_2 = 100 \, \frac{2I_R}{I_D N + I_R} \tag{2}$$

where I_D and I_R are the disc and ring currents, respectively, and N is the collection efficiency.

4. Conclusions

The phenomenon of converting a naturally occurring apricot sap from an apricot tree into a 3D hybrid ORR electro-catalyst composed of N-CF and N-CMS is reported and verified by SEM and TEM. The MNPs initially embedded within the CMS diffused out of the CMS to catalyse for the formation of corrugated hollow N-CF due to the surface destruction caused by the decomposition of melamine during pyrolysis. The 3D integrated N-CMSs and N-CF ORR electro-catalysts prepared using FeNP (N-APG-Fe) or CoNP (N-APG-Co) showed a predominant four-electron transfer pathway for the ORR within the potential region of 0.10–0.70 V. The spherical morphology obtained from non-hazardous apricot sap can be employed in a wide range of areas, such as catalysis applications, absorption studies, and drug delivery.

Supplementary Materials: The following are available online at www.mdpi.com/xxx/s1, Figure S1: EDX analysis of FeMNPC, Figure S2: SEM images of (A) hydrothermally treated apricot sap resin and cobalt acetate (HT-APG-Co), (B) pyrolysed HT-APG-Co at 950 °C with the presence of nitrogen precursor melamine (N-APG-Co), and (C) pyrolysed HT-APG-Co at 950 °C without melamine (APG-Co), Figure S3: SEM images of (A) hydrothermally treated apricot sap resin (HT-APG), (B) pyrolysed HT-APG at 950 °C with the presence of nitrogen precursor melamine (N-APG), and (C) pyrolysed HT-APG at 950 °C without melamine (APG), Figure S4: Pore size distribution of (A) APG-Fe and (B) APG-Co, Figure S5: XPS core level spectra of N-APG-Fe for (A) C1s and (B) N1s, Figure S6: XPS core level spectra of N-APG-Co for (A) C1s and (B) N1s, Figure S7: Cyclic Voltammetry of (A) N-APG, (B) N-APG-Fe, and (C) N-APG-Co at a scan rate of 10 mVS-1 in oxygen-saturated 0.1M KOH solution, Figure S8: (A) Comparison of number of electrons and (B) % HO_2^- of N-APG, N-APG-Co, N-APG-Fe, N-GAL-Fe, and Pt/C catalysts electrodes at 0.4 V applied potential in oxygen-saturated 0.10 M KOH electrolyte at 2000 rpm at a scan rate of 10 mV/s, Figure S9: Rotating ring disc voltammograms of (A) ring current and (B) disc current of catalysts electrodes APG, APG-Co, APG-Fe, GAL-Fe, and Pt/C, pyrolysed without the presence of melamine in oxygen saturated 0.1 M KOH at 2000 rpm at a scan rate of 10 mV/s. (C) Percentage peroxide, and (D) number of electrons of APG, APG-Fe, APG-Co, and Pt/C electrodes at various potential calculated according to RRDE data, Figure S10: RDE polarisation curves of (A) N-APG-Co and (B) N-APG-Fe with a scan rate of 100 mVS^{-1} before and after 6000 potential cycles in an oxygen saturated KOH solution, Figure S11: Rate constants of (A) N-APG, (B) N-APG-Co, (C) N-APG-Fe, and (D) N-GAL-Fe, Table S1: Electro chemical properties of non-doped apricot sap and galactose catalysts, Table S2: Comparison of performance of N-APG-Fe, N-APG-Co, and N-GAL-Fe with other similar carbon-based catalysts.

Acknowledgments: The authors are thankful for the support of the Australian Solar Thermal Research Initiative (ASTRI), ARC Hub for Graphene Enabled Industry Transformation (IH 150100003), the University of Adelaide, School of Chemical Engineering and School of Chemistry. The technical support provided by Adelaide Microscopy and the Micro Analysis Research Facility at Flinders Microscopy (Flinders University) was greatly appreciated.

Author Contributions: Ramesh Karunagaran performed the experiments and analyzed the data, Cameron Shearer conducted the XPS analysis, Campbell Coghlan and Dusan Losic conceived and designed the experiments, and all other authors assisted in writing the manuscript.

Conflicts of Interest: The authors declare no conflict of interest.

References

1. Boudghene, S.A.; Traversa, E. Fuel cells, an alternative to standard sources of energy. *Renew. Sustain. Energy Rev.* **2002**, *6*, 295–304. [CrossRef]

2. Xing, T.; Zheng, Y.; Li, L.H.; Cowie, B.C.; Gunzelmann, D.; Qiao, S.Z.; Huang, S.; Chen, Y. Observation of active sites for oxygen reduction reaction on nitrogen-doped multilayer graphene. *ACS Nano* **2014**, *8*, 6856–6862. [CrossRef] [PubMed]

3. Song, C.; Zhang, J. Electrocatalytic oxygen reduction reaction. In *PEM Fuel Cell Electrocatalysts and Catalyst Layers*; Springer: Berlin, Germany, 2008; pp. 89–134.

4. Holton, O.T.; Stevenson, J.W. The role of platinum in proton exchange membrane fuel cells. *Platin. Met. Rev.* **2013**, *57*, 259–271. [CrossRef]

5. Zhang, S.; Yuan, X.Z.; Hin, J.N.C.; Wang, H.; Friedrich, K.A.; Schulze, M. A review of platinum-based catalyst layer degradation in proton exchange membrane fuel cells. *J. Power Sources* **2009**, *194*, 588–600. [CrossRef]

6. Markovic, N.; Schmidt, T.; Stamenkovic, V.; Ross, P. Oxygen reduction reaction on Pt and Pt bimetallic surfaces: A selective review. *Fuel Cells* **2001**, *1*, 105–116. [CrossRef]

7. Markovic, N.M.; Gasteiger, H.A.; Ross, P.N. Oxygen reduction on platinum low-index single-crystal surfaces in sulfuric acid solution: Rotating ring-Pt(hkl) disk studies. *J. Phys. Chem.* **1995**, *99*, 3411–3415. [CrossRef]

8. Qu, L.; Liu, Y.; Baek, J.B.; Dai, L. Nitrogen-doped graphene as efficient metal-free electrocatalyst for oxygen reduction in fuel cells. *ACS Nano* **2010**, *4*, 1321–1326. [CrossRef] [PubMed]

9. Liang, J.; Zhou, R.F.; Chen, X.M.; Tang, Y.H.; Qiao, S.Z. Fe–N decorated hybrids of CNTs grown on hierarchically porous carbon for high-performance oxygen reduction. *Adv. Mater.* **2014**, *26*, 6074–6079. [CrossRef] [PubMed]

10. Liang, Y.; Li, Y.; Wang, H.; Zhou, J.; Wang, J.; Regier, T.; Dai, H. Co_3O_4 nanocrystals on graphene as a synergistic catalyst for oxygen reduction reaction. *Nat. Mater.* **2011**, *10*, 780–786. [CrossRef] [PubMed]

11. Wu, Z.S.; Yang, S.; Sun, Y.; Parvez, K.; Feng, X.; Müllen, K. 3D Nitrogen-doped graphene aerogel-supported Fe_3O_4 nanoparticles as efficient electrocatalysts for the oxygen reduction reaction. *J. Am. Chem. Soc.* **2012**, *134*, 9082–9085. [CrossRef] [PubMed]

12. Xiang, Z.; Xue, Y.; Cao, D.; Huang, L.; Chen, J.F.; Dai, L. Highly efficient electrocatalysts for oxygen reduction based on 2D covalent organic polymers complexed with non-precious metals. *Angew. Chem. Int. Ed.* **2014**, *53*, 2433–2437. [CrossRef] [PubMed]

13. Bezerra, C.W.; Zhang, L.; Lee, K.; Liu, H.; Marques, A.L.; Marques, E.P.; Wang, H.; Zhang, J. A review of Fe–N/C and Co–N/C catalysts for the oxygen reduction reaction. *Electrochim. Acta* **2008**, *53*, 4937–4951. [CrossRef]

14. Liu, G.; Li, X.; Ganesan, P.; Popov, B.N. Development of non-precious metal oxygen-reduction catalysts for PEM fuel cells based on N-doped ordered porous carbon. *Appl. Catal. B* **2009**, *93*, 156–165. [CrossRef]

15. Nallathambi, V.; Lee, J.W.; Kumaraguru, S.P.; Wu, G.; Popov, B.N. Development of high performance carbon composite catalyst for oxygen reduction reaction in PEM proton exchange membrane fuel cells. *J. Power Sources* **2008**, *183*, 34–42. [CrossRef]

16. Yang, D.S.; Song, M.Y.; Singh, K.P.; Yu, J.S. The role of iron in the preparation and oxygen reduction reaction activity of nitrogen-doped carbon. *Chem. Commun.* **2015**, *51*, 2450–2453. [CrossRef] [PubMed]

17. Wiesener, K. N_4-chelates as electrocatalyst for cathodic oxygen reduction. *Electrochim. Acta* **1986**, *31*, 1073–1078. [CrossRef]

18. Tang, Y.; Allen, B.L.; Kauffman, D.R.; Star, A. Electrocatalytic activity of nitrogen-doped carbon nanotube cups. *J. Am. Chem. Soc.* **2009**, *131*, 13200–13201. [CrossRef] [PubMed]

19. Wang, D.W.; Su, D. Heterogeneous nanocarbon materials for oxygen reduction reaction. *Energy Environ. Sci.* **2014**, *7*, 576–591. [CrossRef]

20. Yang, Z.; Nie, H.; Chen, X.; Xiaohua, C.; Huang, S. Recent progress in doped carbon nanomaterials as effective cathode catalysts for fuel cell oxygen reduction reaction. *J. Power Sources* **2013**, *236*, 238–249. [CrossRef]

21. Vaughan, O. Carbon catalysts: Active sites revealed. *Nat. Nanotechnol.* **2016**, *1*, 361–365. [CrossRef]
22. Yang, T.; Liu, J.; Zhou, R.; Chen, Z.; Xu, H.; Qiao, S.Z.; Monteiro, M.J. N-doped mesoporous carbon spheres as the oxygen reduction reaction catalysts. *J. Mater. Chem. A* **2014**, *2*, 18139–18146. [CrossRef]
23. Liu, H.J.; Bo, S.; Cui, W.; Li, F.; Wang, C.; Xia, Y. Nano-sized cobalt oxide/mesoporous carbon sphere composites as negative electrode material for lithium-ion batteries. *Electrochim. Acta* **2008**, *53*, 6497–6503. [CrossRef]
24. Zhang, W.M.; Hu, J.S.; Guo, Y.G.; Zheng, S.F.; Zhong, L.S.; Song, W.G.; Wan, L.J. Tin-nanoparticles encapsulated in elastic hollow carbon spheres for high-performance anode material in lithium-ion batteries. *Adv. Mater.* **2008**, *20*, 1160–1165. [CrossRef]
25. Du, H.; Jiao, L.; Wang, Q.; Yang, J.; Guo, L.; Si, Y.; Wang, Y.; Yuan, H. Facile carbonaceous microsphere templated synthesis of Co_3O_4 hollow spheres and their electrochemical performance in supercapacitors. *Nano Res.* **2013**, *6*, 87–98. [CrossRef]
26. Zhou, J.; He, J.; Zhang, C.; Wang, T.; Sun, D.; Di, Z.; Wang, D. Mesoporous carbon spheres with uniformly penetrating channels and their use as a supercapacitor electrode material. *Mater. Charact.* **2010**, *61*, 31–38. [CrossRef]
27. Yu, G.; Sun, B.; Pei, Y.; Xie, S.; Yan, S.; Qiao, M.; Fan, K.; Zhang, X.; Zong, B. Fe_xO_y@C spheres as an excellent catalyst for Fischer–Tropsch synthesis. *J. Am. Chem. Soc.* **2009**, *132*, 935–937. [CrossRef] [PubMed]
28. Wen, Z.; Wang, Q.; Zhang, Q.; Li, J. Hollow carbon spheres with wide size distribution as anode catalyst support for direct methanol fuel cells. *Electrochem. Commun.* **2007**, *9*, 1867–1872. [CrossRef]
29. Xiong, K.; Li, J.; Liew, K.; Zhan, X. Preparation and characterization of stable Ru nanoparticles embedded on the ordered mesoporous carbon material for applications in Fischer–Tropsch synthesis. *Appl. Catal. A* **2010**, *389*, 173–178. [CrossRef]
30. Jafri, R.I.; Rajalakshmi, N.; Ramaprabhu, S. Nitrogen doped graphene nanoplatelets as catalyst support for oxygen reduction reaction in proton exchange membrane fuel cell. *J. Mater. Chem.* **2010**, *20*, 7114–7117. [CrossRef]
31. Gong, K.; Du, F.; Xia, Z.; Durstock, M.; Dai, L. Nitrogen-doped carbon nanotube arrays with high electrocatalytic activity for oxygen reduction. *Science* **2009**, *323*, 760–764. [CrossRef] [PubMed]
32. Friedel, B.; Greulich-Weber, S. Preparation of monodisperse, submicrometer carbon spheres by pyrolysis of melamine–formaldehyde resin. *Small* **2006**, *2*, 859–863. [CrossRef] [PubMed]
33. Rybarczyk, M.K.; Lieder, M.; Jablonska, M. N-doped mesoporous carbon nanosheets obtained by pyrolysis of a chitosan–melamine mixture for the oxygen reduction reaction in alkaline media. *RSC Adv.* **2015**, *5*, 44969–44977. [CrossRef]
34. Zhou, X.; Yang, Z.; Nie, H.; Yao, Z.; Zhang, L.; Huang, S. Catalyst-free growth of large scale nitrogen-doped carbon spheres as efficient electrocatalysts for oxygen reduction in alkaline medium. *J. Power Sources* **2011**, *196*, 9970–9974. [CrossRef]
35. Feng, L.; Yang, L.; Huang, Z.; Luo, J.; Li, M.; Wang, D.; Chen, Y. Enhancing electrocatalytic oxygen reduction on nitrogen-doped graphene by active sites implantation. *Sci. Rep.* **2013**, *3*, 3306. [CrossRef] [PubMed]
36. Zhang, L.; Kim, J.; Dy, E.; Ban, S.; Tsay, K.; Kawai, H.; Shi, Z.; Zhang, J. Synthesis of novel mesoporous carbon spheres and their supported Fe-based electrocatalysts for PEM fuel cell oxygen reduction reaction. *Electrochim. Acta* **2013**, *108*, 480–485. [CrossRef]
37. Peng, H.; Mo, Z.; Liao, S.; Liang, H.; Yang, L.; Luo, F.; Song, H.; Zhong, Y.; Zhang, B. High performance Fe- and N-doped carbon catalyst with graphene structure for oxygen reduction. *Sci. Rep.* **2013**, *3*, 1765. [CrossRef]
38. Tang, J.; Liu, J.; Li, C.; Li, Y.; Tade, M.O.; Dai, S.; Yamauchi, Y. Synthesis of nitrogen-doped mesoporous carbon spheres with extra-large pores through assembly of diblock copolymer micelles. *Angew. Chem. Int. Ed.* **2015**, *54*, 588–593. [CrossRef]
39. Chen, P.; Wang, L.K.; Wang, G.; Gao, M.R.; Ge, J.; Yuan, W.J.; Shen, Y.H.; Xie, A.J.; Yu, S.H. Nitrogen-doped nanoporous carbon nanosheets derived from plant biomass: An efficient catalyst for oxygen reduction reaction. *Energy Environ. Sci.* **2014**, *7*, 4095–4103. [CrossRef]
40. Ma, Y.; Zhao, J.; Zhang, L.; Zhao, Y.; Fan, Q.; Hu, Z.; Huang, W. The production of carbon microtubes by the carbonization of catkins and their use in the oxygen reduction reaction. *Carbon* **2011**, *49*, 5292–5297. [CrossRef]

41. Lai, C.; Kolla, P.; Zhao, Y.; Fong, H.; Smirnova, A.L. Lignin-derived electrospun carbon nanofiber mats with supercritically deposited Ag nanoparticles for oxygen reduction reaction in alkaline fuel cells. *Electrochim. Acta* **2014**, *130*, 431–438. [CrossRef]

42. Rana, M.; Arora, G.; Gautam, U.K. N- and S-doped high surface area carbon derived from soya chunks as scalable and efficient electrocatalysts for oxygen reduction. *Sci. Technol. Adv. Mater.* **2015**, *16*, 014803. [CrossRef] [PubMed]

43. Nicolás, E.; Torrecillas, A.; Dell'Amico, J.; Alarcón, J.J. The effect of short-term flooding on the sap flow, gas exchange and hydraulic conductivity of young apricot trees. *Trees* **2005**, *19*, 51–57. [CrossRef]

44. Weaver, D. A gummosis disease of peach trees caused by *Botryosphaeria dothidea*. *Phytopathol* **1974**, *64*, 1429–1432. [CrossRef]

45. Lluveras-Tenorio, A.; Mazurek, J.; Restivo, A.; Colombini, M.P.; Bonaduce, I. Analysis of plant gums and saccharide materials in paint samples: Comparison of GC-MS analytical procedures and databases. *Chem. Cent. J.* **2012**, *6*, 115. [CrossRef] [PubMed]

46. Sun, X.; Li, Y. Colloidal carbon spheres and their core/shell structures with noble-metal nanoparticles. *Angew. Chem. Int. Ed.* **2004**, *43*, 597–601. [CrossRef] [PubMed]

47. Mi, Y.; Hu, W.; Dan, Y.; Liu, Y. Synthesis of carbon micro-spheres by a glucose hydrothermal method. *Mater. Lett.* **2008**, *62*, 1194–1196. [CrossRef]

48. Roberts, A.D.; Li, X.; Zhang, H. Porous carbon spheres and monoliths: Morphology controlling, pore size tuning and their applications as Li-ion battery anode materials. *Chem. Soc. Rev.* **2014**, *43*, 4341–4356. [CrossRef] [PubMed]

49. Wang, J.; Shen, L.; Ding, B.; Nie, P.; Deng, H.; Dou, H.; Zhang, X. Fabrication of porous carbon spheres for high-performance electrochemical capacitors. *RSC Adv.* **2014**, *4*, 7538–7544. [CrossRef]

50. Karunagaran, R.; Tung, T.T.; Shearer, C.; Tran, D.; Coghlan, C.; Doonan, C.; Losic, D. A unique 3D nitrogen-doped carbon composite as high performance oxygen reduction catalysts. *Materials* **2017**, *10*, 921. [CrossRef] [PubMed]

51. Sakaki, T.; Shibata, M.; Miki, T.; Hirosue, H.; Hayashi, N. Reaction model of cellulose decomposition in near-critical water and fermentation of products. *Bioresour. Technol.* **1996**, *58*, 197–202. [CrossRef]

52. Mer, V.K.L. Nucleation in phase transitions. *Ind. Eng. Chem.* **1952**, *44*, 1270–1277. [CrossRef]

53. Terrones, M.; Hsu, W.K.; Kroto, H.W.; Walton, D.R. Nanotubes: A revolution in materials science and electronics. In *Fullerenes and Related Structures*; Springer: Berlin, Germany, 1999; pp. 189–234.

54. Sun, Q.T.M.; Zhang, T.; Wang, G. Rational synthesis of novel π-conjugated poly(1,5-diaminoanthraquinone) for high-performance supercapacitors. *RSC Adv.* **2014**, *4*, 7774–7779. [CrossRef]

55. Terrones, M.; Terrones, H.; Grobert, N.; Hsu, W.; Zhu, Y.; Hare, J.; Kroto, H.; Walton, D.; Kohler-Redlich, P.; Rühle, M. Efficient route to large arrays of CN_x nanofibers by pyrolysis of ferrocene/melamine mixtures. *Appl. Phys. Lett.* **1999**, *75*, 3932–3934. [CrossRef]

56. Terrones, M.; Redlich, P.; Grobert, N.; Trasobares, S.; Hsu, W.K.; Terrones, H.; Zhu, Y.Q.; Hare, J.P.; Reeves, C.L.; Cheetham, A.K. Carbon nitride nanocomposites: Formation of aligned C_xN_y nanofibers. *Adv. Mater.* **1999**, *11*, 655–658. [CrossRef]

57. Lee, K.H.; Han, S.W.; Kwon, K.Y.; Park, J.B. Systematic analysis of palladium–graphene nanocomposites and their catalytic applications in Sonogashira reaction. *J. Colloid Interface Sci.* **2013**, *403*, 127–133. [CrossRef] [PubMed]

58. Gupta, V.; Saleh, T.A. Syntheses of carbon nanotube-metal oxides composites; adsorption and photo-degradation. In *Carbon Nanotubes—From Research to Applications*; InTech: San Francisco, CA, USA, 2011. [CrossRef]

59. Manoj, B.; Kunjomana, A. Study of stacking structure of amorphous carbon by X-ray diffraction technique. *Int. J. Electrochem. Sci.* **2012**, *7*, 3127–3134.

60. Wu, G.; Dai, C.; Wang, D.; Li, D.; Li, N. Nitrogen-doped magnetic onion-like carbon as support for Pt particles in a hybrid cathode catalyst for fuel cells. *J. Mater. Chem.* **2010**, *20*, 3059–3068. [CrossRef]

61. Ferrari, A.C. Raman spectroscopy of graphene and graphite: Disorder, electron-phonon coupling, doping and nonadiabatic effects. *Solid State Commun.* **2007**, *143*, 47–57. [CrossRef]

62. Xie, Z.L.; Huang, X.; Titirici, M.M.; Taubert, A. Mesoporous graphite nanoflakes via ionothermal carbonization of fructose and their use in dye removal. *RSC Adv.* **2014**, *4*, 37423–37430. [CrossRef]

63. Pimenta, M.; Dresselhaus, G.; Dresselhaus, M.S.; Cancado, L.; Jorio, A.; Saito, R. Studying disorder in graphite-based systems by Raman spectroscopy. *Phys. Chem. Chem. Phys.* **2007**, *9*, 1276–1290. [CrossRef] [PubMed]

64. Schwan, J.; Ulrich, S.; Batori, V.; Ehrhardt, H.; Silva, S. Raman spectroscopy on amorphous carbon films. *J. Appl. Phys.* **1996**, *80*, 440–447. [CrossRef]

65. Ghanbarlou, H.; Rowshanzamir, S.; Kazeminasab, B.; Parnian, M.J. Non-precious metal nanoparticles supported on nitrogen-doped graphene as a promising catalyst for oxygen reduction reaction: Synthesis, characterization and electrocatalytic performance. *J. Power Sources* **2015**, *273*, 981–989. [CrossRef]

66. Sing, K.S. Reporting physisorption data for gas/solid systems with special reference to the determination of surface area and porosity (Recommendations 1984). *Pure Appl. Chem.* **1985**, *57*, 603–619. [CrossRef]

67. Khalfaoui, M.; Knani, S.; Hachicha, M.; Lamine, A.B. New theoretical expressions for the five adsorption type isotherms classified by BET based on statistical physics treatment. *J. Colloid Interface Sci.* **2003**, *263*, 350–356. [CrossRef]

68. Chen, J.; Xia, N.; Zhou, T.; Tan, S.; Jiang, F.; Yuan, D. Mesoporous carbon spheres: Synthesis, characterization and supercapacitance. *Int. J. Electrochem. Sci.* **2009**, *4*, 1063–1073.

69. Wang, H.; Hao, Q.; Yang, X.; Lu, L.; Wang, X. Graphene oxide doped polyaniline for supercapacitors. *Electrochem. Commun.* **2009**, *11*, 1158–1161. [CrossRef]

70. Elumalai, E.K.; Kayalvizhi, K.; Silvan, S. Coconut water assisted green synthesis of silver nanoparticles. *J. Pharm. Bioallied Sci.* **2014**, *6*, 241. [CrossRef] [PubMed]

71. Talbot, W.F. Manufacture of Melamine-Aldehyde Condensation Products. U.S. Patent Application No. 2260239 A, 21 October 1941.

72. Permatasari, F.A.; Aimon, A.H.; Iskandar, F.; Ogi, T.; Okuyama, K. Role of C–N Configurations in the photoluminescence of graphene quantum dots synthesized by a hydrothermal route. *Sci. Rep.* **2016**, *6*, 21042. [CrossRef] [PubMed]

73. Lu, M.; Cheng, H.; Yang, Y. A comparison of solid electrolyte interphase (SEI) on the artificial graphite anode of the aged and cycled commercial lithium ion cells. *Electrochim. Acta* **2008**, *53*, 3539–3546. [CrossRef]

74. Martínez, L.; Román, E.; Nevshupa, R. X-Ray Photoelectron spectroscopy for characterization of engineered elastomer surfaces. In *Advanced Aspects of Spectroscopy*; InTech: San Francisco, CA, USA, 2012. [CrossRef]

75. Kelemen, S.R.; Gorbaty, M.L.; Kwiatek, P.J. Quantification of nitrogen forms in coals. *Energeia* **1995**, *6*, 1–3.

76. Zhang, L.S.; Liang, X.Q.; Song, W.G.; Wu, Z.Y. Identification of the nitrogen species on N-doped graphene layers and Pt/NG composite catalyst for direct methanol fuel cell. *Phys. Chem. Chem. Phys.* **2010**, *12*, 12055–12059. [CrossRef] [PubMed]

77. Subramanian, N.P.; Li, X.; Nallathambi, V.; Kumaraguru, S.P.; Colon-Mercado, H.; Wu, G.; Lee, J.W.; Popov, B.N. Nitrogen-modified carbon-based catalysts for oxygen reduction reaction in polymer electrolyte membrane fuel cells. *J. Power Sources* **2009**, *188*, 38–44. [CrossRef]

78. Liu, Y.L.; Xu, X.Y.; Shi, C.X.; Ye, X.W.; Sun, P.C.; Chen, T.H. Iron-nitrogen Co-doped hierachically mesopourous carbon spheres as highly efficient electtocatalysts for oxygen reduction reaction. *RSC Adv.* **2017**, *7*, 8879–8885. [CrossRef]

79. Damjanovic, A.; Genshaw, M.A.; Bockris, J.O. Distinction between intermediates produced in main and side electrodic reactions. *J. Chem. Phys.* **1966**, *45*, 4057–4059. [CrossRef]

80. Darezereshki, E. Synthesis of maghemite (γ-Fe$_2$O$_3$) nanoparticles by wet chemical method at room temperature. *Mater. Lett.* **2010**, *64*, 1471–1472. [CrossRef]

81. Wang, S.; Dou, S.; Tao, L.; Huo, J.; Dai, L. Etched and doped Co$_9$S$_8$/graphene hybrid for oxygen electrocatalysis. *Energy Environ. Sci.* **2016**, *9*, 1320–1326.

Article

Enhanced CO$_2$ Adsorption on Nitrogen-Doped Carbon Materials by Salt and Base Co-Activation Method

Ruiping Wei [1,2], **Xingchao Dai** [1,2] and **Feng Shi** [1,*]

[1]   State Key Laboratory for Oxo Synthesis and Selective Oxidation, Lanzhou Institute of Chemical Physics, Chinese Academy of Sciences, No.18, Tianshui Middle Road, Lanzhou 730000, China; wei828111@163.com (R.W.); daixingchao@licp.cas.cn (X.D.)

[2]   University of Chinese Academy of Sciences, No. 19A, Yuquan Road, Beijing 100049, China

[*]   Correspondence: fshi@licp.cas.cn; Tel.: +86-0931-4968142

Received: 20 February 2019; Accepted: 9 April 2019; Published: 12 April 2019

Abstract: Nitrogen-doped carbon materials with enhanced CO$_2$ adsorption were prepared by the salt and base co-activation method. First, resorcinol-formaldehyde resin was synthesized with a certain salt as an additive and used as a precursor. Next, the resulting precursor was mixed with KOH and subsequently carbonized under ammonia flow to finally obtain the nitrogen-doped carbon materials. A series of samples, with and without the addition of different salts, were prepared, characterized by XRD (X-ray powder diffraction), elemental analysis, BET (N$_2$-adsorption-desorption analysis), XPS (X-ray photoelectron spectroscopy) and SEM (Scanning electron microscopy) and tested for CO$_2$ adsorption. The results showed that the salt and base co-activation method has a remarkable enhancing effect on the CO$_2$ capture capacity. The combination of KCl and KOH was proved to be the best combination, and 167.15 mg CO$_2$ could be adsorbed with 1 g nitrogen-doped carbon at 30 °C under 1 atm pressure. The materials characterizations revealed that the introduction of the base and salt could greatly increase the content of doped nitrogen, the surface area and the amount of formed micropore, which led to enhanced CO$_2$ absorption of the carbon materials.

Keywords: nitrogen-doped carbon materials; carbon dioxide adsorption; salt and base; co-activation method

---

1. Introduction

Terrible scenarios of global warming are attributed to the emission of built-up greenhouse gases. Among these greenhouse gases, carbon dioxide (CO$_2$), released by the combustion of fuels and from certain industrial and resource extraction processes, is one of the main components. Thus, there are many concerns about reducing carbon dioxide in greenhouse gases. As a result, extensive research efforts have been undertaken to develop feasible materials for CO$_2$ capture [1]. Carbon dioxide adsorption especially by porous materials has become a hot research topic because these materials possess many advantages such as low energy requirements, quick and convenient processes of adsorption and desorption compared with chemical absorption [2]. In this context, many porous materials including zeolites [3–5], other inorganic molecular sieves [6–13], metal-organic frameworks [14–19] and carbon-based materials [20–24] have been investigated.

Among them, carbon-based materials are widely accepted as a promising candidate for CO$_2$ adsorption due to their chemical inertness, low cost, high surface area and tunable pore structures. The porous structure and high surface area of carbon materials allow the introduction of several functional groups on the surface to increase the capacity of CO$_2$ adsorption. Various carbon-based materials including metal-carbon composites [25], biowaste derived carbons [26–29] and nitrogen-doped carbons

(NC) [30–41] have been applied in CO_2 capture. Among them, nitrogen doped carbon materials have been reported to exhibit an excellent CO_2 capture capacity and high adsorption selectivity. The incorporation of nitrogen in carbon materials can greatly improve their CO_2 capture capacity by providing basic adsorption sites. In fact, besides nitrogen-doping, the CO_2 adsorption of carbon material could also be remarkably enhanced by base activation [42,43]. For example, nitrogen-free microporous materials [44–47] prepared by alkali etching have been demonstrated to be highly efficient in CO_2 adsorption. It is noteworthy that alkali etching usually led to the formation of a small amount of micropores and, in other words, the pore structure was changed. Thus, it brings a debate on the exact role of doped nitrogen and pore properties for CO_2 adsorption. Recently, it has been reported that the pore structure has a determining effect on CO_2 adsorption at lower temperature and lower pressure, while doped nitrogen plays an important role at higher temperature and higher pressure [48–50]. Therefore, it will be highly desirable to develop a porous carbon material enriched in nitrogen and dominated by micropores.

Based on the above discussions, here, we presented nitrogen-doped carbon materials with high CO_2 capture capacity, which were prepared by the salt and base co-activation method with resorcinol-formaldehyde resin as a precursor. The experimental results showed that the salt and base co-activation method could greatly improve the CO_2 capture capacity of nitrogen-doped carbon material. The characterization analysis revealed an obvious increase of the doped nitrogen content and the amount of the micropores formed in the carbon material prepared by the salt and base co-activation method, which might be the reason for the enhancement of CO_2 adsorption. Therefore, a conclusion could be drawn that CO_2 adsorption was determined by both micropores and the doped nitrogen.

2. Materials and Methods

2.1. Materials Preparation

Precursors of carbon materials were synthesized by a low temperature hydro-thermal method according to the reported references [51–53]. The precursor applied was synthesized as following: Typically, resorcinol (R, 2.20 g, 20 mmol), formaldehyde (F, 3.25 g, 40 mmol, 37 wt % aqueous solution) and 9 mL deionized water were added into a 100 mL Teflon®autoclave. Subsequently, 21.2 mg Na_2CO_3 (1 mol % relative to resorcinol) and 0.25–1.25 g salts (KCl, KNO_3, NaCl, $NaNO_3$, Na_2SO_4) were added into the autoclave. The mixture was stirred for 1 h at room temperature, and then the autoclave was sealed and kept at 80 °C for 24 h and cooled it down to room temperature to provide an R-F resin (R: resorcinol and F: formaldehyde). The wet resin was put into a round-bottom flask and dried at 130 °C in vacuum condition for 3 h and used as the carbon precursor. Precursor without salt additive was prepared through the same process. Next, the synthesized precursors were mixed mechanically with KOH (0.4–2.0 g) and then carbonized at 400–700 °C (a heating rate of 10 °C min^{-1}) for 3 h under ammonia flow (20 mL min^{-1}). The resulting carbon materials were ultrasonically washed with deionized water (about 300 mL) until pH $\approx$ 7.0 and then dried at 80 °C for 6 h to provide the final sample.

2.2. CO_2 Adsorption Measurements

CO_2 adsorption of the carbon materials was measured using a Mettler-Toledo SDTA851 thermogravimetric analyzer according to the reported references [54–56]. In detail, firstly, 10 mg of sample was placed in a porcelain crucible with the volume of 0.1 mL. When the temperature reached 30 °C, the program was started with carbon dioxide (99.9%) as the reaction gas at a flow of 60 mL min^{-1} under 1 atm pressure, and held at that temperature for 50 min. After the completion of the adsorption, the mass of samples after CO_2 adsorption was recorded as m_1. Subsequently, the reaction gas was switched to nitrogen (99.9%) with the same flow rate, and at the same time the temperature was increased to 200 °C at a rate of 10 °C min^{-1} and held for 30 min to ensure the complete removal of

CO_2 that samples' adsorbed. After that, the mass of samples was recorded as m_0, which is used as the true mass of the sample. CO_2 adsorption capacity could be calculated by m_1 and m_0.

2.3. Characterization Techniques

X-ray powder diffraction (XRD) was performed on a Rigaku D/max-2400 X-ray diffractometer (Rigaku, Tokyo, Japan) with Ni-filtered Cu Kα radiation at 40 kV and 100 mA. The XRD patterns were scanned in the 2θ range of 10–80°.

Elemental analysis (C, N, H and O) of the samples was carried out on a Vario EL microanalyzer (Elementar, Hanau, Germany).

X-ray photoelectron spectroscopy (XPS) was performed by using a Thermo Scientific ESCALAB 250 instrument (Thermo Fisher Scientific, Waltham, MA, USA) with a dual Mg/Al anode X-ray source, a hemispherical capacitor analyser and a 5 keV Ar$^+$ ion-gun. All of the spectra were recorded using non-monochromatic Mg Kα (hν = 1253.6 eV) radiation.

The specific surface area (S_{BET}) was calculated using the Brunauer–Emmett–Teller (BET) equation with a relative pressure of 0.05–0.30. The total pore volume (V_{Total}) was obtained at the maximum incremental volume point. Micropore volume was determined from the Dubinin–Radushkevic equation. Mesoporous volume was determined by the subtraction of micropore volume from the total pore volume. Fraction of micropore volume = (micropore volume/total pore volume) * 100. The micropore size distribution was calculated by the Harvath–Kawazoe (H–K) equation based on the N_2/77 K adsorption data.

SEM was performed with a JEOL JSM-6701F (JEOL, Tokyo, Japan) equipped with a cold FEG (Field Emission Gun).

3. Results and Discussion

3.1. CO_2 Adsorption Performance Test

Figure 1 showed the TG curves of CO_2 adsorption and desorption of these samples activated with 1.2 g KOH and different amount of KCl from 0 to 1.25 g with an interval of 0.25g per 2.20 g resorcinol. The CO_2 adsorption-desorption behavior was measured at 30 °C under 1 atm. On the basis of the amount of KCl added, these samples were denoted as NC-KOH, NC-KOH-KCl-0.25, NC-KOH-KCl-0.50, NC-KOH-KCl-0.75, NC-KOH-KCl-1.00 and NC-KOH-KCl-1.25. The unactivated sample was denoted as NC. All these samples were carbonized at 600 °C for 3 h under ammonia flow (20 mL min^{-1}). It can be seen from the Figure 1, all the samples adsorbed CO_2 rapidly at the beginning, then continued with a slower rate and reached an equilibrium in 50 min. During the desorption process, the adsorbed CO_2 is removed rapidly and the mass of samples gradually decreased until a constant value was reached at 200 °C. Figure 1 showed that NC sample had the lowest CO_2 adsorption capacity and a higher CO_2 adsorption capacity was observed in the case of NC-KOH sample, which suggested that the introduction of the base in the carbonization process has a positive effect on the increase of CO_2 adsorption capacity. Similar effect could be also observed by adding the salt in the R-F resin synthesis. Among the tested samples, the samples activated by base and salt exhibited best ability in the CO_2 adsorption, which could be attributed to the synergistic effect of base and salt pretreatment. However, there is no a linear correlation between the CO_2 adsorption capacity of the sample and the amount of the salt added. The CO_2 adsorption capacity of the sample firstly increased then declined, and the maximum (167.15 mg/g) was observed when the sample was activated with 0.75 g KCl and 1.2 g KOH.

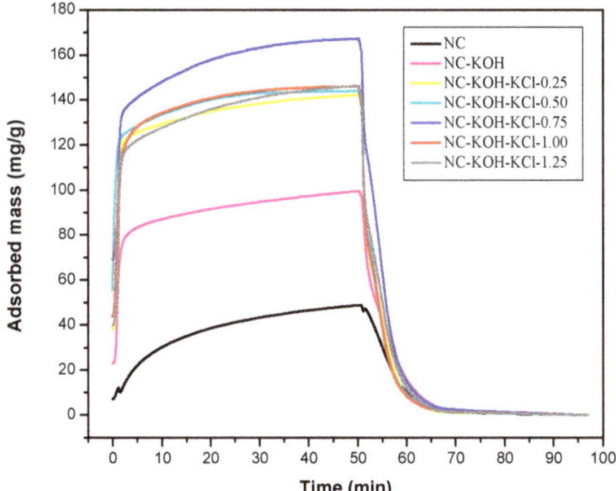

Figure 1. TG curves measured CO_2 adsorption and desorption of samples activated by different amount of KCl and 1.2 g KOH per 2.20 g resorcinol.

After optimizing the amount of KCl, the carbonization temperature of the NC-KOH-KCl-0.75 sample was further optimized in the range of 400–700 °C and the results were shown in the Figure 2. With the increase of the carbonization temperature from 400–600 °C, the CO_2 adsorption capacity of the sample was gradually enhanced, but a drop was observed when the temperature reached at 700 °C. The best CO_2 adsorption performance was obtained when the NC-KOH-KCl-0.75 sample was carbonized at 600 °C.

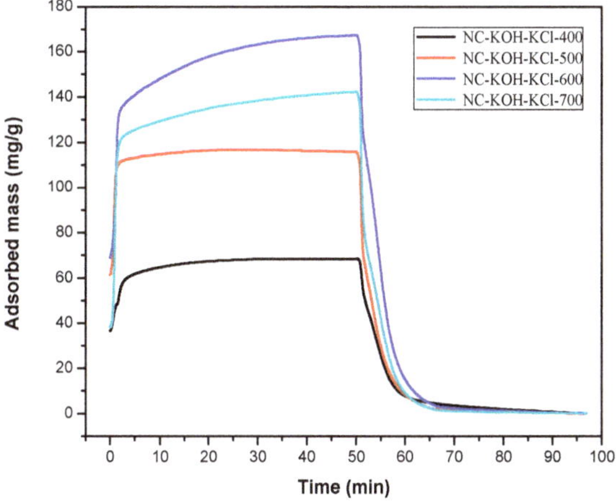

Figure 2. TG curves measured CO_2 adsorption and desorption of samples carbonized at different temperatures.

Following the above results, the effect of the salt kind was investigated (Figure 3). A series of different salts such as KNO_3, $NaNO_3$, KCl, NaCl and Na_2SO_4 were added in the R-F resin synthesis process with the optimized amount of 0.75 g and all the samples were carbonized at 600 °C. The results showed that the kind of the salt added has a great effect on the CO_2 adsorption capacity of the sample. The highest CO_2 adsorption capacity was obtained when the NC-KOH-KCl sample was used, and 167.15 mg CO_2 could be adsorbed with 1 g NC-KOH-KCl sample. Slight or much lower values were observed when the other salts such as KNO_3, $NaNO_3$, $NaCl$ and Na_2SO_4 were used. Among all the samples tested, the CO_2 adsorption capacity of the sample activated by Na_2SO_4 was the lowest.

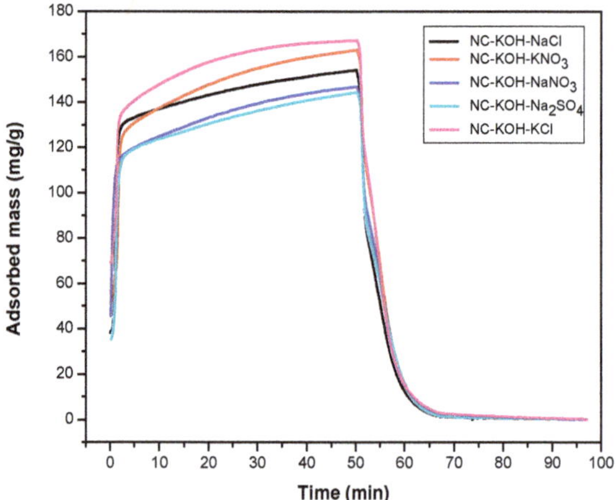

Figure 3. TG curves measured CO_2 adsorption and desorption of samples activated by different salts.

Finally, the amount of KOH added in the carbonization process was optimized in the range of 0–2.0 g and a series of samples activated with 0.75 KCl and different amounts of KOH were prepared. The results in the Figure 4 showed that the CO_2 adsorption capacity of the NC-KCl-KOH-0.4 sample was much higher than that of the sample activated only with KCl, which indicated that the introduction of KOH could greatly increase the CO_2 adsorption capacity of the sample. Further increasing the amount of KOH led to higher CO_2 adsorption capacity, but a sudden drop was observed when 1.6 g KOH was added. The decreased tendency could also be observed when further increasing the amount of KOH to 2.0 g.

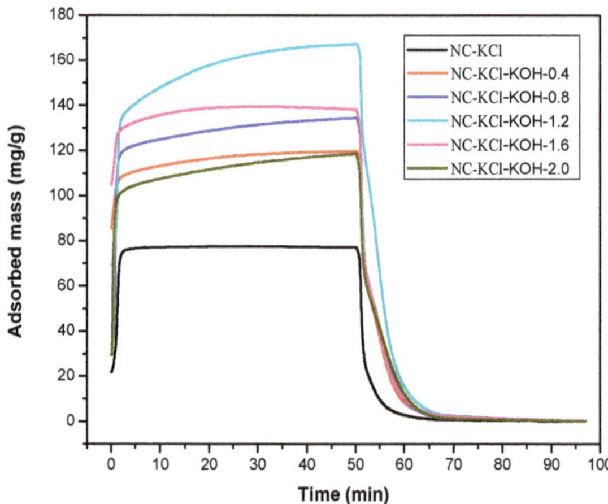

Figure 4. TG curves measured CO_2 adsorption and desorption of samples activated by different amount of KOH and 0.75 g KCl per 2.20 g resorcinol.

In order to better illuminate the effect of salt and base activation on the CO_2 adsorption capacity of nitrogen-doped carbon materials, four typical samples, e.g., NC, NC-KCl, NC-KOH and NC-KCl-KOH were chosen and further compared (Figure 5). In comparison with NC, NC-KCl and NC-KOH both exhibited better CO_2 adsorption capacity, which indicated that the base and salt pretreatment both had a promoted effect on the CO_2 adsorption capacity of the carbon materials, but base is superior to salt by contrast. The highest CO_2 adsorption capacity was obtained when the NC sample was activated by the combination of base and salt, which could be attributed the synergistic effect of base and salt added in the different steps.

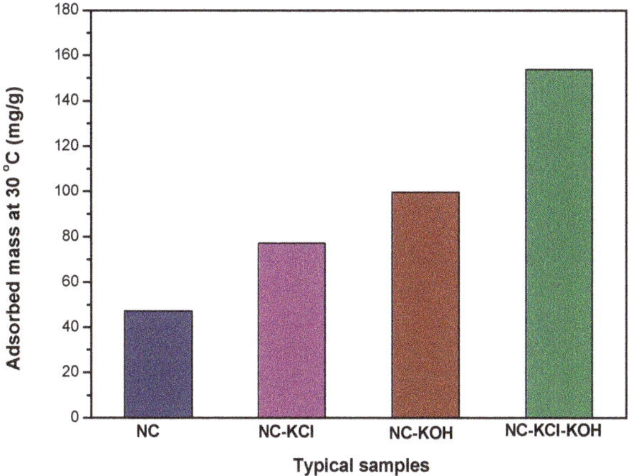

Figure 5. Values of typical samples' CO_2 adsorption measured by TGA at 30 °C under 1 atm pressure.

3.2. Characterization Results and Discussion

In order to explore the relationship of structure and performance, the prepared samples were characterized by elemental analysis and N_2-adsorption-desorption analysis, and the results are shown in Tables 1 and 2, and Figure 6. Obviously, the NC-KOH-KCl-0.75 sample has the highest nitrogen content (12.59 wt %), which implied that the doped nitrogen could promote the CO_2 adsorption (Table 1). The N_2-adsorption-desorption analysis revealed that the NC-KOH-KCl-0.75 sample has the largest specific surface areas and highest fraction of micropore volume to total pore volume, which means that the large specific area and more micropores formation might favors the CO_2 adsorption (Table 2). Thus, the CO_2 adsorption performance of the carbon material could be affected by the content of the doped nitrogen, the specific area and the amount of micropores.

Table 1. The content of N, C, H and O in samples activated by different amount of KCl and 1.2 g KOH per 2.20 g resorcinol.

Samples	N (wt %)	C (wt %)	H (wt %)	O (wt %)
NC	7.78	76.12	1.91	14.19
NC-KOH	11.99	58.22	2.67	27.12
NC-KOH-KCl-0.25	7.05	69.19	1.76	22.00
NC-KOH-KCl-0.50	8.40	70.09	1.88	19.63
NC-KOH-KCl-0.75	12.59	56.23	2.58	28.60
NC-KOH-KCl-1.00	9.43	68.05	1.88	20.64
NC-KOH-KCl-1.25	8.70	66.12	1.74	23.44

Table 2. BET surface area and porosity of samples activated by different amount of KCl and 1.2 g KOH per 2.20 g resorcinol.

Samples	S_{BET} (m^2 g^{-1}) [1]	V_{total} (cm^3 g^{-1}) [2]	V_{Micro} (cm^3 g^{-1}) [3]	V_{Meso} (cm^3 g^{-1}) [4]	F_{Micro} (%) [5]
NC	158	0.190	0.055	0.135	29
NC-KOH	1030	0.659	0.401	0.258	61
NC-KOH-KCl-0.25	911	1.013	0.354	0.659	35
NC-KOH-KCl-0.50	900	0.573	0.349	0.224	61
NC-KOH-KCl-0.75	1034	0.634	0.398	0.236	63
NC-KOH-KCl-1.00	858	0.740	0.361	0.379	49
NC-KOH-KCl-1.25	1006	0.615	0.389	0.226	63

[1] S_{BET} is the specific surface areas determined by the BET method. [2] V_{Total} is the total pore volume. [3] V_{Micro} is the micropore volume. [4] V_{Meso} is the mesoporous volume. [5] F_{Micro} is the fraction of micropore volume to total pore volume.

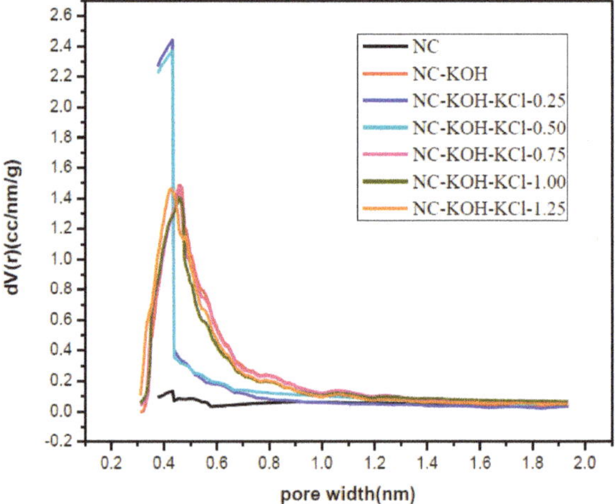

Figure 6. The micropore size distribution of samples activated by different amount of KCl and 1.2 g KOH per 2.20 g resorcinol.

Then, the effect of carbonized temperature on the structure was investigated by the elemental analysis and N_2-adsorption-desorption analysis. It can be seen from the elemental analysis results shown in Table 3 that higher carbonized temperature led to more doped nitrogen in the range of 400–700 °C and the content of doped nitrogen could be increased from 3.5 wt % to 13.08 wt % (Table 3). The NC-KOH-KCl-700 sample has the highest nitrogen content, but its CO_2 adsorption capacity is not the highest, which means that the CO_2 adsorption capacity of the carbon material was not determined by only the content of the doped nitrogen. Further, N_2-adsorption-desorption analysis revealed a good correlation between the micropore volume and the carbonized temperature. In addition, micropore volume enlarged with the increase of the carbonized temperature (Table 4 and Figure 7). The NC-KOH-KCl-600 sample with the best CO_2 adsorption performance has the highest fraction of micropore volume to total pore volume, which is consistent with the above discussions.

Table 3. The content of N, C, H and O in samples carbonized at different temperatures.

Samples	N (wt %)	C (wt %)	H (wt %)	O (wt %)
NC-KOH-KCl-400	3.5	66.00	2.13	28.37
NC-KOH-KCl-500	7.74	69.58	1.84	20.84
NC-KOH-KCl-600	12.59	56.23	2.58	28.60
NC-KOH-KCl-700	13.08	63.95	1.30	21.67

Table 4. BET surface area and porosity of samples carbonized at different temperatures.

Samples	S_{BET} (m^2 g^{-1}) [1]	V_{total} (cm^3 g^{-1}) [2]	V_{Micro} (cm^3 g^{-1}) [3]	V_{Meso} (cm^3 g^{-1}) [4]	F_{Micro} (%) [5]
NC-KOH-KCl-400	167	0.210	0.009	0.201	4
NC-KOH-KCl-500	959	0.671	0.369	0.302	55
NC-KOH-KCl-600	1034	0.634	0.398	0.236	63
NC-KOH-KCl-700	1300	0.812	0.483	0.329	59

[1] S_{BET} is the specific surface areas determined by the BET method. [2] V_{Total} is the total pore volume. [3] V_{Micro} is the micropore volume. [4] V_{Meso} is the mesoporous volume. [5] F_{Micro} is the fraction of micropore volume to total pore volume.

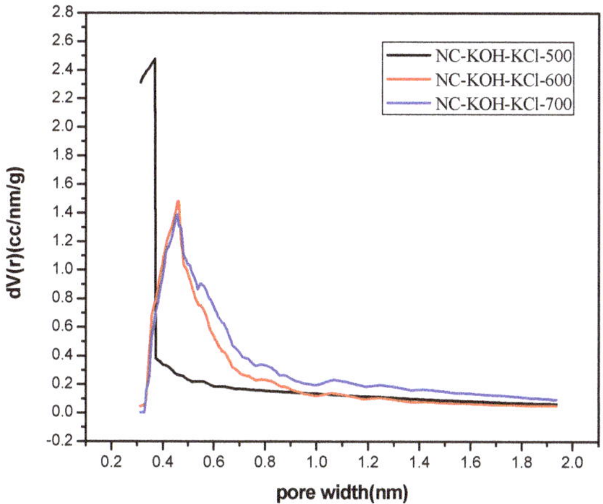

Figure 7. The micropore size distribution of samples carbonized at different temperatures.

Next, the samples activated by KOH and different salts were characterized by elemental analysis and N_2-adsorption-desorption analysis and the results were shown in Tables 5 and 6 and Figure 8. Obviously, these samples co-activated by base and salt have high nitrogen content and all exceeded 12 wt % (Table 5). Especially, for the samples activated by KNO_3, $NaNO_3$ and Na_2SO_4, the nitrogen content above 15 wt % was observed. The N_2-adsorption-desorption analysis revealed that the kind of the activated salt has a great effect on the pore structure of the carbon material. The samples activated by KCl and NaCl exhibited a specific surface area above 1000 m^2 g^{-1} while the smaller specific surface area than 1000 m^2 g^{-1} were observed in the case of other salts (Table 6). Similar phenomena were also observed in the case of total pore volume and micropore volume. It is noteworthy that the NC-KOH-KCl sample exhibited the highest fraction of micropore volume to total pore volume although its micropore volume is not the largest, which suggested that a larger micropore volume did not mean higher CO_2 adsorption capacity.

Table 5. The content of N, C, H and O in samples activated by different salts.

Samples	N (wt %)	C (wt %)	H (wt %)	O (wt %)
NC-KOH-NaCl	12.27	59.51	1.53	26.69
NC-KOH-KNO$_3$	15.45	56.01	1.51	27.03
NC-KOH-NaNO$_3$	15.37	67.18	1.57	15.88
NC-KOH-Na$_2$SO$_4$	15.22	58.14	1.71	24.93
NC-KOH-KCl	12.59	56.23	2.58	28.60

Table 6. BET surface area and porosity of samples activated by different salts.

Samples	S_{BET} (m^2 g^{-1}) [1]	V_{total} (cm^3 g^{-1}) [2]	V_{Micro} (cm^3 g^{-1}) [3]	V_{Meso} (cm^3 g^{-1}) [4]	F_{Micro} (%) [5]
NC-KOH-NaCl	1217	0.775	0.466	0.309	60
NC-KOH-KNO$_3$	874	0.629	0.327	0.302	52
NC-KOH-NaNO$_3$	854	0.620	0.323	0.297	52
NC-KOH-Na$_2$SO$_4$	926	0.602	0.354	0.248	59
NC-KOH-KCl	1034	0.634	0.398	0.236	63

[1] S_{BET} is the specific surface areas determined by the BET method. [2] V_{Total} is the total pore volume. [3] V_{Micro} is the micropore volume. [4] V_{Meso} is the mesoporous volume. [5] F_{Micro} is the fraction of micropore volume to total pore volume.

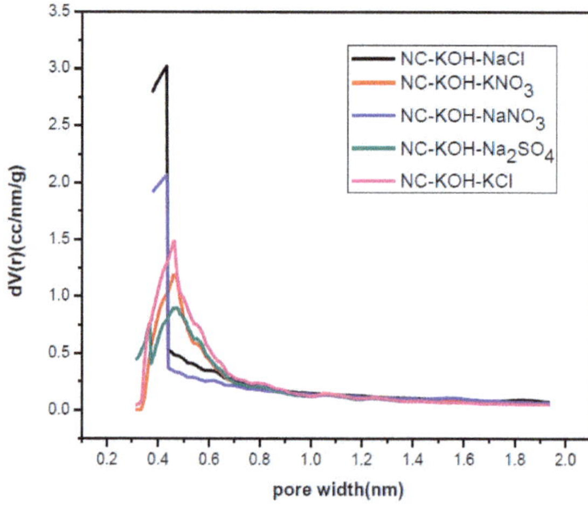

Figure 8. The micropore size distribution of samples activated by different salts.

Furthermore, in order to explore the difference in the structure of the samples activated by different amount of KOH, these samples were characterized by elemental analysis and N_2-adsorption-desorption analysis and the results were shown in Tables 7 and 8 and Figure 9. The elemental analysis showed that the nitrogen content in the NC-KCl sample was 5.28 wt % and the value could be increased to 11.2 wt % by adding 0.4 g KOH (Table 7), which implied that the introduction of KOH could greatly increase the nitrogen content. The addition of more KOH led to higher nitrogen content, but slight -promotion effect was observed if the amount of KOH exceeded 1.2 g. Besides, the promotion effect of KOH was also observed in the specific area. Apart from the NC-KCl-KOH-0.4 sample (Table 8), all the other samples activated by KOH exhibited a larger specific area than the NC-KCl sample, which suggested that the introduction of KOH could increase the specific area of the carbon material, but a certain amount of KOH was required.

Table 7. The content of N, C, H and O in samples activated by different amount of KOH and 0.75g KCl per 2.20 g resorcinol.

Samples	N (wt %)	C (wt %)	H (wt %)	O (wt %)
NC-KCl	5.28	65.28	2.51	26.93
NC-KCl-KOH-0.4	11.2	73.27	1.4	14.13
NC-KCl-KOH-0.8	12.54	61.97	1.39	24.1
NC-KCl-KOH-1.2	12.59	56.23	2.58	28.60
NC-KCl-KOH-1.6	12.65	63.23	1.54	22.58
NC-KCl-KOH-2.0	12.96	58.89	1.4	26.75

Table 8. BET surface area and porosity of samples activated by different amount of KOH and 0.75g KCl per 2.20 g resorcinol.

Samples	S_{BET} (m^2 g^{-1}) [1]	V_{total} (cm^3 g^{-1}) [2]	V_{Micro} (cm^3 g^{-1}) [3]	V_{Meso} (cm^3 g^{-1}) [4]	F_{Micro} (%) [5]
NC-KCl	903	0.686	0.348	0.338	51
NC-KCl-KOH-0.4	702	0.518	0.269	0.249	52
NC-KCl-KOH-0.8	1352	0.758	0.517	0.241	68
NC-KCl-KOH-1.2	1034	0.634	0.398	0.236	63
NC-KCl-KOH-1.6	1159	0.722	0.445	0.277	62
NC-KCl-KOH-2.0	999	0.575	0.377	0.198	66

[1] S_{BET} is the specific surface areasdetermined by the BET method. [2] V_{Total} is the total pore volume. [3] V_{Micro} is the micropore volume. [4] V_{Meso} is the mesoporous volume. [5] F_{Micro} is the fraction of micropore volume to total pore volume.

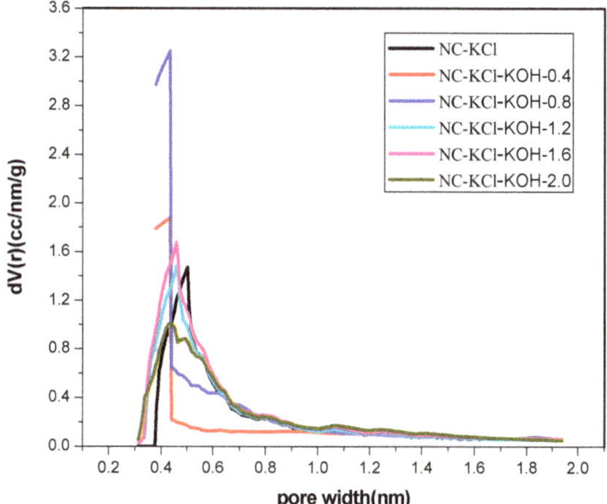

Figure 9. The micropore size distribution of samples activated by different amount of KOH and 0.75 g KCl per 2.20 g resorcinol.

XRD patterns of the sample were shown in the Figure 10 and a typical reflection of amorphous carbon at about 24° was observed in all samples, which could be assigned to hexagonal graphite [47]. Besides, a weak peak appeared at approximately 43° in all the samples but NC-KCl-KOH sample, which could be assigned to rhombohedral graphite [47]. By correlating with the CO_2 adsorption capacity, it's not difficult to make a speculation that the formation of rhombohedral graphite might produce adverse effect for the CO_2 adsorption.

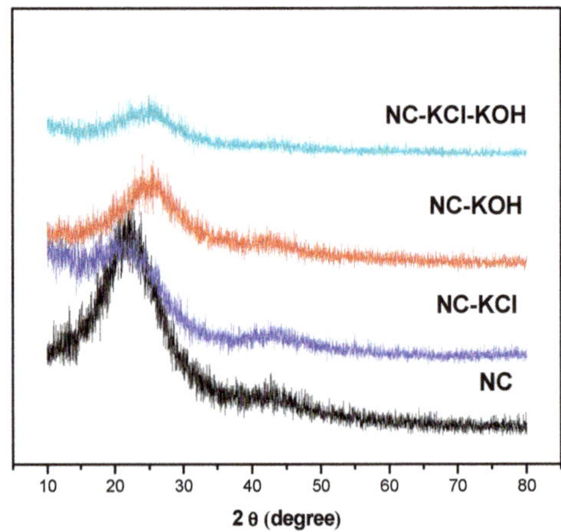

Figure 10. X-ray diffraction patterns of typical samples.

The contents of C, H and N in the sample were determined by elemental analysis, and the content of O was calculated by the subtracting from the content of C, H and N from the total content. The results in Table 9 showed that the activation of the sample by base and salt has a great effect on the content of N. The nitrogen content in the NC sample was 7.78 wt %, and the value could be increased to 11.99 wt % by the KOH activation, which suggested that the introduction of KOH might favors the formation of nitrogen-containing functional groups during ammoxidation process. As is well known, the existence of the doped nitrogen could provide the basic sites to adsorb CO_2 and higher nitrogen content means more CO_2 adsorption sites. Therefore, the higher CO_2 adsorption capacity of the NC-KOH exhibited could be well explained. A similar increase in the nitrogen was observed when the sample was activated by KOH and KCl, which further confirmed the effect of KOH. Considering the enhanced CO_2 adsorption capacity of NC-KOH and NC-KCl-KOH samples in comparison with NC, a speculation could be made that the high nitrogen content in the sample is good for the CO_2 adsorption by providing more basic sites. It is worth noting that the nitrogen content of the sample activated by KCl decreased, but its CO_2 adsorption capacity reversely increased, which implied that the introduction of KCl might increase the CO_2 adsorption capacity by changing the sample's pore structures not increasing the nitrogen content.

Table 9. The content of N, C, H and O in the typical samples determined by elemental analysis.

Samples	N (wt %)	C (wt %)	H (wt %)	O (wt %)
NC	7.78	76.12	1.91	14.19
NC-KCl	5.28	65.28	2.51	26.93
NC-KOH	11.99	58.22	2.67	27.12
NC-KCl-KOH	12.59	56.23	2.58	28.60

Then the porosity and BET specific surface area of typical samples were determined by N_2 adsorption and desorption and the results are shown in Table 10. Obviously, the samples activated by KOH and/or KCl had a larger surface area than the NC sample, and the NC-KCl-KOH sample with the highest CO_2 adsorption capacity exhibited the largest surface area, which suggested that the large surface area might be favorable the CO_2 adsorption. A similar phenomenon could also be observed in the pore volume of the sample and the pore volume of the sample could be increased from 0.19 cm^3 g^{-1} to 0.634 cm^3 g^{-1} by the co-activation of KOH and KCl. However, the NC-KCl sample showed the biggest pore volume although its CO_2 adsorption capacity was lower than that of the NC-KOH and NC-KCl-KOH samples. In order to gain more insights on the correlation of the pore volume and the CO_2 adsorption capacity, the micropore and mesoporous volume as well as the fraction of micropore volume to total pore volume were calculated, respectively. A linear correlation between the CO_2 adsorption capacity and the fraction of micropore volume to total pore volume could be observed, which suggested that the formation of the micropore should be important for the CO_2 adsorption.

Table 10. BET surface area and porosity of typical samples.

Samples	S_{BET} (m^2 g^{-1}) [1]	V_{total} (cm^3 g^{-1}) [2]	V_{Micro} (cm^3 g^{-1}) [3]	V_{Meso} (cm^3 g^{-1}) [4]	F_{Micro} (%) [5]
NC	158	0.190	0.055	0.135	29
NC-KCl	903	0.686	0.348	0.338	51
NC-KOH	1030	0.659	0.401	0.258	61
NC-KCl-KOH	1034	0.634	0.398	0.236	63

[1] S_{BET} is the specific surface areasdetermined by the BET method. [2] V_{Total} is the total pore volume. [3] V_{Micro} is the micropore volume. [4] V_{Meso} is the mesoporous volume. [5] F_{Micro} is the fraction of micropore volume to total pore volume.

Based on the above discussions, the CO_2 adsorption capacity of the sample was determined by the nitrogen content and the pore structure. By contrast, the latter played a more important role.

The micropore size distribution of typical samples was characterized by using the Harvath–Kawazoe (H–K) equation based on the N_2 adsorption and desorption data (Figure 11). Obviously, the micropores in the NC sample were very few, which is consistent with its small micropore volume presented in Table 2. The introduction of KCl in the R-F resin synthesis could promote the formation of more micropores and the micropore size ranged from 0.3–1.9 nm with a peak at 0.50 nm. Similar effect could be also observed when adding KOH in the carbonization process of the material. It's different from the NC-KCl sample that the NC-KOH sample had more micropores with smaller pore size and the peak value shifted left to 0.46 nm. The NC-KCl-KOH sample exhibited a nearly same micropore structure with the NC-KOH sample. It is noteworthy that the size of most micropores in the three samples activated by base and/or salt is smaller than 0.7 nm, and these pores were reported to support the CO_2 adsorption [48].

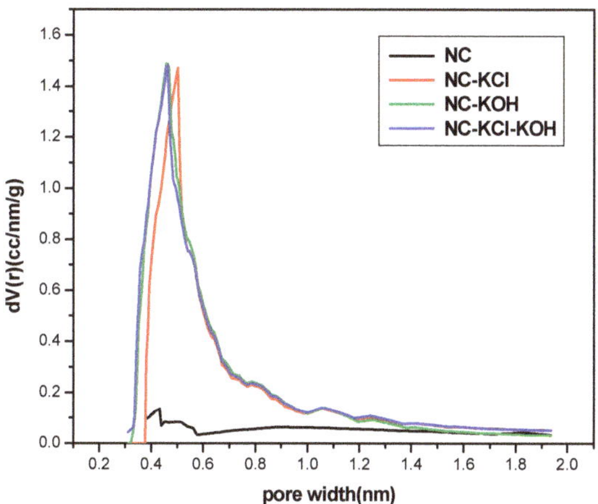

Figure 11. The micropore size distribution of typical samples.

The nitrogen bonding configurations were further studied by XPS and the N 1s spectra of typical samples are shown in Figure 12. Two signal peaks with binding energy at 398.5 and 400.3 eV were observed in all the samples except for NC-KCl-KOH sample. The peak at 398.5 eV could be assigned to pyridinic nitrogen and the peak at 400.3 eV to graphitic nitrogen [49]. It can be found that the ratio of pyridinic nitrogen to graphitic nitrogen was greatly influenced by the pre-treatment activation by comparing the relative intensity of pyridinic and graphitic nitrogen peaks. The salt and base activation pre-treatment led to a decrease in the ratio of pyridinic nitrogen to graphitic nitrogen. In the case of NC-KCl-KOH sample, only the signal peak corresponding to graphitic nitrogen was observed, which suggested that graphitic nitrogen could behave as effective binding sites for CO_2. It has been reported that different kinds of nitrogen functional groups have different degrees of effects on materials' CO_2 adsorption [32], which is also the reason for that the material's nitrogen content obtained by element analysis could not match its CO_2 adsorption performance very well.

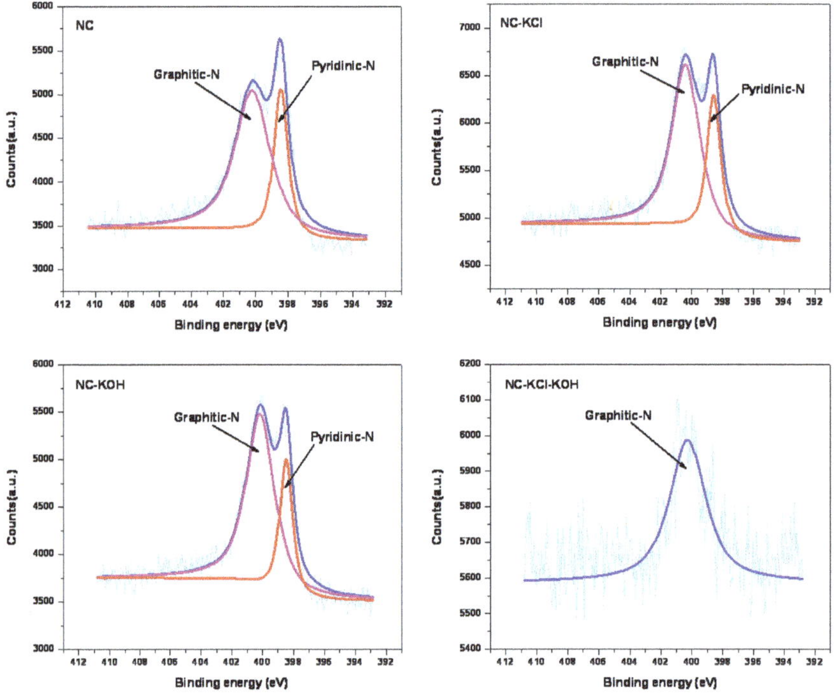

Figure 12. N1s XPS spectra of typical samples.

Finally, the surface morphology of typical samples and their precursors were analyzed by SEM and the results were shown in Figure 13. The morphology of the precursor without KCl activation took on like-lumps feature. When adding 0.75 g KCl in the precursor synthesis, the morphology could be changed to form uniform and close-connected small spheres. However, the surface of the NC and NC-KCl samples both consisted of carbon blocks with different size although the latter contained more carbon blocks with smaller size. Besides irregular carbon blocks, some uniformly small carbon spheres could be also observed when introducing 1.2 g KOH in the carbonization process of the precursor without KCl activation pretreatment, which could be attributed to the etch effect of base. Different from the above case, only uniformly small carbon spheres were obtained when the sample was co-activated by 0.75 KCl and 1.2 g KOH, which suggested a synergistic effect of base and salt activation pretreatment on the regulation of the morphology.

Figure 13. SEM image of typical samples and their precursors.

4. Conclusions

In this work, a series of nitrogen-doped carbon materials with high CO_2 capture capacity were prepared by the ammoxidation of resorcinol-formaldehyde resin precursor with the aid of salt and/or base pretreatment activation. An obvious synergistic effect was observed between base and salt and the combination of 0.75 g KCl and 1.2 g KOH was proven to be the best combination. The sample co-activated by KCl and KOH exhibited the best CO_2 adsorption performance and 1 g typical NC-KCl-KOH sample could adsorb up to 167.15 mg CO_2. The extensive characterization revealed that the introduction of KCl and KOH could increase the doped nitrogen content, change the nitrogen bonding configurations, enlarge the specific surface area and promote the formation of micropores with the size <0.7 nm. Therefore, the CO_2 adsorption capacity of the nitrogen doped carbon material should be co-influenced by the amount and type of doped nitrogen and the pore structure.

Author Contributions: Conceptualization, F.S.; Investigation, R.W.; Writing—original draft preparation, R.W., X.D. and F.S.; Writing—review and editing, R.W., X.D. and F.S.; Funding acquisition, F.S.

Funding: This research was funded by the National Natural Science Foundation of China, grant number (91745106, 21633013).

Acknowledgments: We thank the financially supports by the National Natural Science Foundation of China (91745106, 21633013).

Conflicts of Interest: The authors declare no conflict of interest.

References

1. Samanta, A.; Zhao, A.; Shimizu, G.K.H.; Sarkar, P.; Gupta, R. Post-combustion CO_2 capture using solid sorbents: A review. *Ind. Eng. Chem. Res.* **2011**, *51*, 1438–1463. [CrossRef]
2. Yang, H.; Xu, Z.; Fan, M.; Gupta, R.; Slimane, R.B.; Bland, A.E.; Wright, I. Progress in carbon dioxide separation and capture: A review. *J. Environ. Sci.* **2008**, *20*, 14–27. [CrossRef]

3. Jee, S.E.; Sholl, D.S. Carbon dioxide and methane transport in DDR zeolite insights from molecular simulations into carbon dioxide separations in small pore zeolites. *J. Am. Chem. Soc.* **2009**, *131*, 7896–7904. [CrossRef] [PubMed]

4. Su, F.; Lu, C.; Kuo, S.-C.; Zeng, W. Adsorption of CO_2 on amine-functionalized Y-type zeolites. *Energy Fuels* **2010**, *24*, 1441–1448. [CrossRef]

5. Zukal, A.; Zones, S.I.; Kubů, M.; Davis, T.M.; Čejka, J. Adsorption of carbon dioxide on sodium and potassium forms of STI zeolite. *ChemPlusChem* **2012**, *77*, 675–681. [CrossRef]

6. Sánchez-Zambrano, K.; Lima Duarte, L.; Soares Maia, D.; Vilarrasa-García, E.; Bastos-Neto, M.; Rodríguez-Castellón, E.; Silva de Azevedo, D. CO_2 capture with mesoporous silicas modified with amines by double functionalization: assessment of adsorption/desorption cycles. *Materials* **2018**, *11*, 887. [CrossRef] [PubMed]

7. Zukal, A.; Jagiello, J.; Mayerová, J.; Čejka, J. Thermodynamics of CO_2 adsorption on functionalized SBA-15 silica. NLDFT analysis of surface energetic heterogeneity. *Phys. Chem. Chem. Phys.* **2011**, *13*, 15468. [CrossRef]

8. Yu, J.; Le, Y.; Cheng, B. Fabrication and CO_2 adsorption performance of bimodal porous silica hollow spheres with amine-modified surfaces. *RSC Adv.* **2012**, *2*, 6784. [CrossRef]

9. Vilarrasa-García, E.; Cecilia, J.; Moya, E.; Cavalcante, C.; Azevedo, D.; Rodríguez-Castellón, E. "Low Cost" pore expanded SBA-15 functionalized with amine groups applied to CO_2 adsorption. *Materials* **2015**, *8*, 2495–2513. [CrossRef]

10. Li, Y.; Sun, N.; Li, L.; Zhao, N.; Xiao, F.; Wei, W.; Sun, Y.; Huang, W. Grafting of amines on ethanol-extracted SBA-15 for CO_2 adsorption. *Materials* **2013**, *6*, 981–999. [CrossRef] [PubMed]

11. Koirala, R.; Gunugunuri, K.R.; Pratsinis, S.E.; Smirniotis, P.G. Effect of zirconia doping on the structure and stability of CaO-based sorbents for CO_2 capture during extended operating cycles. *J. Phys. Chem. C* **2011**, *115*, 24804–24812. [CrossRef]

12. Wang, Q.; Tay, H.H.; Zhong, Z.; Luo, J.; Borgna, A. Synthesis of high-temperature CO_2 adsorbents from organo-layered double hydroxides with markedly improved CO_2 capture capacity. *Energy Environ. Sci.* **2012**, *5*, 7526. [CrossRef]

13. Broda, M.; Müller, C.R. Synthesis of highly efficient, Ca-Based, Al_2O_3-stabilized, carbon gel-templated CO_2 sorbents. *Adv. Mater.* **2012**, *24*, 3059–3064. [CrossRef] [PubMed]

14. Millward, A.R.; Yaghi, O.M. Metal-organic frameworks with exceptionally high capacity for storage of carbon dioxide at room temperature. *J. Am. Chem. Soc.* **2005**, *127*, 17998–17999. [CrossRef]

15. Junghans, U.; Kobalz, M.; Erhart, O.; Preißler, H.; Lincke, J.; Möllmer, J.; Krautscheid, H.; Gläser, R. A Series of robust copper-based triazolyl isophthalate MOFs: Impact of linker functionalization on gas sorption and catalytic activity. *Materials* **2017**, *10*, 338. [CrossRef]

16. Torrisi, A.; Bell, R.G.; Mellot-Draznieks, C. Functionalized MOFs for enhanced CO_2 capture. *Cryst. Growth Des.* **2010**, *10*, 2839–2841. [CrossRef]

17. Cao, Y.; Zhang, H.; Song, F.; Huang, T.; Ji, J.; Zhong, Q.; Chu, W.; Xu, Q. UiO-66-NH_2/GO composite: synthesis, characterization and CO_2 adsorption performance. *Materials* **2018**, *11*, 589. [CrossRef]

18. Altintas, C.; Avci, G.; Daglar, H.; NematiVesali Azar, A.; Velioglu, S.; Erucar, I.; Keskin, S. Database for CO_2 separation performances of MOFs based on computational materials screening. *ACS Appl. Mater. Interfaces* **2018**, *10*, 17257–17268. [CrossRef] [PubMed]

19. Allen, A.; Wong-Ng, W.; Cockayne, E.; Culp, J.; Matranga, C. Structural basis of CO_2 adsorption in a flexible metal-organic framework material. *Nanomaterials* **2019**, *9*, 354. [CrossRef] [PubMed]

20. Wahby, A.; Ramos-Fernández, J.M.; Martínez-Escandell, M.; Sepúlveda-Escribano, A.; Silvestre-Albero, J.; Rodríguez-Reinoso, F. High-surface-area carbon molecular sieves for selective CO_2 adsorption. *ChemSusChem* **2010**, *3*, 974–981. [CrossRef]

21. Silvestre-Albero, J.; Wahby, A.; Sepúlveda-Escribano, A.; Martínez-Escandell, M.; Kaneko, K.; Rodríguez-Reinoso, F. Ultrahigh CO_2 adsorption capacity on carbon molecular sieves at room temperature. *Chem. Commun.* **2011**, *47*, 6840. [CrossRef]

22. Maroto-Valer, M.M.; Tang, Z.; Zhang, Y. CO_2 capture by activated and impregnated anthracites. *Fuel Process. Technol.* **2005**, *86*, 1487–1502. [CrossRef]

23. Alghamdi, A.; Alshahrani, A.; Khdary, N.; Alharthi, F.; Alattas, H.; Adil, S. Enhanced CO_2 adsorption by nitrogen-doped graphene oxide sheets (N-GOs) prepared by employing polymeric precursors. *Materials* **2018**, *11*, 578. [CrossRef]

24. Chiang, Y.-C.; Hsu, W.-L.; Lin, S.-Y.; Juang, R.-S. Enhanced CO_2 adsorption on activated carbon fibers grafted with nitrogen-doped carbon nanotubes. *Materials* **2017**, *10*, 511. [CrossRef] [PubMed]

25. Bhagiyalakshmi, M.; Hemalatha, P.; Ganesh, M.; Mei, P.M.; Jang, H.T. A direct synthesis of mesoporous carbon supported MgO sorbent for CO_2 capture. *Fuel* **2011**, *90*, 1662–1667. [CrossRef]

26. Balahmar, N.; Mitchell, A.C.; Mokaya, R. Generalized mechanochemical synthesis of biomass-derived sustainable carbons for high performance CO_2 storage. *Adv. Energy Mater.* **2015**, *5*, 1500867. [CrossRef]

27. Bermúdez, J.; Dominguez, P.; Arenillas, A.; Cot, J.; Weber, J.; Luque, R. CO_2 separation and capture properties of porous carbonaceous materials from leather residues. *Materials* **2013**, *6*, 4641–4653. [CrossRef]

28. Sevilla, M.; Al-Jumialy, A.S.M.; Fuertes, A.B.; Mokaya, R. Optimization of the pore structure of biomass-based carbons in relation to their use for CO_2 capture under low- and high-pressure regimes. *ACS Appl. Mater. Interfaces* **2018**, *10*, 1623–1633. [CrossRef]

29. Yue, L.; Xia, Q.; Wang, L.; Wang, L.; DaCosta, H.; Yang, J.; Hu, X. CO_2 adsorption at nitrogen-doped carbons prepared by K_2CO_3 activation of urea-modified coconut shell. *J. Colloid Interface Sci.* **2018**, *511*, 259–267. [CrossRef]

30. Kou, J.; Sun, L.-B. Fabrication of nitrogen-doped porous carbons for highly efficient CO_2 capture: rational choice of a polymer precursor. *J. Mater. Chem. A* **2016**, *4*, 17299–17307. [CrossRef]

31. Shao, L.; Liu, M.; Huang, J.; Liu, Y.-N. CO_2 capture by nitrogen-doped porous carbons derived from nitrogen-containing hyper-cross-linked polymers. *J. Colloid Interface Sci.* **2018**, *513*, 304–313. [CrossRef]

32. Hao, G.-P.; Li, W.-C.; Qian, D.; Lu, A.-H. Rapid synthesis of nitrogen-doped porous carbon monolith for CO_2 capture. *Adv. Mater.* **2010**, *22*, 853–857. [CrossRef] [PubMed]

33. Plaza, M.G.; Pevida, C.; Arenillas, A.; Rubiera, F.; Pis, J.J. CO_2 capture by adsorption with nitrogen enriched carbons. *Fuel* **2007**, *86*, 2204–2212. [CrossRef]

34. Li, X.; Wang, H.; Robinson, J.T.; Sanchez, H.; Diankov, G.; Dai, H. Simultaneous nitrogen doping and reduction of graphene oxide. *J. Am. Chem. Soc.* **2009**, *131*, 15939–15944. [CrossRef]

35. Pevida, C.; Plaza, M.G.; Arias, B.; Fermoso, J.; Rubiera, F.; Pis, J.J. Surface modification of activated carbons for CO_2 capture. *Appl. Surf. Sci.* **2008**, *254*, 7165–7172. [CrossRef]

36. Plaza, M.G.; Rubiera, F.; Pis, J.J.; Pevida, C. Ammoxidation of carbon materials for CO_2 capture. *Appl. Surf. Sci.* **2010**, *256*, 6843–6849. [CrossRef]

37. Mangun, C.L.; Benak, K.R.; Economy, J.; Foster, K.L. Surface chemistry, pore sizes and adsorption properties of activated carbon fibers and precursors treated with ammonia. *Carbon* **2001**, *39*, 1809–1820. [CrossRef]

38. Li, G.; Wang, Z. Microporous polyimides with uniform pores for adsorption and separation of CO_2 gas and organic vapors. *Macromolecules* **2013**, *46*, 3058–3066. [CrossRef]

39. Park, J.; Nabae, Y.; Hayakawa, T.; Kakimoto, M.-a. Highly selective two-electron oxygen reduction catalyzed by mesoporous nitrogen-doped carbon. *ACS Catal.* **2014**, *4*, 3749–3754. [CrossRef]

40. Tian, Z.; Huang, J.; Zhang, X.; Shao, G.; He, Q.; Cao, S.; Yuan, S. Ultra-microporous N-doped carbon from polycondensed framework precursor for CO_2 adsorption. *Microporous Mesoporous Mater.* **2018**, *257*, 19–26. [CrossRef]

41. Yin, F.; Zhuang, L.; Luo, X.; Chen, S. Simple synthesis of nitrogen-rich polymer network and its further amination with PEI for CO_2 adsorption. *Appl. Surf. Sci.* **2018**, *434*, 514–521. [CrossRef]

42. De Souza, L.K.C.; Wickramaratne, N.P.; Ello, A.S.; Costa, M.J.F.; da Costa, C.E.F.; Jaroniec, M. Enhancement of CO_2 adsorption on phenolic resin-based mesoporous carbons by KOH activation. *Carbon* **2013**, *65*, 334–340. [CrossRef]

43. Yu, J.; Guo, M.; Muhammad, F.; Wang, A.; Yu, G.; Ma, H.; Zhu, G. Simple fabrication of an ordered nitrogen-doped mesoporous carbon with resorcinol-melamine-formaldehyde resin. *Microporous Mesoporous Mater.* **2014**, *190*, 117–127. [CrossRef]

44. Lee, S.-Y.; Park, S.-J. Determination of the optimal pore size for improved CO_2 adsorption in activated carbon fibers. *J. Colloid Interface Sci.* **2013**, *389*, 230–235. [CrossRef] [PubMed]

45. Wickramaratne, N.P.; Jaroniec, M. Importance of small micropores in CO_2 capture by phenolic resin-based activated carbon spheres. *J. Mater. Chem. A* **2013**, *1*, 112–116. [CrossRef]

46. Meng, L.-Y.; Park, S.-J. Effect of heat treatment on CO_2 adsorption of KOH-activated graphite nanofibers. *J. Colloid Interface Sci.* **2010**, *352*, 498–503. [CrossRef] [PubMed]

47. Li, Y.; Ben, T.; Zhang, B.; Fu, Y.; Qiu, S. Ultrahigh gas storage both at low and high pressures in KOH-activated carbonized porous aromatic frameworks. *Sci. Rep.* **2013**, *3*, 2420. [CrossRef]

48. Sánchez-Sánchez, Á.; Suárez-García, F.; Martínez-Alonso, A.; Tascón, J.M.D. Influence of porous texture and surface chemistry on the CO_2 adsorption capacity of porous carbons: acidic and basic site interactions. *ACS Appl. Mater. Interfaces* **2014**, *6*, 21237–21247. [CrossRef]

49. Rehman, A.; Park, S.-J. Comparative study of activation methods to design nitrogen-doped ultra-microporous carbons as efficient contenders for CO_2 capture. *Chem. Eng. J.* **2018**, *352*, 539–548. [CrossRef]

50. Sethia, G.; Sayari, A. Comprehensive study of ultra-microporous nitrogen-doped activated carbon for CO_2 capture. *Carbon* **2015**, *93*, 68–80. [CrossRef]

51. Yang, H.; Cui, X.; Deng, Y.; Shi, F. Highly efficient carbon catalyzed aerobic selective oxidation of benzylic and allylic alcohols under transition-metal and heteroatom free conditions. *RSC Adv.* **2014**, *4*, 59754–59758. [CrossRef]

52. Yang, H.; Cui, X.; Dai, X.; Deng, Y.; Shi, F. Carbon-catalysed reductive hydrogen atom transfer reactions. *Nat. Commun.* **2015**, *6*, 6478. [CrossRef] [PubMed]

53. Liu, J.; Qiao, S.Z.; Liu, H.; Chen, J.; Orpe, A.; Zhao, D.; Lu, G.Q.M. Extension of the stöber method to the preparation of monodisperse resorcinol-formaldehyde resin polymer and carbon spheres. *Angew. Chem. Int. Ed.* **2011**, *50*, 5947–5951. [CrossRef] [PubMed]

54. Azambre, B.; Zenboury, L.; Koch, A.; Weber, J.V. Adsorption and desorption of NO_x on commercial ceria-zirconia ($Ce_xZr_{1-x}O_2$) mixed oxides: a combined TGA, TPD-MS, and DRIFTS study. *J. Phys. Chem. C* **2009**, *113*, 13287–13299. [CrossRef]

55. Ebner, A.D.; Reynolds, S.P.; Ritter, J.A. Nonequilibrium kinetic model that describes the reversible adsorption and desorption behavior of CO_2 in a K-promoted hydrotalcite-like compound. *Ind. Eng. Chem. Res.* **2007**, *46*, 1737–1744. [CrossRef]

56. Xu, X.; Song, C.; Andresen, J.M.; Miller, B.G.; Scaroni, A.W. Novel polyethylenimine-modified mesoporous molecular sieve of MCM-41 type as high-capacity adsorbent for CO_2 capture. *Energy Fuels* **2002**, *16*, 1463–1469. [CrossRef]

Article

Enhanced CO$_2$ Adsorption by Nitrogen-Doped Graphene Oxide Sheets (N-GOs) Prepared by Employing Polymeric Precursors

Abdulaziz Ali Alghamdi [1,*], Abdullah Fhead Alshahrani [1], Nezar H. Khdary [2], Fahad A. Alharthi [1], Hussain Ali Alattas [1] and Syed Farooq Adil [1,*]

[1]   Department of Chemistry, College of Science, King Saud University, P.O. 2455, Riyadh 11451, Saudi Arabia; a.t.alshahrani@hotmail.com (A.F.A.); fharthi@ksu.edu.sa (F.A.A.); Attashussain@gmail.com (H.A.A.)

[2]   King Abdulaziz City for Science and Technology (KACST) P.O. Box 6086, Riyadh 11442, Saudi Arabia; nkhdary@kacst.edu.sa

*   Correspondence: aalghamdia@ksu.edu.sa (A.A.A.); sfadil@ksu.edu.sa (S.F.A.)

Received: 9 March 2018; Accepted: 6 April 2018; Published: 10 April 2018

Abstract: Nitrogen-doped graphene oxide sheets (N-GOs) are prepared by employing N-containing polymers such as polypyrrole, polyaniline, and copolymer (polypyrrole-polyaniline) doped with acids such as HCl, H$_2$SO$_4$, and C$_6$H$_5$-SO$_3$-K, which are activated using different concentrations of KOH and carbonized at 650 °C; characterized using SEM, TEM, BET, TGA-DSC, XRD, and XPS; and employed for the removal of environmental pollutant CO$_2$. The porosity of the N-GOs obtained were found to be in the range 1–3.5 nm when the KOH employed was in the ratio of 1:4, and the XRD confirmed the formation of the layered like structure. However, when the KOH employed was in the ratio of 1:2, the pore diameter was found to be in the range of 50–200 nm. The SEM and TEM analysis reveal the porosity and sheet-like structure of the products obtained. The nitrogen-doped graphene oxide sheets (N-GOs) prepared by employing polypyrrole doped with C$_6$H$_5$-SO$_3$-K were found to possess a high surface area of 2870 m^2/g. The N-GOs displayed excellent CO$_2$ capture property with the N-GOs; PPy/Ar-1 displayed ~1.36 mmol/g. The precursor employed, the dopant used, and the activation process were found to affect the adsorption property of the N-GOs obtained. The preparation procedure is simple and favourable for the synthesis of N-GOs for their application as adsorbents in greenhouse gas removal and capture.

Keywords: nitrogen-doped graphene oxide; polypyrrole; polyaniline; CO$_2$; adsorption studies

---

1. Introduction

CO$_2$ emission is a growing problem, which is mainly caused due to the burning of fossil fuel for the production of energy, which is used to drive the present day factories and automobiles. According to the report published by International Monetary Fund (IMF), there are 35.9 billion tonnes of carbon dioxide emissions worldwide, which is a dangerously alarming level, effecting human lives and ecological systems. Apart from its impact on the environment, studies have revealed that CO$_2$ emissions have an impact on the gross domestic product (GDP) as well [1].

In order to overcome this problem, efforts are being made throughout the world to curb pollutions by replacing fossil fuels with other sources of sustainable energy. Apart from this, there are efforts such as afforestation and reforestation for the consumption of CO$_2$ from the environment [2]. Other than this, there are attempts being made to enable adsorption of CO$_2$ into materials, which is also known as CO$_2$ capture and sequestration (CCS) technology, which can be employed further for the generation of electricity [3–7].

One of the most commonly used technique is entrapment of CO_2, which is carried out by adsorption techniques using various adsorbents such as MOFs [8], mesoporous material such as MgO [9], metal decorated phosphorene [10], silica mesospheres [11,12], nanostructrual copolymer, and ionic liquid [13], etc. Carbon-based material such as carbon nanotubes [14], graphene [15], carbon spheres [16], activated carbon, and activated carbon fibers [17] has been extensively used for the entrapment CO_2. Among these, graphene and graphene oxide has been extensively used for this study due to its superior surface area among the various carbon compounds used [18,19]. These were further modified by impregnating it with various metal salts such as Fe_3O_4 [20], Mn_3O_4 [21], Cu [22], Ca [23], and polymeric additives such as PEI [24] to improve the capture and entrapment of CO_2. Further modification is carried out using nitrogen doping of the graphene sheets, wherein the nitrogen-doped graphene was employed for the capture of CO_2, and it was found to possess better adsorption and selectivity towards CO_2 [25].

Nitrogen-doped graphene is synthesized by employing various methods such as graphene exposed to ammonia vapors at elevated temperatures, graphene, graphene mixed with melamine as a source of nitrogen doping on the graphene sheets, the hydrazine steaming process, and the arc discharge method [26–29]. However, based on the information obtained from the previous literature, it can be understood that it is the need of the hour to find easier methods for the synthesis of this important material.

Hence, in continuation of our work on the synthesis of graphene and graphene-based composites for various applications [30–32], in this study, we prepared the N-doped graphene oxide sheets by employing 'N'-containing polymers such as polypyrole, polyaniline, and copolymers (polypyrrole-anlline) doped with acids such as HCl, H_2SO_4, and C_6H_5-SO_3-K, which were activated using different concentrations of KOH. The as-prepared carbonaceous materials were characterized using SEM, TEM, BET, XRD and XPS and tested for CO_2 capture property.

2. Materials and Methods

All materials were obtained from commercial sources such as potassium salt benzene sulphonic acid (97%) and Sulphuric acid (98%) were procured from BDH, Poole, UK, while the polymeric precursors, i.e., monomers, such as Pyrrole (98%) and Aniline (99%), were procured from Alfa Aesar, Karlsruhe, Germany. Hydrochloric acid (~36%), Nitric acid (68%), and Methanol (99.5%) were procured from Fisher Chemical, Loughborough, UK. Ammonium persulphate (98%) and Pottassium hydroxide pellets (85%) were procured from Alfa Aesar, Karlsruhe, Germany. Solvents such as Acetone (99.5%) were procured from Panreac, Barcelona, Spain, and Isopropanol (+99%) was produced from WiNLAB, Middlesex, UK. All solvents were used as received without further purification.

The surface areas, pore size, and pore volumes of the prepared nitrogen-doped graphene oxide (N-GOs) were determined through nitrogen adsorption at 77 K, using Automated gas sorption system Micromeritics analyzer (Gemini VII, 2390 Surface Area and Porosity, Micromeritics, Norcross, GA, USA). Before analysis, each sample was degassed at 150 °C (under N_2 flow) for 1 h to eliminate moisture and gasses. The specific surface area (SBET) was calculated by Brunauer, Emmett, and Teller (BET) method using adsorption isotherm in the range of $0.05 \leq p/p_o \leq 0.30$. The average pore width and micro-pore volume were measured by Dubinin-Radushkevich (DR) equation from the N_2 adsorption isotherm. (DR) equation is expressed by:

$$\frac{W}{W_o} = \exp\left[-\frac{BT^2}{\beta}\ln^2\left(\frac{P_o}{P}\right)\right]$$

in which

W = volume of the pores that has been filled at p/p_o (cm^3/g)

W_o = total volume of the micropore system (cm^3/g)

β = structural constant related to the width of the Gaussian pore distribution (K^{-2})

T = temperature at which the isotherm has been taken (K)

B = similarity constant, depending solely on the adsorbate (-)

p_o/p = inverse of the relative pressure of the adsorbate (-)

The total pore volume was estimated by Barrett, Joyner, and Halenda (BJH) model from the quantity of N_2 adsorbed at relative pressure (p/p_o) of 0.99 by the software of the instrument. The meso-pore volume was calculated by subtracting the micro-pore volume from the total pore volume:

$$V_{Meso} = V_t - V_{Mic}$$

in which V_{Meso} is the mesopore volume, V_t is the total pore volume, and V_{Mic} is the micropore volume.

Scanning electron microscopy (SEM) images were obtained using JSM-6380-LA (JEOL, Tokyo, Japan), which was used for morphology analysis showing the surface texture, pore structure, and pore distribution of the prepared samples. Transmission electron microscopy (TEM) images were recorded on JEM-1011, transmission electron microscope (JEOL, Tokyo, Japan).

Carbon dioxide uptake and the heat of adsorption of the treated nitrogen-doped graphene oxide sheets (N-GOs) were measured by Thermo Gravimetric Analyser (TGA)/Differential Scanning Calorimetry (DSC) SDT-Q600, (TA instruments, New Castle, DE, USA) in the temperature range of 25–1000 °C at a heating rate of 10 °C min^{-1}, using 99.9999% purity Carbon dioxide, additionally purified by a molecular sieve filter. Before analysis, each sample of 5–15 mg was cleaned up at 25–120 °C to eliminate moisture and gasses (under He flow), and the samples' heating rate was 10 °C min^{-1}, using isotherm 120 °C for 30 min. The carbon dioxide uptake experiments were performed at 50 °C, using uptake isotherm at 50 °C, and the flow rate of carbon dioxide was maintained at 100 mL/min for 30 min. The results are calculated as mg/g measurement, which is converted to mmol/g and presented as such.

2.1. Experimental

2.1.1. Polymer Preparation

Polyaniline (PANI) and Polypyrrole (PPy)

0.04 mol of monomer (aniline, pyrrole) was taken in a round bottomed flask along with 0.06 mol of desired dopants (HCl, H_2SO_4, C_6H_5-SO_3-K), and the mixture of reaction was stirred for 30 min at 0–5 °C. In another beaker, 13.7 g Ammonium persulphate (APS) was dissolved in 30 mL water and stirred in beaker for 30 min at 0–5 °C, which was added to the previous reaction mixture while stirring at 0–5 °C; the solid product formed wass filtrated and washed with water and methanol.

Poly(Aniline-Co-Pyrrole) Copolymer

0.02 mol of aniline monomer and 0.02 mol of pyrrole monomer were taken in a round bottomed flask along with 0.06 mol of desired dopants (HCl, H_2SO_4, C_6H_5-SO_3-K), and the mixture of reaction was stirred for 30 min at 0–5 C. In another beaker, 13.7 g of APS was dissolved in 30 mL water and stirred in beaker for 30 min at 0–5 °C, which was added to the round bottomed flask containing the reaction mixture while stirring at 0–5 °C; the solid product formed was filtrated and washed with water and methanol.

Carbonization/Activation of Prepared Polymers

The carbonization/chemical activation of the prepared polymers was performed taking 2 g of the polymer and mixing it with KOH in weight ratios of 2 and 4, i.e., KOH/Polymer weight ratio of 2 or 4. The mixture was carbonized at 650 °C attained at a ramp rate of 3 °C/min for 2 h using Carbolite furnace under nitrogen atmosphere. The weight of the pyrolysed sample was noted. The pyrolyzed samples were then thoroughly washed with 10 wt % HCl to remove any inorganic salts, and then with distilled water until neutral pH was attained. They were then dried in an oven at 120 °C. The dry weight of the washed sample was noted as weight of N-GOs obtained. The samples prepared were denoted as mentioned in Table 1.

Table 1. List of N-GOs samples prepared and denotation method employed in the manuscript.

Polymers	Polymer:KOH					
	1:2 [3]			1:4 [3]		
	HCl [2]	H_2SO_4 [2]	C_6H_5-SO_3-K [2]	HCl [2]	H_2SO_4 [2]	C_6H_5-SO_3-K [2]
ppy [1]	PPy/HCl-1	PPy/H_2SO_4-1	PPy/Ar-1	PPy/HCl-2	PPy/H_2SO_4-2	PPy/Ar-2
PANI [1]	PANI/HCl-1	PANI/H_2SO_4-1	PANI/Ar-1	PANI/HCl-2	PANI/H_2SO_4-2	PANI/Ar-2
Copolymer [1]	Co-P/HCl-1	Co-P/H_2SO_4-1	Co-P/Ar-1	Co-P/HCl-2	Co-P/H_2SO_4-2	Co-P/Ar-2

[1.] Polymer used to prepare the nitrogen-doped graphene oxide sheets; [2.] Dopant used when polymer was prepared; [3.] Ratio of KOH used along with polymer before carbonization.

2.2. Characterization

Surface area measurements, pore size distribution, and various other surface related parameters were carried out using the BET method based on N_2 physisorption capacity at 77 K using Gemini VII 2390 V1.03 apparatus (Micromeritics, Norcross, GA, USA) with the instrument operating in single-point and multi-point modes. Prior to analysis, the samples were degassed for 3 h at 150 °C. The activated carbon prepared was also subjected to scanning electron microscopy to find out the surface porosity and morphology.

3. Results and Discussion

3.1. Textural Properties

The N-GOs obtained after carbonization were weighed, and it was observed that the amount of N-GOs obtained after the acid wash varied based on the polymer used and the dopant added; however, there was no particular pattern observed with regard to it. The amount of polymer used for carbonization was 2 grams, and the highest amount of N-GOs, i.e., about 0.82 g, was obtained when the polypyrrole doped with HCl was carbonized, while the least amount was obtained from the carbonization of copolymer, i.e., Poly(aniline-co-pyrrole) doped with H_2SO_4. The amount of carbonaceous material obtained from the various polymers and dopants are illustrated graphically in Figure 1.

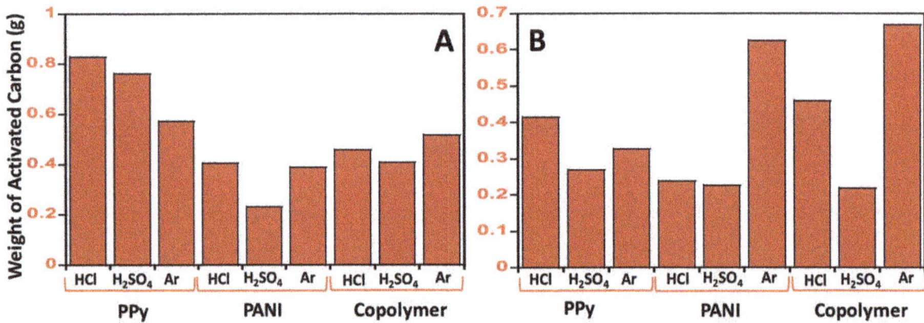

Figure 1. Graphical illustration of amount of activated carbon from pyrolysis of various polymers: (**A**) amount of N-GOs obtained when carbonization and activation were carried using polymer: KOH at a ratio of 1:2; (**B**) amount of N-GOs obtained when carbonization and activation were carried using polymer: KOH in the ratio of 1:4.

The surface area analysis revealed that the N-GOs prepared were found to possess surface area varying from 4.75 m^2/g to ~3000 m^2/g. It was found that the N-GOs obtained from the carbonization of polypyrrole, polyaniline, and co-polymer, which were mixed with KOH at a ratio of 1:2 (Figure 2A), yielded surface area in the range of ~1063–1556 m^2/g, while the N-GOs obtained from the carbonization of polypyrrole, polyaniline, and co-polymer, which were mixed with KOH at a ratio of 1:4 (Figure 2B), yielded a wide range of surface area ranging from 4.75 m^2/g to ~3000 m^2/g; among these the samples, PPy/HCl-2, PPy/H$_2$SO$_4$-2, and PPy/Ar-2 yielded samples with surface areas of 2870 m^2/g, 2134 m^2/g, and 2943 m^2/g surface area, respectively. These samples of N-GOs, i.e., PPy/HCl-2, PPy/H$_2$SO$_4$-2, and PPy/Ar-2, were obtained when the polymer polypyrole doped with HCl, H$_2$SO$_4$, and C$_6$H$_5$-SO$_3$-K was mixed with KOH during the carbonization step in the ratio 1:4, while the same polymer doped with HCl, H$_2$SO$_4$, and C$_6$H$_5$-SO$_3$-K was mixed with KOH during the carbonization step in the ratio 1:2. Then, the surface area for the N-GOs obtained was found to be 1491 m^2/g, 1374 m^2/g, and 1397 m^2/g, respectively. This indicates that the amount of KOH mixed with the polymer before carbonization plays an important role in the surface area of the N-GOs obtained. In the case of PPy, the surface area increased up to 64% upon employing KOH in the ratio 1:4 from 1:2; however, in the N-GOs obtained from PANI and Co-poly, the surface area decreased drastically.

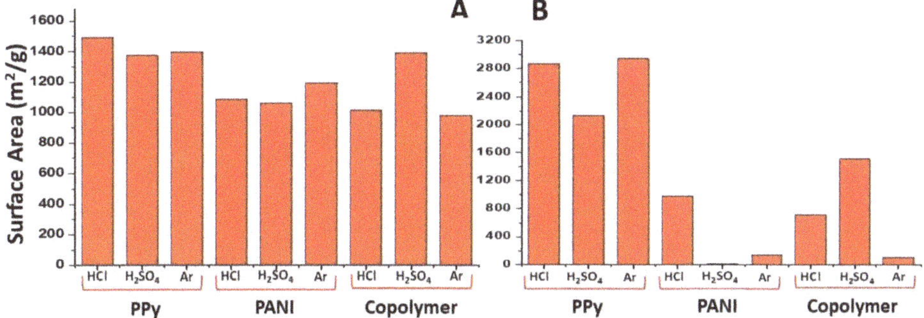

Figure 2. Graphical illustration of surface area of the activated carbon obtained from pyrolysis of various polymers: (**A**) amount of N-GOs obtained when carbonization and activation were carried using polymer: KOH at a ratio of 1:2; (**B**) amount of N-GOs obtained when carbonization and activation were carried using polymer: KOH at a ratio of 1:4.

The nitrogen adsorption-desorption isotherms for the N-GOs with higher surface areas of more than 1400, i.e., PPy/HCl-1, PPy/H$_2$SO$_4$-1, PPy/Ar-1, Co-P /H$_2$SO$_4$-1, PPy/HCl-2, PPy/H$_2$SO$_4$-2, PPy/Ar-2, and Co-P/H$_2$SO$_4$-2, were selected for further textural and surface morphological evaluations. The N$_2$ adsorption/desorption isotherms were obtained and plotted in Figure 3. The different polymeric precursors, and the amount of KOH and dopants employed for the synthesis were found to have an effect on the textural and surface morphological of the N-GOs obtained, which was ascertained from the different isotherms obtained. The N$_2$ adsorption/desorption isotherms curves obtained were found to increase in the low relative pressure, i.e., p/p$_o$ < 0.2, indicating the presence of pore structures; the isotherm was found to match to Type I adsorption curves, indicating the presence of mono layers.

However, in the case of N-GOs obtained by employing KOH at a ratio of 1:4, as the relative pressure increases the knee of the isotherm becomes more open than the one obtained in the case of N-GOs obtained by employing KOH at a ratio of 1:2, and a regular increase was observed, indicating the increase in the amount of nitrogen adsorbed with the increase in relative pressure, signifying the formation of mesopores with Type IV isotherm. Hence, it can be said that the N-GOs that were obtained display a combination of the Type I and Type IV isotherms, signifying the formation of microporous and mesoporous N-GOs. Furthermore, the occurrence of hysteresis loop of the H4 type suggests that the pores are random with irregular structures.

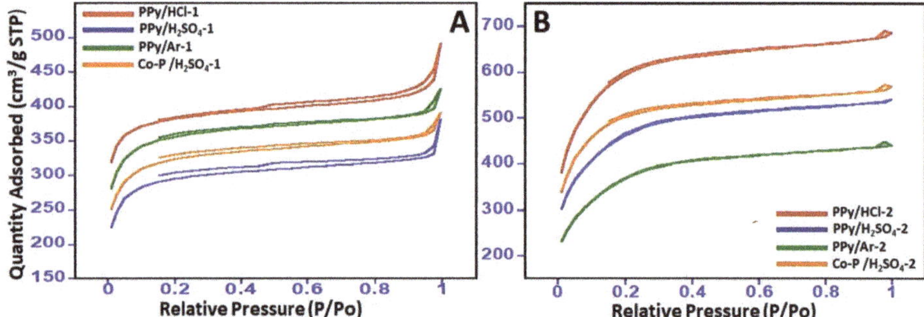

Figure 3. Adsorption/desorption isotherms obtained for (**A**) N-GOs obtained when carbonization and activation were carried using polymer: KOH at a ratio of 1:2; (**B**) N-GOs obtained when carbonization and activation were carried using polymer: KOH at a ratio of 1:4.

3.2. Morphological and Microscopic Analysis

The pore size distribution study was carried out using the BET and the data obtained is presented graphically in Figure 4. Figure 4A shows the microstructure of the inside of N-GOs prepared from the polymer substrates employing 1:2 KOH in the pre-carbonization step, which indicates that the porous structure obtained ranges from 50–200 nm in diameter, with volume ranging from <0.005–0.01 cm^3/g. However, when the N-GOs prepared from the polymer substrates employing 1:4 KOH in the pre-carbonization step were evaluated, they were found to possess diameter ranging from 1–3.5 nm, while their pore volume ranged from 0.04–0.05 cm^3/g. From the values obtained, it is clear that the KOH treatment pre-carbonization plays an important role in the surface morphology and porosity of the N-GOs obtained, with the KOH at a ratio of 1:2; the pore diameter ranges from 50–200 nm, and the pore volume denotes that the pores formed are wide and shallow, while with the 1:4 KOH ratio, the pores appear to be long cylindrical capillaries with pore diameter ranging from 1–3.5 nm (Figure 4B).

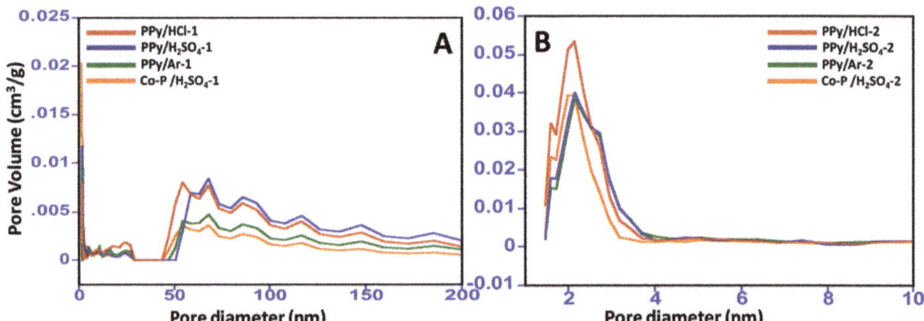

Figure 4. Pore size distribution for (**A**) N-GOs obtained when carbonization and activation were carried using polymer: KOH at a ratio of 1:2 and (**B**) N-GOs obtained when carbonization and activation were carried using polymer: KOH at a ratio of 1:4.

The SEM analysis of the obtained N-GOs was carried out to understand the surface morphology, while the TEM analysis helps one understand the formation of sheet-like structure as desired. The SEM and TEM images that were obtained are given in Figures 5 and 6, respectively.

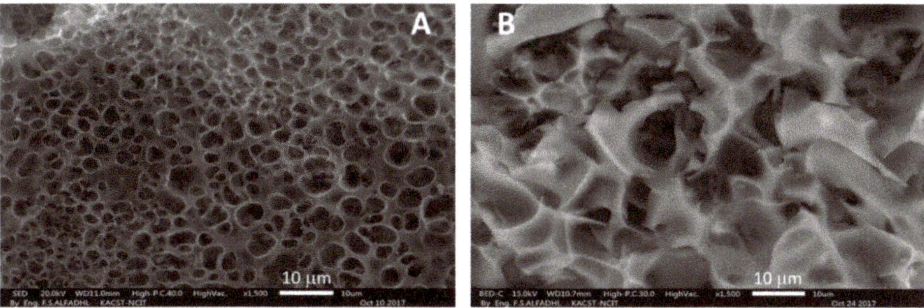

Figure 5. SEM micrograms obtained for N-GOs (**A**) PPy/Ar-1 (**B**) PPy/Ar-2.

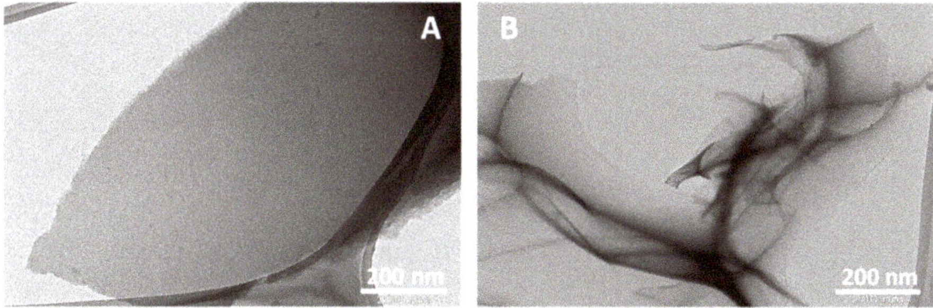

Figure 6. TEM micrograms obtained for graphene sheets-like structure of (**A**) PPy/Ar-1 (**B**) PPy/Ar-2.

The SEM micrograms obtained indicated that a porous structure was well-formed with rigid borders in case of N-GOs formed from the PPy/Ar-1 obtained when polymer: KOH was at a ratio of 1:2, which is in accordance with the data obtained from the pore size distribution graph obtained as given in Figure 5A. However, in case of the N-GOs formed from the PPy/Ar-2 obtained when polymer: KOH was at a ratio of 1:4, the surface appears to be rugged, and porosity is not very evident, as shown in Figure 5B. Hence, it can be said that the pore size and pore size distribution data of N-GOs is in good agreement with that of SEM images.

The TEM analysis of the samples revealed that a sheet-like structure was formed that resembled the exfoliated graphene sheets. However, from the images it is evident that the N-GOs formed from the PPy/Ar-1 obtained when polymer: KOH was at a ratio of 1:2 appear to be thicker than the ones obtained when the polymer: KOH was at a ratio of 1:4, i.e., PPy/Ar-2, which indicates the activation process, effects the thickness of the sheet formed. The porous structure obtained from the TEM images yielded that it is in the range of 1–3 nm (Figure 6A,B).

3.3. XRD Spectral Analysis

The composition of amorphous carbon is generally made of organized graphite-like microcrystals, non-organized carbon, and single-reticular-plane carbon [33]. The N-GOs PPy/Ar-1 and PPy/Ar-2 that were prepared by different activation methods and were subjected to XRD analysis to confirm the composition and the diffraction patterns obtained are shown in Figure 7. The absence of sharp peaks along with very broad diffraction peaks reveals that the N-GOs formed are amorphous in nature.

The two obvious diffraction peaks represent, respectively, the diffraction-characteristic peaks of the microcrystalline (002) and (10l) crystal face of the turbostratic graphite structure of the PPy/Ar-1 and PPy/Ar-2. The slight difference in the diffraction pattern between the two N-GOs PPy/Ar-1 and PPy/Ar-2 can be attributed to the extent of disorderliness prevailing in the graphitic structure of the N-GOs prepared. The varying intensity of the diffraction spectrum suggests that the difference between the surface of the pores in the PPy/Ar-1 are more occupied with the dopants employed, while PPy/Ar-2 appears to be more naked and exposed due to the employment of higher amount of KOH during the activation process [34].

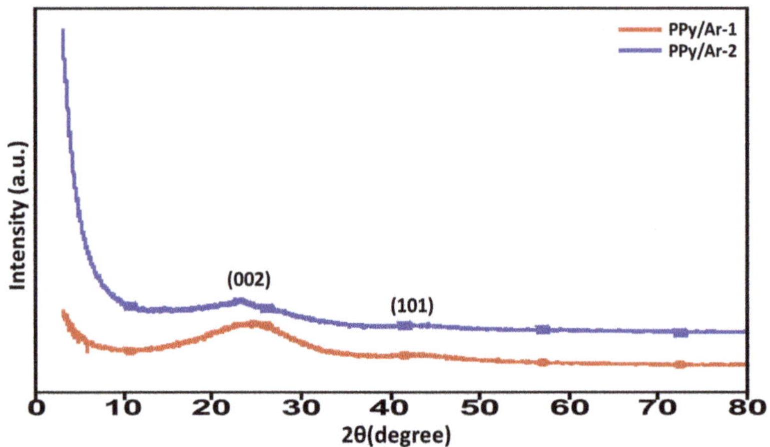

Figure 7. XRD diffractograms obtained for N-GOs PPy/Ar-1 and PPy/Ar-2.

3.4. X-ray Photoelectron Spectroscopy (XPS)

X-ray photoelectron spectroscopy (XPS) was also conducted to study the chemical compositions of the N-GOs, i.e., PPy/Ar-1 and PPy/Ar-2 were prepared by employing the polymeric precursors, and the results are shown in Figure 8. From the results obtained, it is found that the N-GOs PPy/Ar-1 and PPy/Ar-2 obtained are found to be mainly composed of C, N, and O elements observed in the survey spectrum. The C 1s spectrum of PPy/Ar-1 (thick line) and PPy/Ar-2 (dotted line) yielded a broad peak at 284 eV, which can be attributed to the graphitic carbon and C–N [35]. As for N 1s spectrum of PPy/Ar-1 (thick line) and PPy/Ar-2 (dotted line), it yielded a peak that can be separated into two peaks with binding energy of 398.1 eV and 400.02 eV. The major peak located at 400.02 eV indicates the presence of pyrrolic-N (398–402 eV), while the peak at 398.1 can be assigned to pyridinic-N (397.1–399.3 eV). However, the intensities in the peaks vary, indicating the varying presence of the pyrrolic-N and pyridinic-N, which indicates that the percentage of presence of pyrrolic-N and pyridinic-N is found to be greater in PPy/Ar-1 (thick line) than in PPy/Ar-2 (dotted line) [36,37]. The high-resolution O 1s spectra (Figure 6) in case of PPy/Ar-1 (thick line) reveal the presence of three peaks corresponding to C=O groups (530.8 eV), C–OH, and/or C–O–C groups (532.6 eV), and chemisorbed oxygen and/or water (535.6 eV), while the PPy/Ar-2 (dotted line) yielded two peaks corresponding to C=O groups (530.7 eV), C–OH, and/or C–O–C groups (532.9 eV). A variation in the intensities of the O 1s peaks obtained can be observed in this case, just like in the N 1s spectra. Moreover, the percentage composition calculated revealed that PPy/Ar-1 possesses more N atoms on the surface, while the percentage of O atoms is greater in the PPy/Ar-2 sample. The elemental composition calculated from the XPS spectras of PPy/Ar-1 and PPy/Ar-2 is given in Table 2.

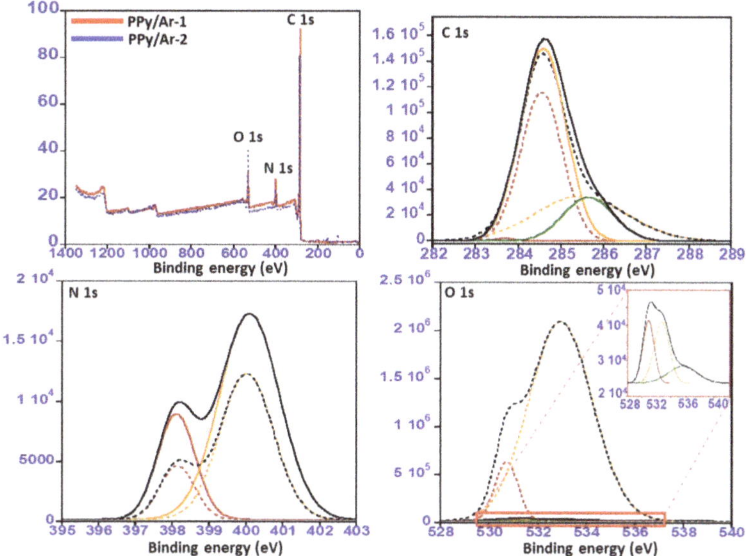

Figure 8. XPS spectrum obtained for N-GOs PPy/Ar-1 (thick line) and PPy/Ar-2 (dotted line).

Table 2. Elemental composition of N-GOs PPy/Ar-1 and PPy/Ar-2.

N-GOs	Elements		
	C	N	O
PPy/Ar-1	80.80	8.08	11.11
PPy/Ar-2	80.80	5.05	14.14

3.5. Thermal Stability

The thermal stability of the synthesized N-GOs was evaluated by employing thermal analysis by heating the samples in N_2 atmosphere from 25 °C to 900 °C, and it is found that different samples displayed differing thermal stability (Figure 9). The best thermal stability is displayed by PPy/Ar-1 and PPy/Ar-2, which displays 32.8% and 32.4% weight loss, respectively, while PPy/HCl displays least thermal stability with a weight loss of ~52%.

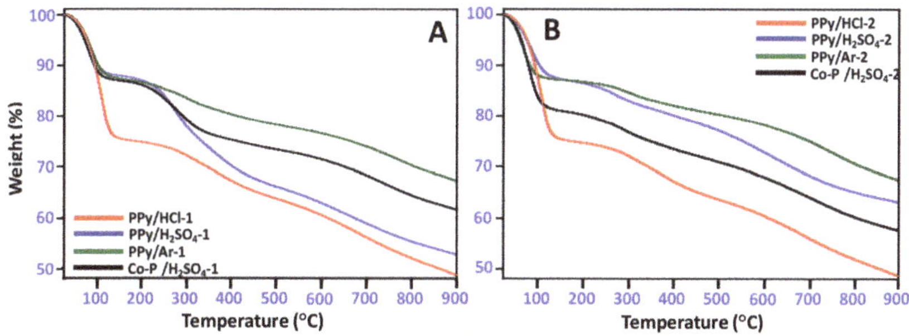

Figure 9. TGA thermograms obtained for (**A**) N-GOs obtained when carbonization and activation were carried using polymer: KOH in the ratio of 1:2; (**B**) N-GOs obtained when carbonization and activation were carried using polymer: KOH in the ratio of 1:4.

3.6. CO$_2$ Adsorption Properties

The preliminary CO$_2$ adsorption-desorption behavior measured at 50 °C and 1.0 atm for the N-GOs samples prepared is shown in Figure 7. Firstly, the N-GOs tested were activated at 120 °C by He flow for 20 min, then cooled down to 50 °C. CO$_2$ was then passed over until no further weight gain was observed and a complete adsorption–desorption cycle was completed. Each N-GO was found to gradually adsorb CO$_2$ over the first 25 min, which then continued at a slower rate until equilibrium was apparently achieved, which lasted for 60 min for all samples. A graphical representation of the CO$_2$ capture pattern exhibited by the various N-GOs is displayed in Figure 10.

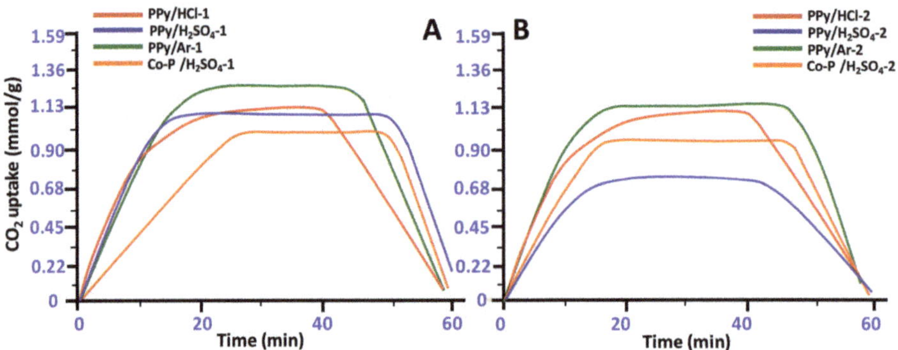

Figure 10. CO$_2$ adsorption isotherms measured by TGA-DSC for (**A**) amount of N-GOs obtained when carbonization and activation were carried using polymer: KOH at a ratio of 1:2 and (**B**) amount of N-GOs obtained when carbonization and activation were carried using polymer: KOH at a ratio of 1:4.

It is observed that other than the polymeric precursor and the dopant employed, the activation process plays an important role in the CO$_2$ adsorption performance of the N-GOs. Among the various polymeric precursors employed, the N-GOs obtained from the PPy yielded best surface area and pore size distribution. The N-GOs obtained from PPy precursor display the best CO$_2$ adsorption efficiency, while the ones obtained from employing the other N-GOs from Co-PPy-PANI also display optimum CO$_2$ adsorption capability. Among the N-GOs from PPy, the ones doped with HCl and H$_2$SO$_4$ display a lower CO$_2$ adsorption capability than the ones obtained from C$_6$H$_5$-SO$_3$-K-doped, indicating the role played by the dopant. The N-GOs obtained with varying ratios of KOH effected the CO$_2$ adsorption performance of the N-GOs obtained. In case of N-GOs obtained from PPy, when the activation is carried out employing KOH in the ratio 1:2, the N-GOs obtained are found to adsorb CO$_2$ better than those of the N-GOs obtained when KOH was used with the ratio of 1:4. However, the adsorption pattern obtained is found to be contrary to the surface area of the N-GOs, which is found to be higher in case of N-GOs obtained employing KOH in the ratio 1:4 than those obtained in case of 1:2. Hence, the adsorption and capture of CO$_2$ can be attributed to the porosity of the N-GOs obtained, which in case of polymer: KOH in the ratio of 1:2 was found to possess a wide range of porosity (i.e., 50–200 nm), while in the case of polymer: KOH in the ratio of 1:4, it was found to be narrow (i.e., 1–3.5 nm). The activation process with higher amount of KOH renders the pore more naked and exposed, which reduces the adsorption of CO$_2$. Apart from the pore morphology, it is also observed that polymeric precursor and the dopant also play an important role.

The heat of adsorption integrated from the TGA-DSC heat flow curves provides information about the interaction of the CO$_2$ molecules and the active sites in the N-GOs. The heat of adsorption is found to be in the range of 10–96 kJ/mol, which indicates that the N-GOs obtained displayed varying interactions between the CO$_2$ molecules and the active sites in the N-GOs, leading to the fluctuating adsorption of the CO$_2$. Among the N-GOs prepared, the PPy/Ar-1 yielded the highest heat of adsorption, i.e., 96.04 kJ/mol, which is higher and more stable (chemical adsorption) than the heat of adsorption for PPy/Ar-2 (94.13 kJ/mol).

In case of N-GOs, PPy/Ar-1 was found to possess a pore size of 50–200 nm, while PPy/Ar-2 was found to possess a narrow pore size of 1–3.5 nm. PPy/Ar-1 and PPy/Ar-2 possess pyrrolic-N and pyridinic-N along with C=O groups, C–OH, and/or C–O–C groups. However, the percentage composition varies in the N-GOs prepared, which can be attributed to the difference in CO_2 adsorption. Of the N-GOs prepared, PPy/Ar-1 possesses a higher percentage of pyrrolic-N and pyridinic-N and wider pore size that could possibly enhance the interaction of the CO_2, which is indicated by a high heat of adsorption of 96.04 kJ/mol, which greatly promotes CO_2 adsorption. However, PPy/Ar-2 displays lower CO_2 adsorption, which can be attributed to the narrow pore size of 1–3.5 nm and the higher percentage of O on the surface of the graphene-like sheets, which could probably hinder the interaction of CO_2 molecule with the pyrrolic-N and pyridinic-N, which is evident by the lower heat of adsorption 94.13 kJ/mol.

A series of absorbent materials that were previously studied and reported in literature for their adsorption capacities is compiled in the Table 3. When the saturated CO_2 adsorption capacity of the as-prepared N-GOs is compared to previously reported adsorbent materials, it was found that the as-prepared N-GOs performed better than a few absorbent materials like Cu-propyl ethylenediamine-silica and propyl ethylenediamine-silica composites, while many of the adsorbent materials such as carbonized porous aromatic framework, activated carbon, and mesoporous carbon were found to be better than the as-prepared N-GOs.

Table 3. CO_2 adsorption capacity comparison of various porous materials with N-GOs PPy/Ar-1 and PPy/Ar-2.

Materials	Capacity (mmol g^{-1})	References
Carbonized porous aromatic framework (PAF)	4.5	[38]
Activated carbon-phloroglucinol-500 °C	4.37	[39]
Microporous carbon ultrafine fibers	2.92	[40]
N-containing porous carbon monoliths	2.9	[41]
Porous carbon nanosheets	2.88	[42]
Alkali-modified activated Carbon	2.46	[43]
Mesoporous carbons	2.27	[44]
Isoreticular zeolitic imidazolate frameworks	2.2	[45]
Mesoporous carbons	2.14	[44]
Commercially activated carbons including BPL, Maxsorb, and Norit R1	<2.00	[46]
Soft-templated mesoporous carbons	1.49	[47]
KOH-activated graphite nanofibers	1.35	[48]
PPy/Ar-1	1.28	This work
PPy/Ar-2	1.18	This work
Cu-propyl ethylenediamine-silica composites	0.58	[49]
Propyl ethylenediamine-silica composites	0.45	[49]

4. Conclusions

In conclusion, polymeric precursors such as polypyrrole (PPy), polyaniline (PANI), and copolymer (PPy-PANI), along with various dopants such as HCl, H_2SO_4, and C_6H_5-SO_3-K, were employed for the preparation of N-doped graphene oxide (N-GOs). Among the polymeric precursors employed, the polypyrrole precursor doped with C_6H_5-SO_3-K yielded the desired N-GOs. The porosity and surface area varied upon the varying use of KOH. The N-doped graphene (NDG) obtained from the use of 1:2, polymer: KOH, i.e., PPy/Ar-1, yielded a porosity in the range of 50–150 nm, while the surface area was found to be 1400 m^2/g; however, when 1:4, polymer: KOH, i.e., PPy/Ar-2, the porosity was found in the range of 1–3.5 nm, while the surface area was found to be ~3000 m^2/g. Among the N-GOs prepared, the N-GOs obtained from employing PPy doped with C_6H_5-SO_3-K was found to display the best adsorption property. The N-GOs obtained from PPy doped with C_6H_5-SO_3-K activated by employing KOH at a ratio of 1:2, polymer: KOH, i.e., PPy/Ar-1, yielded a 1.3 mmol/g adsorption of CO_2; however, when activated with a ratio of 1:4, polymer: KOH, i.e., PPy/Ar-2, yielded a 1.2 mmol/g adsorption of CO_2. Upon comparison of the obtained adsorption CO_2 values with the previously reported ones, it was found that the CO_2 adsorption values obtained were for the as-prepared material, i.e., N-GOs were slightly lower than some of the materials employed earlier, which suggests that these materials can be studied further by fine tuning them to enhance their adsorption properties. This study also revealed a facile synthesis of nitrogen-doped graphene oxide and can be extended to various other polymers, which can be obtained from recyclable material and can be useful for tackling the two environmental issues of the recycling of polymeric waste and air pollution.

Acknowledgments: The authors extend their appreciation to the Deanship of Scientific Research at King Saud University for funding this work through the research group No. RGP-1438-040.

Author Contributions: A.A.A. and N.H.K. played an important role in the project design and execution, while A.F.A. carried out the experimental work. F.A.A., H.A.A., and S.F.A. played a crucial role in the characterization and elucidation of the results. A.A.A. and S.F.A. compiled the data and prepared the manuscript. All authors read and approved the final manuscript.

Conflicts of Interest: The authors declare that they have no conflict of interests.

References

1. Mitić, P.; Munitlak Ivanović, O.; Zdravković, A. A cointegration analysis of real GDP and CO_2 emissions in transitional countries. *Sustainability* **2017**, *9*, 568. [CrossRef]
2. Di Vita, G.; Pilato, M.; Pecorino, B.; Brun, F.; D'Amico, M. A review of the role of vegetal ecosystems in CO_2 capture. *Sustainability* **2017**, *9*, 1840. [CrossRef]
3. Rossi, F.; Nicolini, A.; Palombo, M.; Castellani, B.; Morini, E.; Filipponi, M. An innovative configuration for CO_2 capture by high temperature fuel cells. *Sustainability* **2014**, *6*, 6687–6695. [CrossRef]
4. Blumberg, T.; Sorgenfrei, M.; Tsatsaronis, G. Design and assessment of an igcc concept with CO_2 capture for the co-generation of electricity and substitute natural gas. *Sustainability* **2015**, *7*, 16213–16225. [CrossRef]
5. Al Sadat, W.I.; Archer, L.A. The O_2-assisted Al/CO_2 electrochemical cell: A system for CO_2 capture/conversion and electric power generation. *Sci. Adv.* **2016**, *2*, e1600968. [CrossRef] [PubMed]
6. Atrens, A.D.; Gurgenci, H.; Rudolph, V. Electricity generation using a carbon-dioxide thermosiphon. *Geothermics* **2010**, *39*, 161–169. [CrossRef]
7. Randolph, J.B.; Saar, M.O.; Bielicki, J. Geothermal energy production at geologic CO_2 sequestration sites: Impact of thermal drawdown on reservoir pressure. *Energy Procedia* **2013**, *37*, 6625–6635. [CrossRef]
8. Crake, A.; Christoforidis, K.C.; Kafizas, A.; Zafeiratos, S.; Petit, C. CO_2 capture and photocatalytic reduction using bifunctional TiO_2/MOF nanocomposites under UV-vis irradiation. *App. Catal. B Environ.* **2017**, *210*, 131–140. [CrossRef]
9. Hiremath, V.; Shavi, R.; Seo, J.G. Mesoporous magnesium oxide nanoparticles derived via complexation-combustion for enhanced performance in carbon dioxide capture. *J. Colloid Interface Sci.* **2017**, *498*, 55–63. [CrossRef] [PubMed]
10. Kuang, A.; Kuang, M.; Yuan, H.; Wang, G.; Chen, H.; Yang, X. Acidic gases (CO_2, NO_2 and SO_2) capture and dissociation on metal decorated phosphorene. *Appl. Surf. Sci.* **2017**, *410*, 505–512. [CrossRef]
11. Minju, N.; Nair, B.N.; Mohamed, A.P.; Ananthakumar, S. Surface engineered silica mesospheres—A promising adsorbent for CO_2 capture. *Sep. Purif. Technol.* **2017**, *181*, 192–200. [CrossRef]
12. Panek, R.; Wdowin, M.; Franus, W.; Czarna, D.; Stevens, L.; Deng, H.; Liu, J.; Sun, C.; Liu, H.; Snape, C.E. Fly ash-derived MCM-41 as a low-cost silica support for polyethyleneimine in post-combustion CO_2 capture. *J. CO_2 Util.* **2017**, *22*, 81–90. [CrossRef]
13. Lim, J.Y.; Kim, J.K.; Lee, C.S.; Lee, J.M.; Kim, J.H. Hybrid membranes of nanostructrual copolymer and ionic liquid for carbon dioxide capture. *Chem. Eng. J.* **2017**, *322*, 254–262. [CrossRef]
14. Rahimi, M.; Singh, J.K.; Müller-Plathe, F. CO_2 adsorption on charged carbon nanotube arrays: A possible functional material for electric swing adsorption. *J. Phys. Chem. C* **2015**, *119*, 15232–15239. [CrossRef]

15. Takeuchi, K.; Yamamoto, S.; Hamamoto, Y.; Shiozawa, Y.; Tashima, K.; Fukidome, H.; Koitaya, T.; Mukai, K.; Yoshimoto, S.; Suemitsu, M.; et al. Adsorption of CO_2 on graphene: A combined TPD, XPS, and vdW-DF study. *J. Phys. Chem. C* **2017**, *121*, 2807–2814. [CrossRef]

16. Wickramaratne, N.P.; Jaroniec, M. Activated carbon spheres for CO_2 adsorption. *ACS Appl. Mater. Interfaces* **2013**, *5*, 1849–1855. [CrossRef] [PubMed]

17. Cazorla-Amorós, D.; Alcaniz-Monge, J.; Linares-Solano, A. Characterization of activated carbon fibers by CO_2 adsorption. *Langmuir* **1996**, *12*, 2820–2824. [CrossRef]

18. Garcia-Gallastegui, A.; Iruretagoyena, D.; Gouvea, V.; Mokhtar, M.; Asiri, A.M.; Basahel, S.N.; Al-Thabaiti, S.A.; Alyoubi, A.O.; Chadwick, D.; Shaffer, M.S. Graphene oxide as support for layered double hydroxides: Enhancing the CO_2 adsorption capacity. *Chem. Mater.* **2012**, *24*, 4531–4539. [CrossRef]

19. Li, X.; Cheng, Y.; Zhang, H.; Wang, S.; Jiang, Z.; Guo, R.; Wu, H. Efficient CO_2 capture by functionalized graphene oxide nanosheets as fillers to fabricate multi-permselective mixed matrix membranes. *ACS Appl. Mater. Interfaces* **2015**, *7*, 5528–5537. [CrossRef] [PubMed]

20. Mishra, A.; Ramaprabhu, S. Enhanced CO_2 capture in Fe_3O_4-graphene nanocomposite by physicochemical adsorption. *J. Appl. Phys.* **2014**, *116*, 064306. [CrossRef]

21. Zhou, D.; Liu, Q.; Cheng, Q.; Zhao, Y.; Cui, Y.; Wang, T.; Han, B. Graphene-manganese oxide hybrid porous material and its application in carbon dioxide adsorption. *Chin. Sci. Bull.* **2012**, *57*, 3059–3064. [CrossRef]

22. Alves, D.C.; Silva, R.; Voiry, D.; Asefa, T.; Chhowalla, M. Copper nanoparticles stabilized by reduced graphene oxide for CO_2 reduction reaction. *Mater. Renew. Sustain. Energy* **2015**, *4*, 2. [CrossRef]

23. Cazorla, C.; Shevlin, S.; Guo, Z. Calcium-based functionalization of carbon materials for CO_2 capture: A first-principles computational study. *J. Phys. Chem. C* **2011**, *115*, 10990–10995. [CrossRef]

24. Yang, S.; Zhan, L.; Xu, X.; Wang, Y.; Ling, L.; Feng, X. Graphene-based porous silica sheets impregnated with polyethyleneimine for superior CO_2 capture. *Adv. Mater.* **2013**, *25*, 2130–2134. [CrossRef] [PubMed]

25. Kemp, K.C.; Chandra, V.; Saleh, M.; Kim, K.S. Reversible CO_2 adsorption by an activated nitrogen doped graphene/polyaniline material. *Nanotechnology* **2013**, *24*, 235703. [CrossRef] [PubMed]

26. Deng, D.; Pan, X.; Yu, L.; Cui, Y.; Jiang, Y.; Qi, J.; Li, W.-X.; Fu, Q.; Ma, X.; Xue, Q. Toward N-doped graphene via solvothermal synthesis. *Chem. Mater.* **2011**, *23*, 1188–1193. [CrossRef]

27. Wei, D.; Liu, Y.; Wang, Y.; Zhang, H.; Huang, L.; Yu, G. Synthesis of N-doped graphene by chemical vapor deposition and its electrical properties. *Nano Lett.* **2009**, *9*, 1752–1758. [CrossRef] [PubMed]

28. Li, N.; Wang, Z.; Zhao, K.; Shi, Z.; Gu, Z.; Xu, S. Large scale synthesis of N-doped multi-layered graphene sheets by simple arc-discharge method. *Carbon* **2010**, *48*, 255–259. [CrossRef]

29. Tang, J.; Yang, J.; Zhou, X.; Chen, G.; Xie, J. Toward N-doped graphene nanosheets via hydrazine steaming process. *Mater. Lett.* **2014**, *131*, 340–343. [CrossRef]

30. Khan, M.; Tahir, M.N.; Adil, S.F.; Khan, H.U.; Siddiqui, M.R.H.; Al-warthan, A.A.; Tremel, W. Graphene based metal and metal oxide nanocomposites: Synthesis, properties and their applications. *J. Mater. Chem. A* **2015**, *3*, 18753–18808. [CrossRef]

31. Khan, M.; Al-Marri, A.H.; Khan, M.; Shaik, M.R.; Mohri, N.; Adil, S.F.; Kuniyil, M.; Alkhathlan, H.Z.; Al-Warthan, A.; Tremel, W.; et al. Green approach for the effective reduction of graphene oxide using *Salvadora persica* L. Root (Miswak) extract. *Nanoscale Res. Lett.* **2015**, *10*, 281. [CrossRef] [PubMed]

32. Al-Marri, A.H.; Khan, M.; Khan, M.; Adil, S.F.; Al-Warthan, A.; Alkhathlan, H.Z.; Tremel, W.; Labis, J.P.; Siddiqui, M.R.H.; Tahir, M.N. *Pulicaria glutinosa* extract: A toolbox to synthesize highly reduced graphene oxide-silver nanocomposites. *Int. J. Mol. Sci.* **2015**, *16*, 1131–1142. [CrossRef] [PubMed]

33. Kim, B.-J.; Lee, H.-M.; Kim, H.-G.; An, K.-H.; Kang, H.-R. Comparative studies of porous carbon nanofibers by various activation methods. *Carbon Lett.* **2013**, *14*, 180–185.

34. Wang, X.; Li, H.; Liu, H.; Hou, X. As-synthesized mesoporous silica MSU-1 modified with tetraethylenepentamine for CO_2 adsorption. *Microporous Mesoporous Mater.* **2011**, *142*, 564–569. [CrossRef]

35. Biniak, S.; Szymański, G.; Siedlewski, J.; Światkowski, A. The characterization of activated carbons with oxygen and nitrogen surface groups. *Carbon* **1997**, *35*, 1799–1810. [CrossRef]

36. Deng, Y.; Tang, L.; Zeng, G.; Zhu, Z.; Yan, M.; Zhou, Y.; Wang, J.; Liu, Y.; Wang, J. Insight into highly efficient simultaneous photocatalytic removal of Cr(VI) and 2, 4-diclorophenol under visible light irradiation by phosphorus doped porous ultrathin g-C_3N_4 nanosheets from aqueous media: Performance and reaction mechanism. *Appl. Catal. B Environ.* **2017**, *203*, 343–354. [CrossRef]

37. Putri, L.K.; Ong, W.-J.; Chang, W.S.; Chai, S.-P. Heteroatom doped graphene in photocatalysis: A review. *Appl. Surf. Sci.* **2015**, *358*, 2–14. [CrossRef]

38. Ben, T.; Li, Y.; Zhu, L.; Zhang, D.; Cao, D.; Xiang, Z.; Yao, X.; Qiu, S. Selective adsorption of carbon dioxide by carbonized porous aromatic framework (PAF). *Energy Environ. Sci.* **2012**, *5*, 8370–8376. [CrossRef]

39. De Souza, L.K.; Wickramaratne, N.P.; Ello, A.S.; Costa, M.J.; da Costa, C.E.; Jaroniec, M. Enhancement of CO_2 adsorption on phenolic resin-based mesoporous carbons by KOH activation. *Carbon* **2013**, *65*, 334–340. [CrossRef]

40. Nan, D.; Liu, J.; Ma, W. Electrospun phenolic resin-based carbon ultrafine fibers with abundant ultra-small micropores for CO_2 adsorption. *Chem. Eng. J.* **2015**, *276*, 44–50. [CrossRef]

41. Hao, G.-P.; Li, W.-C.; Qian, D.; Wang, G.-H.; Zhang, W.-P.; Zhang, T.; Wang, A.-Q.; Schüth, F.; Bongard, H.-J.; Lu, A.-H. Structurally designed synthesis of mechanically stable poly (benzoxazine-co-resol)-based porous carbon monoliths and their application as high-performance CO_2 capture sorbents. *J. Am. Chem. Soc.* **2011**, *133*, 11378–11388. [CrossRef] [PubMed]

42. Hao, G.-P.; Jin, Z.-Y.; Sun, Q.; Zhang, X.-Q.; Zhang, J.-T.; Lu, A.-H. Porous carbon nanosheets with precisely tunable thickness and selective CO_2 adsorption properties. *Energy Environ. Sci.* **2013**, *6*, 3740–3747. [CrossRef]

43. Acar, B.; Başar, M.S.; Eropak, B.M.; Caglayan, B.S.; Aksoylu, A.E. CO_2 adsorption over modified ac samples: A new methodology for determining selectivity. *Catal Today* **2018**, *301*, 112–124. [CrossRef]

44. Nelson, K.M.; Mahurin, S.M.; Mayes, R.T.; Williamson, B.; Teague, C.M.; Binder, A.J.; Baggetto, L.; Veith, G.M.; Dai, S. Preparation and CO_2 adsorption properties of soft-templated mesoporous carbons derived from chestnut tannin precursors. *Microporous Mesoporous Mater.* **2016**, *222*, 94–103. [CrossRef]

45. Banerjee, R.; Phan, A.; Wang, B.; Knobler, C.; Furukawa, H.; O'keeffe, M.; Yaghi, O.M. High-throughput synthesis of zeolitic imidazolate frameworks and application to CO_2 capture. *Science* **2008**, *319*, 939–943. [CrossRef] [PubMed]

46. Himeno, S.; Komatsu, T.; Fujita, S. High-pressure adsorption equilibria of methane and carbon dioxide on several activated carbons. *J. Chem. Eng. Data* **2005**, *50*, 369–376. [CrossRef]

47. Saha, D.; Deng, S. Adsorption equilibrium and kinetics of CO_2, CH_4, N_2O, and NH_3 on ordered mesoporous carbon. *J. Colloid Interface Sci.* **2010**, *345*, 402–409. [CrossRef] [PubMed]

48. Meng, L.-Y.; Park, S.-J. Effect of heat treatment on CO_2 adsorption of KOH-activated graphite nanofibers. *J. Colloid Interface Sci.* **2010**, *352*, 498–503. [CrossRef] [PubMed]

49. Khdary, N.H.; Ghanem, M.A. Metal-organic-silica nanocomposites: Copper, silver nanoparticles-ethylenediamine-silica gel and their CO_2 adsorption behaviour. *J. Mater. Chem.* **2012**, *22*, 12032–12038. [CrossRef]

 *materials*

Article

An Enhanced Carbon Capture and Storage Process (e-CCS) Applied to Shallow Reservoirs Using Nanofluids Based on Nitrogen-Rich Carbon Nanospheres

Elizabeth Rodriguez Acevedo [1,2,*], **Farid B. Cortés** [1,*], **Camilo A. Franco** [1], **Francisco Carrasco-Marín** [2], **Agustín F. Pérez-Cadenas** [2], **Vanessa Fierro** [3], **Alain Celzard** [3], **Sébastien Schaefer** [3] and **Agustin Cardona Molina** [4]

[1] Grupo de Investigación en Fenómenos de Superficie–Michael Polanyi, Facultad de Minas, Universidad Nacional de Colombia-Sede Medellín, Medellín 050034, Colombia
[2] Research Group in Carbon Materials, Faculty of Sciences, University of Granada, Granada 18071, Spain
[3] Bio-Sourced Materials Research Group, Institut Jean Lamour, UMR CNRS–Université de Lorraine, Epinal 88051, France
[4] Grupo de Investigación en Yacimientos de Hidrocarburos, Facultad de Minas, Universidad Nacional de Colombia-Sede Medellin, Medellín 050034, Colombia
* Correspondence: ecrodrig@unal.edu.co (E.R.A.); fbcortes@unal.edu.co (F.B.C.); Tel.: +57-301-3995162 (E.R.A.); +57-318-3474625 (F.B.C.)

Received: 1 June 2019; Accepted: 22 June 2019; Published: 28 June 2019

Abstract: The implementation of carbon capture and storage process (CCS) has been unsuccessful to date, mainly due to the technical issues and high costs associated with two main stages: (1) CO_2 separation from flue gas and (2) CO_2 injection in deep geological deposits, more than 300 m, where CO_2 is in supercritical conditions. This study proposes, for the first time, an enhanced CCS process (e-CCS), in which the stage of CO_2 separation is removed and the flue gas is injected directly in shallow reservoirs located at less than 300 m, where the adsorptive phenomena control CO_2 storage. Nitrogen-rich carbon nanospheres were used as modifying agents of the reservoir porous texture to improve both the CO_2 adsorption capacity and selectivity. For this purpose, sandstone was impregnated with a nanofluid and CO_2 adsorption was evaluated at different pressures (atmospheric pressure and from 3×10^{-3} MPa to 3.0 MPa) and temperatures (0, 25, and 50 °C). As a main result, a mass fraction of only 20% of nanomaterials increased both the surface area and the molecular interactions, so that the increase of adsorption capacity at shallow reservoir conditions (50 °C and 3.0 MPa) was more than 677 times (from 0.00125 to 0.9 mmol g^{-1}).

Keywords: adsorption; carbon capture and storage process (CCS); carbon dioxide; nanofluids; nanoparticles and shallow reservoirs

---

1. Introduction

In recent decades, climate changes have generated negative consequences such as loss of sea ice, accelerated rise of sea level, extinction of some species, drought and population displacements, among others [1–5]. These changes are caused by human activities, mainly through the release of greenhouse gases [4–7]. The anthropogenic emissions of carbon dioxide, CO_2, from fossil fuel combustion and industrial processes contributed to about 78% of the total greenhouse gases emissions [8–10], and CO_2 emissions have increased by 46% since pre-industrial times [11,12]. The huge CO_2 daily emissions and their growth increase its responsibility on global climate change [12–16]. Raw material industries such as chemical, petrochemical, iron, steel, cement, and others, contribute with the most to CO_2 emissions worldwide [17,18]. The advantage of CO_2 industrial emissions is that they are fixed sources so that they can be controlled in-situ. A typical flue gas from coal-fired boilers may contain 12–14 vol% CO_2, 8–10 vol% H_2O, 3–5 vol% O_2, and 72–77% N_2 [19–21]. The method most frequently used at industrial levels for CO_2 capture is absorption. However, this method has significant limitations related to solvent regeneration, composition of flue gas, CO_2 concentration, corrosion, high-energy consumption, etc. [15,16,18,22]. Other methods are cryogenic distillation, membranes, and adsorption by porous solids. Hence, the current methods are not enough, and CO_2 emissions continue to increase [12,17,23]. For this reason, it is necessary to broaden the portfolio of methods to reduce CO_2 emissions significantly in the years to come.

The Intergovernmental Panel on Climate Change (IPCC) promotes the carbon capture and storage process (CCS), which allows the geological storage of CO_2 emitted by the industry over the long term [10,18,24]. The CCS process might decrease CO_2 emissions by approximately 22% in 2035 [25,26]. The CCS process has three main stages: 1) CO_2 capture and separation from flue gas, 2) CO_2 transport to the storage site, and 3) CO_2 injection into deep geological storage sites, 300 to 4000 m, and 800 m on average [22]. In this case, CO_2 capture and storage are mainly due to the inter-particle volume filling. Figure 1a presents a scheme of the CCS process.

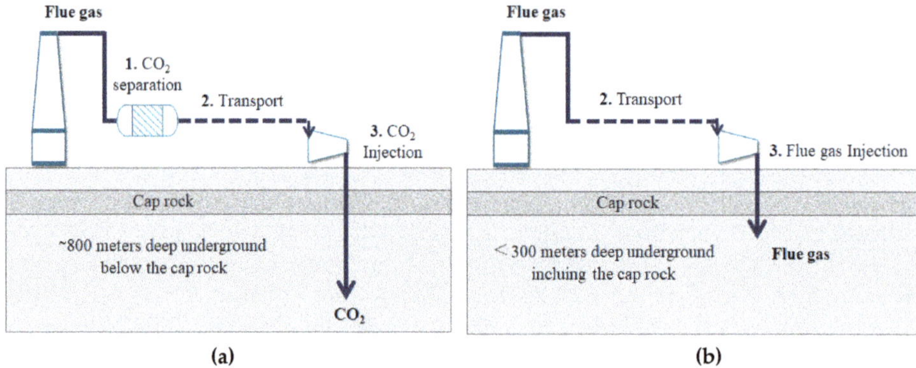

Figure 1. Configurations of the carbon capture and storage (CCS) process: (a) conventional CCS process; (b) proposed enhanced CSS (e-CCS) process.

It is estimated that 50% of the CCS projects use sandstone deposits for storage because they have technical advantages and high availability [24]. Oil and gas deposits are indeed mostly composed of sandstone, which provides a natural storage structure, with porous texture and upper and lateral seals that allow long-term storage. Carbon capture and storage have been coupled with other processes such as enhanced oil recovery (EOR) or enhanced coal bed methane production (ECBM) [24].

However, its in situ industrial implementation has been unsuccessful mainly due to the technical and economic costs associated with the separation and storage stages. In general, the cost of CO_2 capture is 70–80% of the total CCS costs [26–28]. For this reason, the current research focuses mainly on increasing the CO_2/flue gas separation capacity in order to decrease energy consumption, or on the use of waste energy, among others. This study proposes an alternative to minimize the technical and economic cost for the viability of the CCS process. Figure 1b presents the configuration of an enhanced CCS process (e-CCS), in which the stage of CO_2 capture/separation is removed, and the flue gas is injected directly into shallow deposits, at depths less than 300 m. In this case, CO_2 remains in a gaseous state, and the adsorption process controls the capture and storage. The density of CO_2 is very different in gaseous or supercritical conditions, which affects the amount of CO_2 stored in the e-CCS process. For this, it is necessary to add a modifying agent to the surface of the porous medium in order to improve the adsorption capacity and the CO_2 selectivity, since the adsorption (capture/storage) process is done underground. In addition, the modifying agent should not affect the naturally porous structure of the deposit to avoid operational problems.

Nanostructured materials constitute a vast and active field of research in various areas, due to their intrinsic characteristics that can be adjusted depending on the foreseen application, such as porosity, structure, molecular affinity, high surface-area-to-volume ratio, high surface activity, dispersion capacity, and optical and electronic properties, etc. [29]. Nanoparticles have been used to increase the recovery of oil and gas by modifying the physicochemical properties of the reservoirs [30,31], and to modify reservoirs' wettability, asphaltenes adsorption, catalysis and stabilization of fines in reservoirs, among others [30,32–35]. The obtained results demonstrate their effectiveness in achieving this objective without obstruction of the porous media. In this way, nanospheres can be an option for the modification of the surface in shallow reservoirs for the e-CCS process. Although nanoparticles have been evaluated for the conventional CCS process in the form of a nanofluid used to improve CO_2 transport in saline aquifers [36], they have not been used for rock modification and improvement of the CO_2 storage capacity, such as in the case of the proposed e-CCS process, to the best of our knowledge.

Many adsorbents for CO_2 capture have been reported, such as carbon materials, metal organic frameworks, zeolites, alkali metal carbonates, etc. [16,37–39], but not widely at the nanoscale level, with spherical structures allowing their application in geological deposits. Many carbon nanostructures have been evaluated for CO_2 capture, among them nanofibers, nanosheets, and nanotubes, leading to adsorption capacities ranging from 0.26 to 4.15 mmol g^{-1} under atmospheric conditions [16,40–45]. However, these materials are not applicable to reservoirs due to their structure and dimensions that might affect the nature of the reservoir's porous structure. Carbon nanospheres, thus, appear to be the best choice for the e-CCS process. Wang et al. [16] analyzed different adsorbents and concluded that carbon materials are one of the best options, mainly due to their low cost, high surface area, adjustable porous texture, and easy surface functionalization [16,37–39]. Chen et al. [46] reported hollow carbon nanospheres with a CO_2 adsorption capacity of 3.65 mmol g^{-1} under atmospheric conditions [46], which is similar to what was reported for other nanomaterials such as carbon fibers, carbon nanosheets, carbon nanotubes, metal organic frameworks, zeolites, alkali metal carbonates, etc [16,37–39]. For CO_2 adsorption applications, a high surface area is essential as well as basic functionalities, due to the acidic character of the CO_2 molecule. Therefore, high nitrogen content in the adsorbent allows obtaining a basic nature and enhances the adsorption capacity and selectivity. Some authors suggested increasing the nitrogen content by impregnating the materials with amines like the materials commonly called supported amine material, which consist of a porous support onto which an amine is attached or immobilized [16,37–39,47]. However, treatment with amines can obstruct the microporosity of the nanomaterial, thereby decreasing its adsorption capacity and increasing its final cost. Also, in some cases, like samples containing grafted primary and tertiary monoamines, the material could be deactivated in the presence of oxygen-containing gases [47]. Thus, it is desirable to incorporate nitrogen groups during the material's synthesis. Hence, amine or amino acid functional groups are grafted on the surface of the support during the synthesis process instead of physically dispersed in the pores after synthesizing the support material [48].

The main objective of this manuscript was, therefore, to experimentally study the possibility of improving the CCS process through nanotechnology. For this, carbon nanospheres with different structures were synthesized, characterized, and used to impregnate sandstone using different mass ratios of nanoparticles to sandstone. The CO_2 adsorption process, as well as its thermodynamic parameters, were evaluated under atmospheric and high-pressure conditions.

2. Materials and Methods

Two different carbon nanostructures were synthesized using either a sol-gel method or a solvothermal method. The synthesized nanostructures were labeled and synthesized as follows:

(1) CN.LYS: Carbon nanospheres obtained from a sol-gel method, using resorcinol/formaldehyde as carbon precursor and L-lysine as catalyst and nitrogen precursor.

(2) CN.MEL: Carbon nanostructures obtained from a solvothermal method, using carbon tetrachloride as carbon precursor and melamine as nitrogen precursor.

Both CN.LYS and CN.MEL were characterized in order to select the best material, considering the nanometer size, adsorption capacity, lower technical and economic cost, and method of synthesis. Ottawa sandstone was used as porous medium and was impregnated with the best nanomaterial at different percentages. The performances of the materials were evaluated by CO_2 adsorption at atmospheric pressure and at 0 °C, and by varying pressure and temperature conditions. The detailed procedures are presented below.

2.1. Materials and Reagents

For the synthesis processes, the following reagents were used, all from Sigma–Adrich, St. Louis, USA: carbon tetrachloride ($\geq$ 99.9%), melamine (99%), formaldehyde (37%), resorcinol ($\geq$99%), L-lysine (>98%), sodium dodecylbenzene sulfonate (SDBS), and deionized water.

For cleaning, drying, and carbonization, the following chemicals were used, all from Sigma–Adrich again except N_2: acetone (99.9%), ethanol (99.5%), hydrochloric acid (37%,), *tert*-butanol ($\geq$99.5%), and N_2 (high purity, grade 5.0). Clean Ottawa sandstone and sandstone from a real reservoir were used as porous media. The real sandstone was obtained from a Colombian oil field, which allows evaluating the real rock that might be used to implement the e-CCS process in depleted oil fields.

2.2. Synthesis of Nanomaterials

2.2.1. CN.LYS Synthesis

The process was adapted from Yong–Rong et al. [48], changing the lysine concentration, the reaction time and the resorcinol/water molar ratio. A solution (S1) of resorcinol/formaldehyde in a 1:2 molar ratio, and deionized water was stirred at 25 °C for 1 h. In parallel, a solution (S2) of L-lysine and deionized water was stirred at 60 °C for 1 h. The molar ratio of resorcinol/L-lysine was 1:0.16. Subsequently, the solutions S1 and S2 were mixed at 60 °C for 1 h to obtain the solution S3. The latter was maintained at 25 °C for 20 h to benefit from the natural precipitation of the nanomaterial. Finally, the obtained polymer was dried at 120 °C for 12 h and carbonized under N_2 flowing at 60 mL min^{-1}, using a tubular furnace. The temperature was increased up to 800 °C at a rate of 1 °C min^{-1}, and the final temperature was held for 5 h. The employed molar ratios of resorcinol/water (for S1) were 1:2778 (without dilution), 1:5556 (dilution 1), and 1:11112 (dilution 2) for obtaining the CN.LYS1, CN.LYS2, and CN.LYS3 materials, respectively. Different molar ratios of resorcinol to water were used to reduce particle size.

2.2.2. CN.MEL Synthesis

The CN.MEL synthesis was adapted from Bai et al. [49] by dissolving 2 g of melamine in 120 mL of carbon tetrachloride [49]. This solution was put in a stainless-steel autoclave (Parr Instrument, Illinois, USA) with a capacity of 200 mL and introduced into an oven (Thermo Fisher Scientific, Massachusetts, USA) at 250 °C for 24 h. The synthesis was carried out under auto-generated pressure. Subsequently, the carbonaceous material that formed was separated from the solution, and was cleaned with acetone, ethanol, and finally 0.1 mol L^{-1} HCl. A mixture of nanospheres and fibers (formed by aggregation of nanospheres) was obtained (Gel.MEL1).

To obtain N-rich carbon spheres, the Gel.MEL1 was coated with a mixture of resorcinol and formaldehyde in a 1:2 molar ratio. Initially, the gel was stirred with sodium dodecylbenzene sulfonate (SDBS) at 25 °C for 18 h to promote the subsequent interaction with resorcinol/formaldehyde. After 18 h, the gel/SDBS was mixed with resorcinol/formaldehyde (in proportions of 0.37 g of nanoparticles per 1 g of resorcinol) and put in a stainless-steel autoclave at 130 °C for 24 h. In order to maintain its porous texture, the gel was dried by freeze-drying. First, the material was impregnated with *tert*-butanol for three days. Then, the impregnated material was frozen at −5 °C and lyophilized for two days until the *tert*-butanol was entirely removed (Gel.MEL2). Finally, the lyophilized gel was carbonized under N_2 flow (60 mL min^{-1}) in a tubular furnace (Thermo Fisher Scientific, Massachusetts, USA). The temperature was increased at a heating rate of 1 °C min^{-1} up to 700 °C, and the latter temperature was held for 6 h.

2.2.3. Impregnation of Sandstones

The sandstone was impregnated to decorate the rock surface with the nanoparticles and improve the surface area and molecular interactions. Ottawa sandstone (SS) was impregnated with CN.LYS2 at mass fractions of 0.01, 0.1, 1, 5, 10, and 20% by immersion and soaking [50]. Initially, a nanofluid composed of nanoparticles and deionized water was sonicated at 40 °C for 4 h. Subsequently, the SS was introduced in the nanofluid at 60 °C for either 6 h at 600 rpm or for 24 h without stirring.

The latter method better mimics the reservoir conditions in which the porous medium might be impregnated. Finally, the impregnated material was dried at 110 °C for 12 h. The same procedure was followed for impregnating the sandstone from a real reservoir (RS) but using only a mass fraction of 10 and 20% of CN.LYS2 to RS and 24 h of soaking.

2.3. *Characterization of the Nanomaterials*

The following procedures allowed characterizing the physicochemical characteristics of the materials, essential for a good understanding of the results. For e-CCS application, nanoparticles must have a nanometer size, a spherical shape, a high surface area, and a high nitrogen content.

2.3.1. Size and Structure of Nanomaterials

Different techniques were used to evaluate the particle size distribution: scanning electron microscopy (SEM) was used to obtain the dry particle size, size distribution, and morphology of CN.MEL, whereas transmission electron microscopy (TEM) was used to analyze the dry particle size, size distribution, and surface characteristics of CN.LYS. The CN.LYS showed a different porous structure after dilutions, which was observed by N_2 adsorption (at −196 °C). Further, TEM was used to characterize the structure after dilutions to analyze the causes of pore structure modification. Scanning electron microscopy analysis was also carried out to analyze the distribution of carbon nanoparticles on sandstone after impregnation. The observations were carried out by means of a JSM-7100 emission electron microscope (JEOL, Nieuw-Vennep, The Netherlands), a GEMINI-LEO1530 VP FE-SEM emission electron microscope (Carl Zeiss, Cambridge, UK), and a Tecnai F20 Super Twin TMP transmission electron microscope (FEI, Hillsboro, USA).

Dynamic light scattering (DLS) was carried out by means of a NanoPlus-3 zeta/nanoparticle analyzer (Micromeritics, Norcross, USA) at 25 °C in a glass cell (capacity of 0.9 mL), which was used to obtain the mean particle size of nanoparticles dispersed in a fluid, which hydrate and interact with each other. The mean particle size was calculated from the diffusional properties of the particle indicating the size of the hydrated and solvated particle. For this purpose, a nanoparticle solution, 10 mg L^{-1}, was dispersed in water or ethanol and sonicated for 6 h before analysis. Particles suspended in a liquid have a Brownian motion due to the random collisions with solvent molecules. This motion causes the particles to diffuse through the medium. The diffusion coefficient, D, is inversely proportional to the particle size or hydrodynamic diameter, d, according to the Stokes–Einstein equation:

$$D = \frac{k_B\, T}{3\, \pi\, \eta\, d} \tag{1}$$

where, k_B is Boltzmann's constant, T is the absolute temperature, and η is the viscosity.

2.3.2. Porous Structure of Nanomaterials and Sandstone

All materials were characterized by N_2 and CO_2 adsorption at $-196\ °C$ and $0\ °C$, respectively, using a 3-Flex manometric adsorption equipment (Micromeritics, Norcross, USA). The total adsorbed volume ($V_{0.95}$) was taken as the physiosorbed volume of N_2 at a relative pressure $P/P_0 = 0.95$. The Brunauer–Emmett–Teller (BET) model was applied to obtain the BET area (A_{BET}). Micropore volume (V_{mic}), average pore size (L_0), and CO_2 adsorption energy ($E_{ads}-_{CO2}$) were obtained by application of the Dubinin–Radushkevich equation. The mesopore volume (V_{meso}) was obtained through the Barrett–Joyner–Halenda (BJH) model.

2.3.3. Chemical Composition of Nanomaterials and Sandstone

The chemical characterization was carried out by carbon, hydrogen, oxygen and nitrogen (CHON) analysis for nanomaterials, and by Fourier transform infrared spectroscopy (FTIR) for sandstone. An IRAffinity-1S FTIR spectrometer (Shimadzu, Columbia, USA)was operated at room temperature using potassium bromide in a KBr-to-material ratio of 30:1 (% w/w). The impregnation percentages of nanoparticles on sandstone were corroborated by thermogravimetric analysis (TGA) (TA Instruments, New Castle, USA). For this, the sample was burned under an air atmosphere at $10\ °C\ min^{-1}$ up to $800\ °C$.

2.3.4. Rheological Analysis of CN.LYS Synthesis Solutions

The stability of the synthesis solutions was evaluated at different concentrations of CN.LYS (CN-LYS1, CN.LYS2, and CN.LYS3). For this purpose, a Kinexus Ultra+ rheometer (Malvern Panalytical, Malvern, UK) utilizing a concentric cylinder sensor equipped with a Peltier cell for temperature control was used. The tests were first carried out by varying the shear rate from 1 to $250\ s^{-1}$ in order to define the adequate shear rate ($50\ s^{-1}$). The test conditions were carried out to mimic the real synthesis conditions in terms of temperature and stirring. The process started at $60\ °C$ for 1 h, after which the temperature was controlled to simulate the natural cooling process at 57, 45, 37, and $34\ °C$ until the viscosity reached a constant value. Only 30 mL were needed for the test, while the current reaction was carried out in 1.8 L.

2.3.5. Dispersion of Nanoparticles in Solution

The electrophoretic light scattering (ELS) technique was used to evaluate the surface charge of the particles and their dispersion stability at $25\ °C$ in a NanoPlus-3 zeta/nanoparticle analyzer (Micromeritics, Norcross, USA). In this test, several nanoparticle suspensions were prepared at $10\ mg\ L^{-1}$, with a pH adjusted between 2 to 12 by adding solutions of $0.1\ mol\ L^{-1}$ HCl or $0.01\ mol\ L^{-1}$ NaOH, and then subjected to analysis. The zeta potential was calculated using the Smoluchowski equation, derived from the calculation of the Doppler effect.

$$\zeta = \eta\, U\, /\, \varepsilon \tag{2}$$

$$U = \frac{V}{E} \tag{3}$$

$$\Delta v = 2V\, n\, \sin\!\left(\frac{\theta}{2}\right)/\lambda \tag{4}$$

where ζ is the zeta potential, η is the viscosity of the fluid (water), U is the electrophoretic mobility, ε is the permittivity, V represents the speed of movement of the particles, E is the electric field, Δv is the Doppler effect, n is the index of refraction, θ is the angle of detection, and λ is the wavelength of the incident light.

2.4. Adsorption Tests at High Pressure

The adsorption tests carried out below atmospheric pressure were described in sub-Section 2.3.2. At high pressure, the CO_2 adsorption capacity was evaluated in two different conditions: (i) under pure CO_2 in a manometric device (up to 3.0 MPa) and (ii) under a CO_2/N_2 flow in a gravimetric device (up to 2.6 MPa). The details of each protocol are presented below.

2.4.1. Adsorption at High Pressure for Pure CO_2–Manometric Device

The carbon nanospheres (CN.LYS2), sandstone, and impregnated sandstone (at mass fractions of 10 and 20%) were investigated in High Pressure Volume Analyzer, HPVAII-200 (Micromeritics, Norcross, USA) at 0 °C, 25 °C, and 50 °C and at pressures from 3×10^{-3} up to 3.0 MPa. In order to have enough total surface area for adsorption and to minimize measurement errors, the amount of each material inside the sample holder was around 0.5 g for nanoparticles, 1.5 g for impregnated sandstone, and 14 g for sandstone. The contribution of the empty sample holder was systematically measured and subtracted to all data to improve accuracy. The isosteric heat of adsorption, Q_{ST}, was calculated using the isosteric method with the Microactive software (from Micromeritics, Norcross, USA) from three adsorption isotherms at 0, 25, and 50 °C, based on the Clausius–Clapeyron equation [51]:

$$-\frac{Q_{ST}}{R} = \frac{\partial \ln (P)}{\partial (1/T)} \tag{5}$$

where R is the universal gas constant (8.314 J mol^{-1} K^{-1}), P is the absolute pressure, and T is the temperature.

The excess adsorbed CO_2 amount (N_{exc}, g$_{CO2}$·g$_{adsorbent}^{-1}$) was equal to the absolute adsorbed CO_2 amount (N_{ads}, g$_{CO2}$·g$_{adsorbent}^{-1}$) minus the product of gas density in the bulk phase by the volume of the adsorbed phase. The values provided by the HPVA device were obtained on an excess basis, and therefore, the absolute amounts had to be determined as follows [52]:

$$N_{ads} = N_{exc} \left(1 + \frac{P + M_{CO2}}{Z \, \rho_{liq} \, R \, T}\right) \tag{6}$$

where M_{CO2} is the molecular weight of CO_2 (44.013 g mol^{-1}), Z is the compressibility factor at the considered pressure and temperature, and ρ_{liq} is the density of liquid CO_2 (1032×10^3 g m^{-3}).

The isotherms were fitted with the Sips and Toth models, which take into account multilayer adsorption. Table 1 presents the equations for each model [53–55]. K_S and K_T represent adsorption equilibrium constants for the Sips and Toth models, respectively, and the n and t parameters indicate the heterogeneity of the system for the Sips and Toth models, respectively. The heterogeneity may originate from the solid structure, from the solid energy properties, or from the adsorbate [54]. The n or t parameters are usually greater than unity, and when they are the unit, the models assume the Langmuir equation [54].

$$N_{ads} = N_m \frac{(K_S \, P)^{1/n}}{1 + (K_S \, P)^{1/n}} \tag{7}$$

$$N_{ads} = N_m \frac{(K_T \, P)^{1/t}}{1 + (K_T \, P)^{1/t}} \tag{8}$$

Table 1. Models for adsorption isotherms.

Model	Equations	Parameters
Sips	(7)	N_{ads} (mmol g^{-1}) is the adsorbed amount, N_m (mmol g^{-1}) is the adsorption capacity at equilibrium, P (kPa) is the equilibrium pressure, and K_S and n are the Sips adsorption equilibrium constants, related to the affinity and the heterogeneity of the system, respectively.
Toth	(8)	N_{ads}, N_m, and P have the same meaning as above, and K_T and t are the Toth adsorption equilibrium constants, related to the affinity and the heterogeneity of the system, respectively.

2.4.2. Adsorption at High Pressure for CO_2 and N_2–Gravimetric Device

The CN.LYS2, sandstone, and impregnated sandstone CO_2 isotherms (with a mass fraction of 20% of nanoparticles) were investigated using a HP TGA 750 thermogravimetric analyzer (TA Instruments, New Castle, USA) at 50 °C and high pressure from 0.03 to 3.0 MPa for CO_2 and N_2. This device was equipped with a magnetic levitation top-loading balance, which made it possible to achieve high accuracy and reduce the volume of the system. The amount of each material put inside the sample holder was around 15 mg for nanoparticles, 40 mg for sandstone, and 40 mg for impregnated sandstone, to have enough total surface area for adsorption. The contribution of the buoyancy effect was manually subtracted from the data using blank tests carried out in the same conditions but with an empty sample holder. From the adsorption results of each N_2 and CO_2 isotherm, it was possible to predict the selectivity by applying the ideal adsorbed solution theory (IAST), which allows estimating the competitive adsorption of the compounds in a mixture of gases as the flue gas. Based on the literature, a model flue gas comprising 80% N_2 and 20% CO_2 was selected. The IAST was implemented in a Python routine (package-pyIAST from Simon et al. [56]).

3. Results and Discussion

The results are divided into two main sections: (a) materials characteristics (nanoparticles and sandstone) and (b) study of the interaction between CO_2/nanoparticles/sandstone by adsorption isotherms under different operation conditions (T, P).

3.1. Materials Characteristics

The morphology of the carbon materials obtained from melamine and carbon tetrachloride was very heterogeneous. Figure 2a,b presents SEM images of two different zones wherein more or less agglomerated nanospheres can be observed (Figure 2b). When the images are observed at lower magnification, it can be noticed that some areas have fiber and block morphologies (Figure 2c), while other areas have nanospheres/microspheres morphologies (Figure 2d).

After coating with resorcinol/formaldehyde, the hydrodynamic diameter of the CN.MEL particles was higher than the limit of detection of the equipment (10 μm). For the e-CCS process, this material might, thus, induce technical problems, due to the possible obstruction of the naturally porous structure of the rock. The ultimate analysis (Table 2) shows that this material had a nitrogen content close to 50% before coating and carbonization (Gel.MEL1), but of only 9.2% after coating with resorcinol/formaldehyde (Gel.MEL2) and 2.2% after carbonization (CN.MEL). Therefore, CN.LYS and CN.MEL materials exhibited similar nitrogen contents. Some N-rich carbon materials, reported in the literature, have nitrogen content close to those obtained in this work [16]. The oxygen content was measured independently from carbon, hydrogen and nitrogen content.

(a)   (b)

(c)   (d)

Figure 2. SEM images at 5 kV of carbon nanospheres from melamine (CN.MEL) before resorcinol/formaldehyde coating and final pyrolysis: (**a**,**b**) nanospheres and build fibers and blocks; (**c**) area with structures in the form of fibers and blocks, and (**d**) distribution of nanospheres, fibers, and blocks.

Table 2. Ultimate analysis of nanoparticles synthesized with melamine (CN.MEL) and L-lysine (CN.LYS).

	C (Mass Fraction %)	H (Mass Fraction %)	N (Mass Fraction %)	O (Mass Fraction %)
Gel.MEL1	22.8	4.7	49.1	7.3
Gel.MEL2	55.9	5.0	9.2	31.0
CN.MEL	85.6	2.3	2.2	10.2
Gel.LYS1	59.5	6.7	5.1	31.9
Gel.LYS2	61.7	6.7	5.0	31.6
Gel.LYS3	62.7	6.4	5.0	28.3
CN.LYS1	88.6	1.7	1.7	9.9
CN.LYS2	91.1	1.7	1.9	12.0
CN.LYS3	91.9	2.1	2.2	12.8

The Gel.MEL1 (49.1% of nitrogen) was submitted to CO_2 and N_2 adsorption (at 0 °C and −196 °C, respectively) but it was impossible to obtain the corresponding isotherms, possibly because of its too low surface area. It is well known that the pyrolysis step induces a significant increase of narrow porosity by elimination of volatile species. However, when the Gel.MEL1 is directly pyrolyzed, most of the material undergoes decompositions and the yield is very low (< 5%).

The adsorption and desorption isotherms (N_2 at −196 °C and CO_2 at 0 °C) for nanomaterials synthesized with melamine (CN.MEL) and L-lysine (CN.LYS) are presented in Figure 3, and the textural parameters obtained from adsorption isotherms are presented in Table 3.

The micropore and the mesopore fractions of the CN.MEL and CN.LYS2 materials were similar (67–64% and 33–36%, respectively) although their total pore volumes (at $P/P_0 = 0.95$) were rather different. Both CN.LYS1 and CN.LYS3 exhibited higher micropore fractions (72.7 and 92.3%, respectively), but had lower values of A_{BET} and total pore volumes than those of CN.MEL and CN.LYS2 materials (Figure 3). The mesoporous volumen of CN.MEL is evidenced by the hysteresis loop in the range of relative pressures between 0.46 and 0.66. In the CN.LYS series, only CN.LYS2 was also mesoporous, but with a different structure as deduced from the different shape of the hysteresis loop, occurring at higher relative pressure. The CN.LYS1 texture was moderately mesoporous (28.6%). After the first dilution, the CN.LYS2 texture was a little more mesoporous (36.6%), but after the third dilution, CN.LYS3 had a predominantly microporous texture as evidenced by the type Ia of its N_2 isotherm. The adsorption capacity of CO_2 at 0 °C was as expected according to the range reported in the literature, but the synthesis process reported in this study was easier than many others reported so far [16,43,57]. The experimental development considered important parameters for a possible industrial application, such as operation at atmospheric pressure, relatively low temperature (60 °C), and relatively low time (1 h) before carbonization. The carbon nanospheres are usually synthesized by methods that demand greater energy and time [29].

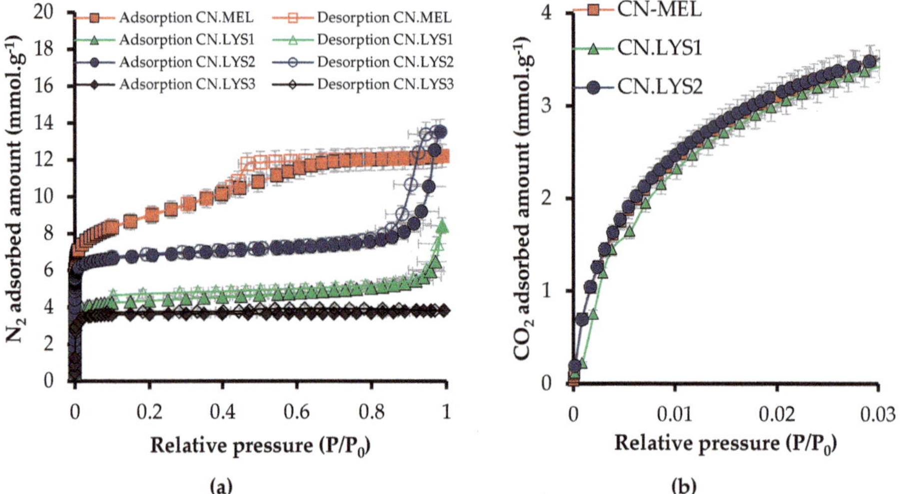

Figure 3. Adsorption (full symbols) and desorption (empty symbols) isotherms at atmospheric pressure for nanoparticles synthesized with melamine (CN.MEL) and L-lysine (CN.LYS). (a) N_2 at −196 °C and (b) CO_2 at 0 °C.

Table 3. Parameters obtained from adsorption isotherms (N_2 at −196 °C and CO_2 at 0 °C) for nanomaterials synthesized with melamine (CN.MEL) and L-lysine (CN.LYS).

	A_{BET} $(m^2\ g^{-1})$	$V_{0.95}$ $(cm^3\ g^{-1})$	$V_{mic\text{-}N2}$ $(cm^3\ g^{-1})$	$V_{mic\text{-}CO2}$ $(cm^3\ g^{-1})$	V_{mes} $(cm^3\ g^{-1})$	L_0 (nm)	$E_{ads.CO2}$ $(kJ\ mol^{-1})$
CN.MEL	713	0.42	0.28 (66.6%)	0.26	0.14	0.77	30.9
CN.LYS1	385	0.22	0.16 (72.7%)	0.18	0.06	0.84	32.4
CN.LYS2	612	0.36	0.23 (63.9%)	0.25	0.13	0.56	31.6
CN.LYS3	320	0.13	0.12 (92.3%)	-	0.01	0.62	-

In order to provide explanations of the aforementioned trends, Figure 4 presents TEM pictures of Gel.LYS (materials before carbonization) and CN.LYS (materials after carbonization).

Figure 4a,b present CN.LYS1 before and after carbonization, respectively. Here a mixture of larger and smaller nanospheres can be seen. Figure 4c,d present smaller particles (approximately 50 nm) for CN.LYS2 before and after carbonization. These particles are more transparent than those of the CN.LYS1 and CN.LYS3 materials, due to the more mesoporous texture of the CN.LYS2 material. Figure 4e,f present CN.LYS3 before and after carbonization. It can be seen that a gel was formed around the spheres. The L-lysine acts as a catalyst in the reaction, and therefore, if its amount is limited for the third dilution process, it might be the reason for the formation of a gel coating the spheres instead of producing more nanospheres, which obstructs the porous structure of CN.LYS3. This would affect the results presented in Figure 3 and Table 3 for CN.LYS3. The reproducibility for CN.LYS is significant, the variation of size and CO_2 capacity is less than 1%.

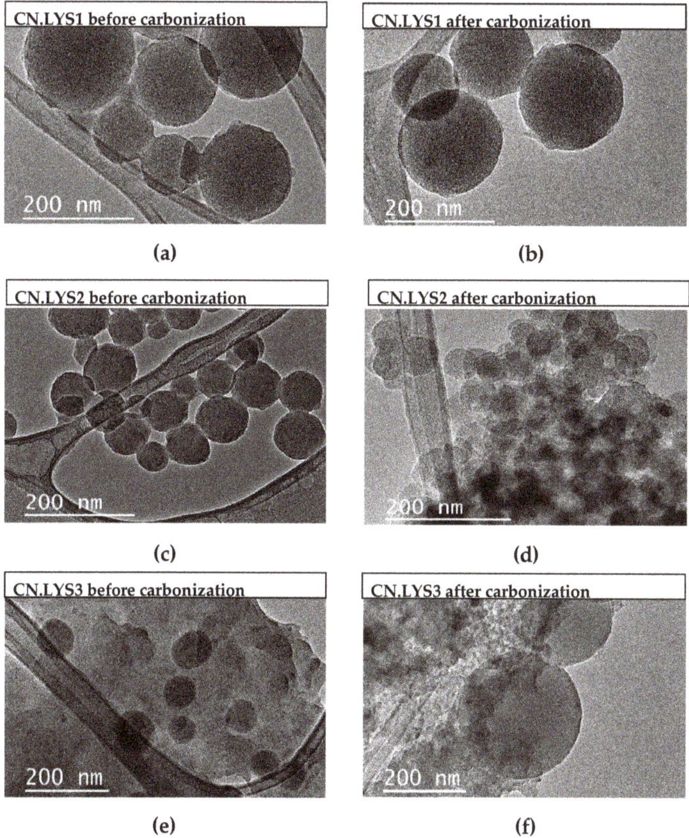

Figure 4. TEM images of carbon nanospheres synthesized with L-lysine. (**a**) Gel.LYS1, (**b**) CN.LYS1, (**c**) Gel.LYS2, (**d**) CN.LYS2, (**e**) Gel.LYS3, and (**f**) CN.LYS3.

Figure 5 presents the rheological behavior of the synthesis solution/colloid/suspension (Sol.LYS1 without dilution, Sol.LYS2 for dilution 1, and Sol.LYS3 for dilution 2). The changes in temperature and the stirring were chosen to reproduce the conditions of synthesis (60 °C for 1 h and after cooling at room temperature). The cooling stage was controlled in the rheometer. The viscosity has changed during synthesis because the reactive system underwent polymerization and condensation-related changes to form the nanospheres, so that the system passed from the solution to the colloid and the suspension. The differences in viscosity were related to the concentration of reagents in the solutions. The third solution, Sol.LYS3, presented fewer changes during the synthesis process because the system had a lower concentration of reagents as presented in Figure 4.

Based on the particle size analysis, porous texture, nitrogen content, and adsorbed amount of CO_2 (at 0 °C and up to 1 bar), the CN.LYS material was selected to perform the adsorption tests in conditions closer to those of the reservoir. Despite the high CO_2 adsorption capacities exhibited by the CN.MEL material, its size and shape would not allow its application in real reservoir conditions. Besides, its synthesis process was more complex, with a higher number of steps and a higher consumption of energy and time, which were not compensated by significantly better physicochemical properties regarding the CN.LYS.

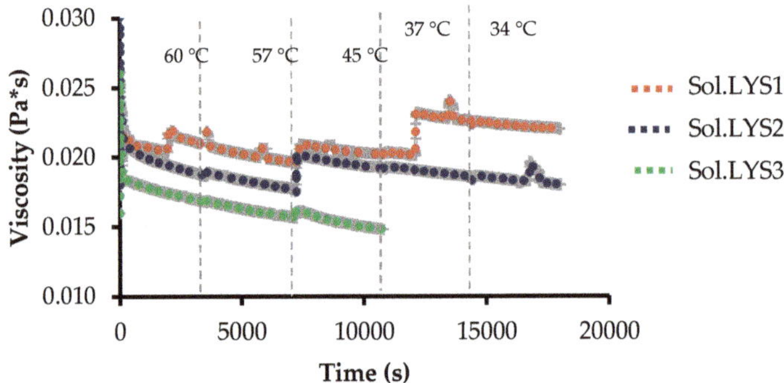

Figure 5. Rheological analysis of synthesis solutions with L-Lysine (Sol.LYS) in the same thermal conditions but with different resorcinol/water molar ratios: 1:2778 (Sol.LYS1), 1:5556 (Sol.LYS2), and 1:11112 (Sol.LYS3).

To analyze the behavior of the nanoparticles in aqueous medium, Table 4 presents the mean particle size of nanomaterials (CN.LYS) and Figure 6 presents their zeta potentials. According to the Stokes–Einstein equation, the diffusion coefficient is inversely proportional to particle size or hydrodynamic diameter; therefore, it is possible to analyze whether the nanoparticles could interact to form aggregates. Precipitation is only possible if the aggregates are big enough. Another important concept is zeta potential; if zeta potential is high (negative or positive), the particles are stable due to the high electrostatic repulsion between them. On the contrary, a low zeta potential (approaching zero) increases the probability of particles colliding, and thus forming aggregates. The hydrodynamic diameter was calculated for nanoparticles in water (at pH 5.8) and ethanol (at pH 7). Aggregate size was less in ethanol (Table 4) because the pH affects the behavior in solution (Figure 6). However, the results for nanoparticles suspended in water was close those obtained for nanoparticles suspended in ethanol. For an industrial application and injection into the porous medium, the most economical way is suspension in water.

Table 4. Mean particle size of nanomaterials in suspension, synthesized with L-lysine.

Material	d_p 50 (nm) in Water (pH 5.8)	d_p 50 (nm) in Ethanol (pH 7)
Gel.LYS1	579.4	361.5
Gel.LYS2	314.2	274.2
Gel.LYS3	1083.4	1957.1
CN.LYS1	801.1	785.9
CN.LYS2	242.6	239.9
CN.LYS3	2828.7	2587.4

For CN.LYS1 and CN.LYS3, a pH higher than 7 was better for rocks impregnation because the zeta potential was farther from zero. For CN.LYS2, a pH higher than 7 or lower than 4.7 was better for rocks impregnation. Gel.LYS2 presented the highest values of zeta potential at pH below 4, increasing the natural precipitation time of nanoparticles after the synthesis process.

Figure 7 presents SEM images of the sandstone surface before (Figure 7a) and after (Figure 7b,c) the impregnation step. Impregnation was achieved in water because of its lower cost and non-hazardous nature, making it ideal for industrial applications. The distribution of CN.LYS2 particles was homogeneously distributed on the surface (Figure 7b). The size of the aggregates was between 100 and 200 nm (Figure 7b). After one year, the sandstone continued to be impregnated, without showing any disintegration of nanoparticles from the sandstone surface. By thermogravimetric analysis, variations of less than 5% of the percentage of impregnation were obtained. This can also be related to the impregnation method without stirring, which might produce zones of lower nanoparticle concentration.

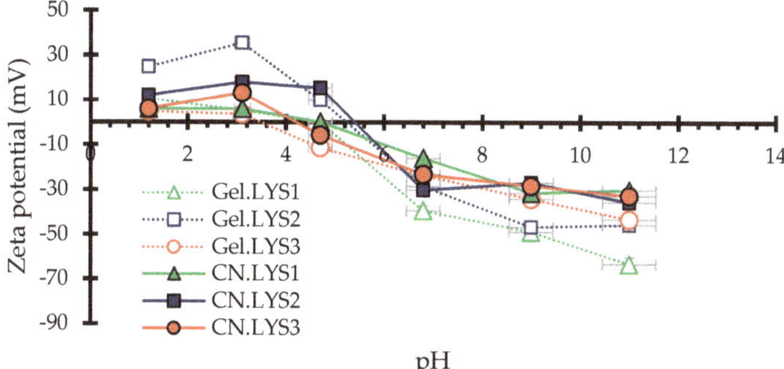

Figure 6. Zeta potential for carbon nanoparticles synthesized with L-lysine.

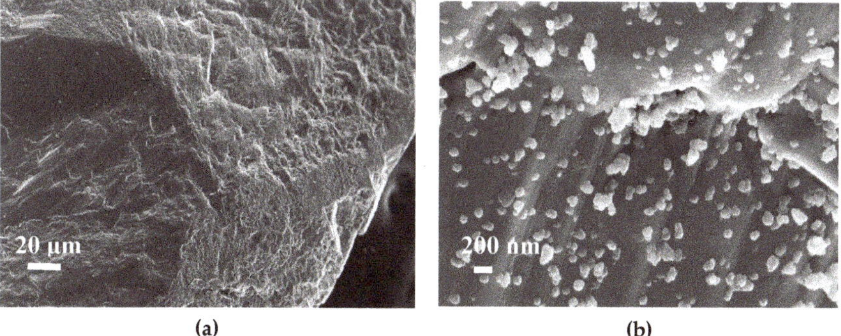

Figure 7. SEM images of (**a**) sandstone and (**b**) sandstone impregnated with a mass fraction of 20% of CN.LYS2.

The sandstone presented an A_{BET} of 0.4 m^2 g^{-1}, and its CO_2 adsorption capacity could not be measured using conventional methods (<0.0013 mmol g^{-1} at 0 °C and atmospheric pressure). Sandstone is mainly composed of silica, which has an acidic character as the CO_2 molecule. Consequently, if the specific area of the sandstone is low, its CO_2 adsorption capacity is even lower than that which might be expected for this specific area.

The sandstone was impregnated at a low nanoparticle concentration to evaluate its economic feasibility at the industrial level. The sandstone impregnated with mass fractions of 0.1 and 0.01% did not show a significant increase in its surface properties, unlike samples with higher mass fractions as shown in Table 5. The textural parameters of the impregnated sandstones were indeed improved as the percentage of nanoparticles on their surface increased. At a mass fraction of 20%, A_{BET} and $V_{0.95}$ increased by factors as high as 225 and 670, respectively.

Figure 8 shows adsorption isotherms of CO_2 at atmospheric pressure and 0 °C, for raw sandstone and sandstone impregnated at mass fractions of 0.01, 0.1, 1, 5, 10, and 20% of CN.LYS2. In addition, it shows the slope changes related to the affinity between the adsorbent medium and the adsorbate. The materials did not have significant affinity with CO_2 at mass fractions of 0.01 and 0.1% and without CN.LYS2. The affinity and adsorption capacity increased with the percentage of nanoparticles in the system. The latter presented a different behavior above a mass fraction of 1%. Indeed, at a mass fraction of 1%, the adsorption capacity was increased by a factor 21 with respect to raw sandstone, although the value was still low, 0.03 mmol g^{-1}. At a mass fraction of 20%, the value was far higher, 0.63 mmol g^{-1}, corresponding to an increment factor of 499. Different materials have been reported in the literature [16,43] with specific surface modifications to increase the adsorption capacity of CO_2, but the value of adsorption capacity was similar to that of sandstone by adding a mass fraction of 10 or 20% of nanoparticles.

Table 5. Parameters obtained from adsorption isotherms (N_2 at −196 °C and CO_2 at 0 °C) for sandstone impregnated with mass fractions of 1, 5, 10, and 20 % of CN.LYS2.

	A_{BET} (m^2 g^{-1})	$V_{0.95}$ (cm^3 g^{-1})	V_{mic-N2} (cm^3 g^{-1})	$V_{mic-CO2}$ (cm^3 g^{-1})	V_{mes} (cm^3 g^{-1})	L_0 (nm)
SS-1	2	0.003	0.002	0.002	0.001	0.56
SS-5	20	0.016	0.01	0.012	0.006	0.53
SS-10	49	0.035	0.021	0.023	0.014	0.51
SS-20	99	0.067	0.042	0.044	0.025	0.52

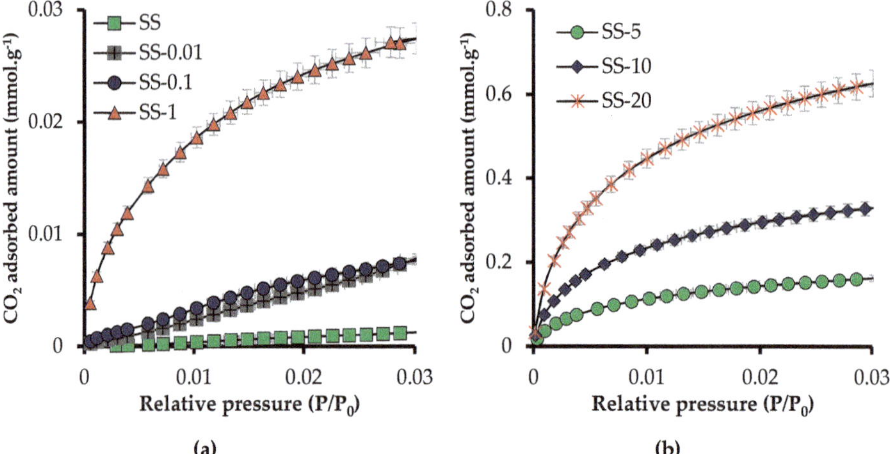

Figure 8. Adsorption isotherms of CO_2 at atmospheric pressure and 0 °C. (**a**) Mass fractions ≤1% and (**b**) mass fractions ≥5%.

The effect of the impregnation method on the CO_2 adsorption capacity at atmospheric pressure and 0°C was then evaluated. As explained before, immersion and soaking were achieved using two different sets of conditions: (i) 6 h and 600 rpm and (ii) 24 h without stirring. Increasing the soaking time by 18 h improved the adsorption capacity by more than 25% (Table 6). The values presented in Figure 8, adsorption isotherms of CO_2 at atmospheric pressure and 0 °C, thus correspond to Conditions 2.

Table 6. Relationship between soaking time and adsorption capacity of CO_2 at atmospheric pressure, 0 °C and mass fractions of 5, 10, and 20%. Conditions 1: 6 h and 600 rpm; Conditions 2: 24 h without stirring.

	5%	10%	20%
Conditions 1 (mmol g^{-1})	0.12	0.27	0.49
Conditions 2 (mmol g^{-1})	0.16	0.33	0.63
Increment (%) from conditions 1 to 2	40.5	25.4	27.2

To evaluate the possible synergistic behavior between NC.LYS2 and sandstone, the theoretical and experimental values of the CO_2 adsorption capacity are presented in Table 7. Theoretical values were calculated by assuming a linear relationship and taking into account the CO_2 adsorption capacities and the percentages of each solid. The difference between theoretical and experimental values ranged from 5 to 10%, which corresponds to the experimental error given the inaccuracy in the measurement of the very low CO_2 adsorption capacity of the sandstone. The differences could also be related to the segregation of nanoparticles during the impregnation process, the nanoparticles not being homogeneously distributed on the surface of the sandstone.

Table 7. CO_2 adsorption capacity at atmospheric pressure, 0°C, and mass fractions of 1, 5, 10, and 20 %. Theoretical and experimental values.

	1%	5%	10%	20%
Theoretical N_{ads} (mmol g^{-1})	0.036	0.175	0.349	0.697
Experimental N_{ads} (mmol g^{-1})	0.027	0.162	0.333	0.627
Relative difference (%)	23.8	7.4	4.7	10.1

3.2. High-Pressure Adsorption Tests

3.2.1. Pure CO_2 Adsorption at High Pressure–Manometric Measurement Method

The e-CCS process requires evaluating the behavior of the materials at high pressure (up to 3 MPa) and at the temperature of a hypothetical reservoir (50 °C). Figure 9a–c present the absolute CO_2 adsorbed amount and the excess amount for CN.LYS2 at 0, 25, and 50 °C. The difference between excess and absolute amounts appeared above 1 MPa, and represented 8.3% at 0 °C, 6.7% at 25 °C, and 5.9% at 50 °C. This difference was lower when the temperature increased. Such a trend is consistent with the fact that, at similar pressure, the density of the bulk phase decreases when the temperature increases. The high-pressure intrinsic CO_2 adsorption capacity of sandstone without impregnation had a negligible effect on the measurement. Therefore, the evaluation was only carried out for sandstone impregnated at mass fractions of 10% and 20% (Figure 9d,e). The difference between excess and absolute adsorbed amounts of impregnated sandstone was similar to that of CN.LYS2, stranded out after 1 MPa, and represented less than 10%, on average.

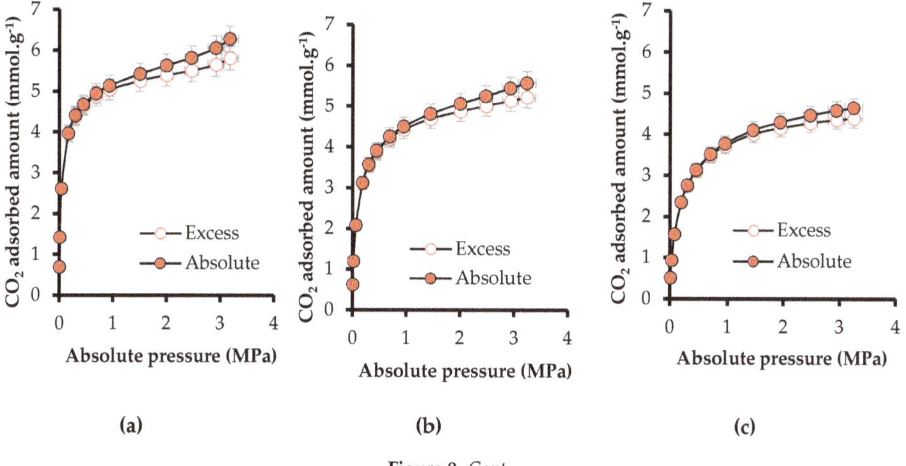

Figure 9. *Cont.*

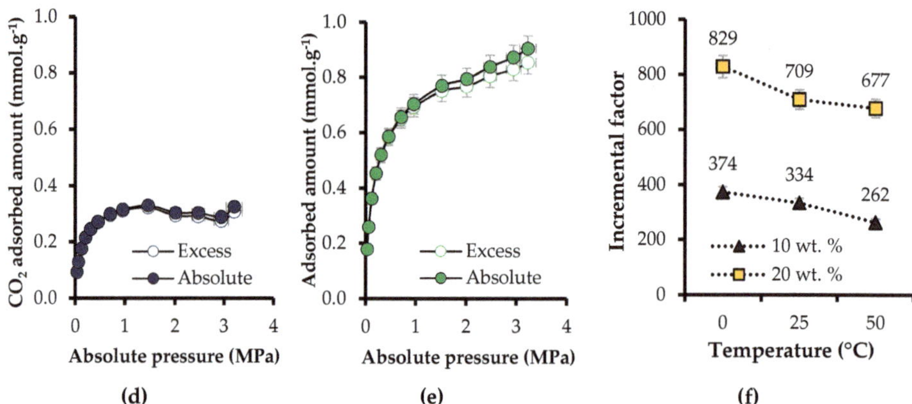

Figure 9. Adsorption isotherms of CO_2 at high pressure (3×10^{-3} up to 3.0 MPa) of CN.LYS2 at (**a**) 0 °C, (**b**) 25 °C, and (**c**) 50 °C; (**d**) sandstone impregnated with a mass fraction of 10% at 50 °C, (**e**) sandstone impregnated with a mass fraction of 20% at 50 °C. (**f**) Relationship between the impregnation percentages (mass fractions of 10 and 20%) and the adsorption capacity of CO_2 at 3 MPa and 0, 25, and 50 °C.

To observe the pressure effect on the adsorption capacity of CN.LYS2, the N_{ads} at atmospheric pressure and 0 °C (3.48 mmol g^{-1}) was compared to N_{ads} at 3 MPa and 0 °C (5.80 mmol g^{-1}). The increase of pressure produced an increase of N_{ads} of 66.6%, indicating physisorption as the main adsorption mechanism. At 50 °C, as expected, the adsorption capacity decreased by 20% due to the exothermic character of adsorption. The obtained adsorption capacity is competitive compared to other results reported for nanomaterials under similar conditions [43,44,58]. The effect of pressure on the impregnated sandstone was also observed by comparing N_{ads} at atmospheric pressure and 0 °C (0.34 mmol g^{-1} for SS-10 and 0.63 mmol g^{-1} for SS-20) to N_{ads} at 3 MPa and 0 °C (0.47 mmol g^{-1} for SS-10 and 1.04 mmol g^{-1} for SS-20). The corresponding increases were 38.2% (SS-10) and 66.0% (SS-20), respectively. The maximum N_{ads} under reservoir conditions was 0.85 mmol g^{-1} at a mass fraction of 20% of CN.LYS2.

Figure 9e presents incremental factors comparing N_{ads} for sandstone without impregnation to N_{ads} after impregnation with mass fractions of 10 and 20% at 3 MPa and 0, 25, and 50 °C. In the conditions of a shallow reservoir (50 °C and 3.0 MPa), the incremental factor was 677 for SS-20. For the e-CCS process, the pressure could increase up 6.0 MPa so that CO_2 is still in vapor phase, which allows a higher adsorbed amount. Appendix A presents the fit of the Sips and Toth models to the CO_2 adsorption isotherms at high pressure and at 0, 25, and 50°C for CN.LYS2, SS-10, and SS-20 (Figures A1–A3, respectively). As expected, the Toth and Sips models led to very good fits: $R^2 > 0.99$ for CN.LYS2, SS-10, and SS-20, and $R^2 > 0.75$ for SS-10 at 0 °C. In the latter case, the concavity of the isotherm was mainly due to the very rapid increase of the bulk density as a function of pressure with respect to the increase of the density of the adsorbed phase at pressure higher than 1.5 MPa.

Figure 10a presents the isosteric heat of adsorption (Q_{st}) of CN.LYS2 and of sandstone impregnated with mass fractions of 10 and 20%, as a function of N_{ads} expressed in mmol per gram of total adsorbent material. N_{ads} for SS-10 and SS-20 was small, and thus, Figure 10b presents Q_{st} as a function of N_{ads} in mmol per gram of carbon adsorbent material. The values of Q_{st} varied from 25 to 33 kJ mol^{-1}, which indicates a strong interaction in the adsorption system (high affinity). The interactions with the carbon porous structure could be increased by nitrogen-containing groups that are present onto the carbon surface. Nitrogen groups can indeed promote interactions between CO_2 and the substrate. The values of Q_{st} compare favorably with those reported in the literature for different materials doped or not with nitrogen (20–25 kJ mol^{-1}, on average) [45,59–61].

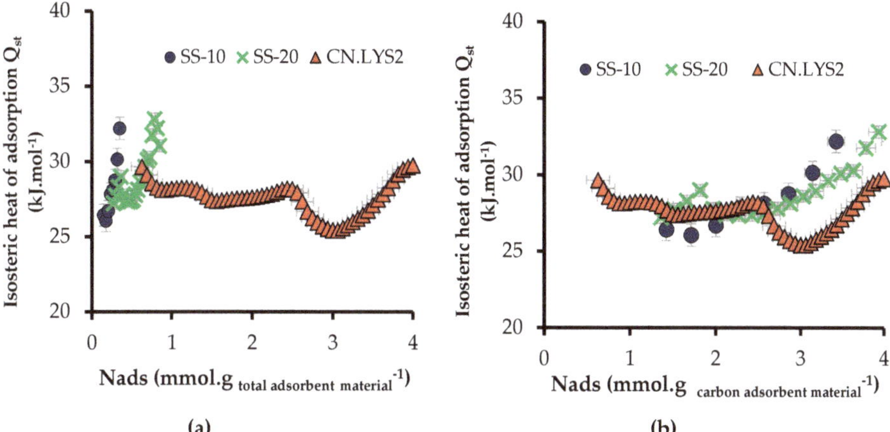

Figure 10. Isosteric heat of adsorption of CN.LYS2 and sandstone impregnated with mass fractions of 10 and 20%, as a function of the adsorbed CO_2 amount expressed: (**a**) in mmol per total amount of adsorbent material (sandstone and CN.LYS2) and (**b**) in mmol per amount of carbon adsorbent material.

The isosteric heat of adsorption of CN.LYS2 presented two distinct behaviors, whether N_{ads} was either (i) lower or (ii) higher than 3 mmol $g_{carbonadsorbentmaterial}^{-1}$. For case (i), Q_{st} decreased with N_{ads} due to the increasing distance between the last adsorbed layer and the carbon surface, thus decreasing the molecular interactions (Figure 10b). For case (ii), i.e., when N_{ads} was higher than 3 mmol $g_{carbonadsorbentmaterial}^{-1}$, Q_{st} increased due to the increased interactions between adsorbate molecules [61]. For SS-10 and SS-20, the value of Q_{st} also increased with N_{ads}. Because the interactions between sandstone (silica) and CO_2 were weak and because the contribution of the surface area associated to the carbon material was lower than for CN.LYS2 alone, it can be assumed again that the interactions between the adsorbed gas layers prevailed [61].

3.2.2. CO_2 and N_2 Adsorption at High Pressure–Gravimetric Measurement Method

Using the PyIAST application and using the pure-components adsorption isotherms obtained by experimental tests, it is possible to characterize the behavior of each component and to predict the adsorbed amount of each component present in a mixture [56]. It is, thus, possible to obtain the adsorbed amount at a constant temperature by varying the concentration of the components (at constant pressure) or the pressure (at constant concentrations). Initially, it is necessary to use the "isothermal interpolator" to generate data points that follow a given isothermal model, avoiding the search for an appropriate analytical model and examining the quality of its fit to the data (i.e., Langmuir, Freundlich or Toth, among others) [56]. The isothermal interpolator is a tool included in the PyIAST package. After that, PyIAST takes the interpolated data for its calculations. Calculations are done using a predesigned routine [56].

In the present case, the experimental data of each pure component (CO_2 and N_2) were obtained by HP-TGA at 50 °C, between 0.1 and 2.5 MPa, and with a flow of CO_2 or N_2 (50 mL min^{-1} up to 1.0 MPa and 70 mL min^{-1} up to 2.5 MPa). The prediction was calculated at 50 °C by varying: (1) the system pressure from 0.1 to 2.5 MPa at constant CO_2 concentration of 20% and (2) the concentration of CO_2, from 5 to 100% at a constant pressure of 2.5 MPa. The evaluated materials were CN.LYS2 and sandstone from a real reservoir (RS) impregnated with a mass fraction of 20% of CN.LYS2. It is important to mention that it might be possible to obtain considerable adsorbed quantities for a cleaning and adsorption balance of more than 12–24 h, because the nanomaterials (main adsorbent) are micro/mesoporous. A longer cleaning time thus allows eliminating adsorbed gases and moisture. In the same way, an adequate equilibrium time allows for greater diffusion of the gas into the porous structure and greater interactions with the material, which would allow a higher adsorbed amount. Therefore, to analyze the selectivity, a shorter time was used (cleaning and adsorption equilibrium time of 2 h at each pressure).

It was not possible to obtain the CO_2 isotherm for RS because the latter had a too low surface area, so the maximum value for RS at 50 °C and 2.5 MPa was 0.033 mmol g^{-1}. For RS, the adsorbed amount of N_2 was 3.19 mmol g^{-1} at 50 °C and 2.5 MPa. Figure 11a,b present the CO_2 and N_2 adsorption isotherms under continuous gas flow for CN.LYS2 and RS-20. The increment factor of N_{ads} for RS-20 with respect to RS was 19 (0.66 mmol g^{-1} at 50 °C and 2.5 MPa). In addition, the theoretical value of N_{ads} for RS with a mass fraction of 20% of CN.LYS was 0.78 mmol g^{-1} but the experimental value was 0.66 mmol g^{-1} (i.e., 85% of the theoretical one).

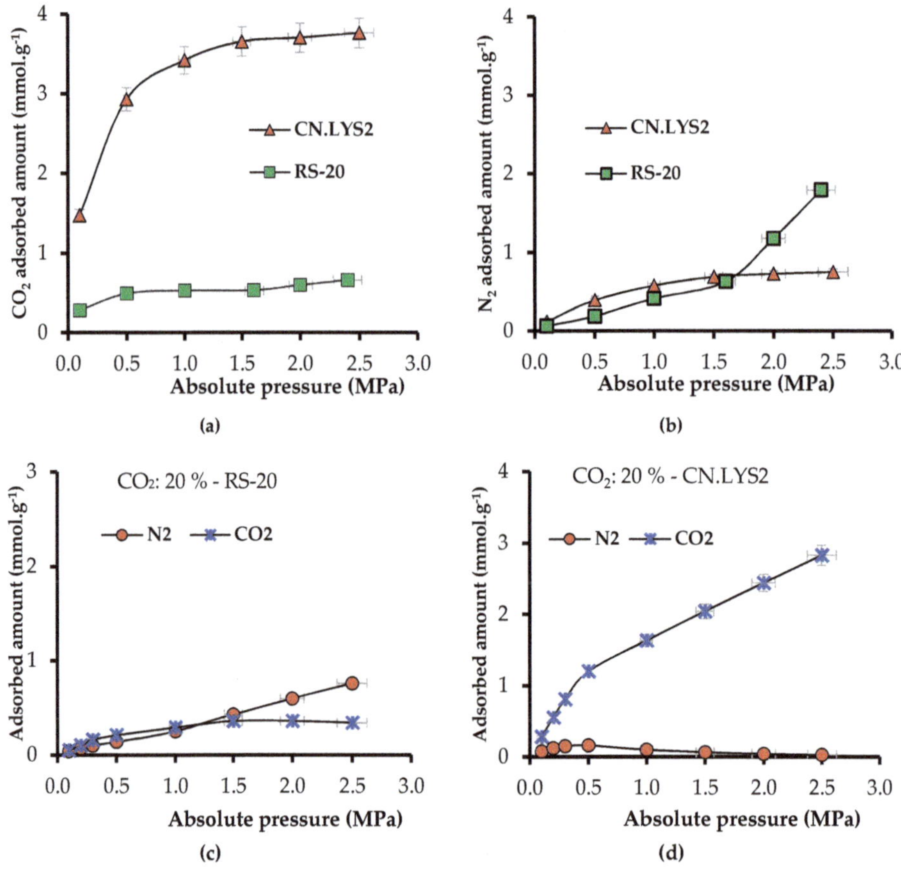

Figure 11. *Cont.*

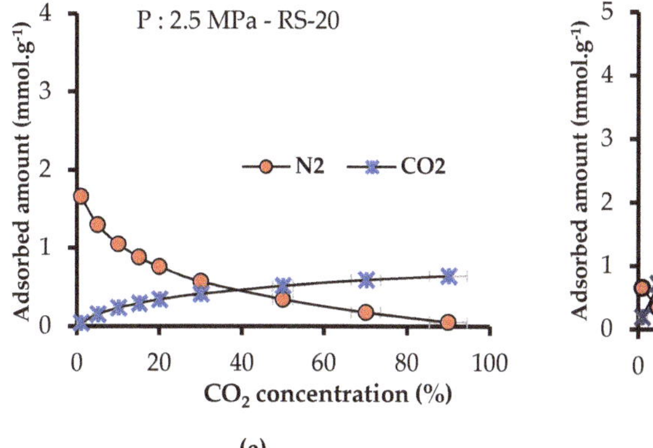
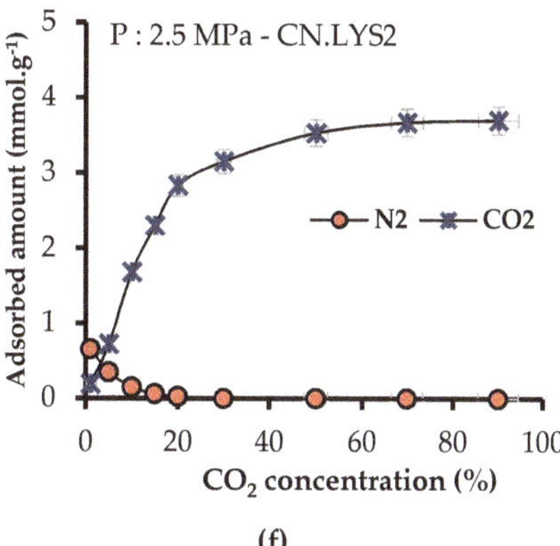

(e)   (f)

Figure 11. Adsorption isotherm of CN.LYS2 and RS-20 at 50 °C between 0.1 and 2.5 MPa for (**a**) pure CO_2 and (**b**) pure N_2. Adsorption isotherm simulating a N_2/CO_2 mixture at 50 °C, constant CO_2 concentration of 20% and varying the pressure for (**c**) RS-20 and (**d**) CN.LYS2; and varying the CO_2 concentration at constant pressure of 2.5 MPa for (**e**) RS-20 and (**f**) CN.LYS2.

Figure 11c,d present the simulated adsorbed amount for a N_2/CO_2 mixture at constant CO_2 concentration of 20% and at 50 °C by varying the pressure for RS-20 and CN.LYS2, respectively. For RS-20, the affinity for CO_2 was higher (P < 1 MPa) compared to RS without impregnation, but the N_{ads} for N_2 was superior (123.5%) to that for CO_2 (Figure 11c). For RS-20, the CO_2 isotherm obtained a form that would be impossible to obtain without impregnation due to the lack of surface area and molecular interactions. The isotherm for CN.LYS2 showed a higher affinity for CO_2 and a correspondingly higher N_{ads} than for N_2, which, above 0.5 MPa, decreased considerably (Figure 11d). Similar conditions occurred when the CO_2 concentration was varied at 2.5 MPa and 50 °C (Figure 11e,f), CN.LYS2 had more affinity for CO_2 than RS-20, and CN.LYS2 had a considerably higher N_{ads} for CO_2 than for N_2. Figure 11e shows the increment of N_{ads} only for CO_2 while the CO_2 concentration was increased in RS-20; at low concentrations of CO_2 (< 30%), the N_{ads} for N_2 was higher. For CN.LYS2 at low concentrations of CO_2 (1%), the slope increment was significant, indicating a higher affinity for CO_2 at low concentrations.

From Figure 11, it can be concluded that the pyIAST is a useful tool because it is possible to simulate the behavior of adsorption systems from some pure gas adsorption data.

4. Conclusions

To the best of our knowledge, this is the first study using nanoparticles to modify the CO_2 adsorption capacities of a reservoir in a carbon capture and storage (CCS) process. Moreover, this is the first research proposing a possible new configuration of the CCS process in which the storage is performed in shallow reservoirs (less than 300 m). We called it enhanced CCS (e-CCS), for which the main advantage is that the CO_2 capture/separation step is removed, and the flue gas is injected directly into shallow deposits, where the CO_2 is gaseous and where the adsorption phenomena control capture and storage.

Nitrogen-rich carbon nanospheres allowed increasing the adsorption capacity by 67,700% with a mass fraction of only 20% under realistic reservoir conditions (50 °C and 3 MPa). This was possible thanks to the higher surface area and to the favorable chemical composition, which promoted the capture and storage of CO_2. These N-doped carbon nanospheres, synthesized by a simple process, had competitive CO_2 capture performances compared to other special materials reported in the literature. Therefore, this research opens an interesting line of research that would expand knowledge in the field of carbon nanospheres for application in the adsorption and geological storage of CO_2.

135

Author Contributions: Conceptualization, F.B.C., C.A.F., F.C.-M., V.F., A.C., A.C.M. and E.R.A.; methodology, F.B.C., C.A.F., F.C.-M., V.F. and E.R.A.; software, F.B.C., E.R.A.; validation, F.B.C., F.C.-M., V.F., S.S. and E.R.A.; formal analysis, F.B.C., C.A.F., F.C.-M., V.F., A.C., S.S. and E.R.A.; investigation, S.S. and E.R.A.; resources, F.B.C., C.A.F., F.C.-M., A.F.P.-C., V.F., A.C., S.S., A.C.M. and E.R.A.; Writing—Original Draft preparation, E.R.A.; Writing—Review and Editing, F.B.C., C.A.F., F.C.-M., A.F.P.-C., V.F., A.C., S.S. and E.R.A.; visualization, E.R.A.; supervision, F.B.C., F.C.-M., and V.F.

Funding: The authors thank COLCIENCIAS for financing the doctoral studies of Elizabeth Rodriguez Acevedo through the call 647-2014. The authors thank COLCIENCIAS, Agencia Nacional de Hidrocarburos-ANH provided by agreement 272-2017 for the support provided and Universidad Nacional de Colombia for the support provided in the agreement 272-2017. The authors thank to Spanish Ministry of Science, Innovation and Universities, FEDER funds, contract number RTI2018-099224-B-I00. The authors also thank to ERASMUS+ program (agreement F NANCY43) and ENLAZAMUNDOS-SAPIENCIA for the support of academic internships. French authors acknowledge FEDER funds, through TALiSMAN project, for the financial support.

Acknowledgments: The authors thank the Universidad Nacional de Colombia, the University of Granada, and the University of Lorraine-Institut Jean Lamour for their logistical and financial support. The authors also thank Philippe Gadonneix and Saray Perez–Robles for their technical support in the experimental tests.

Conflicts of Interest: The authors declare no conflict of interest.

Appendix A

This section presents the adsorption isotherms fitted by the Sips and Toth models, for CO_2 at high pressure and 0, 25, and 50 °C. This information is related to Section 3.2.1. Pure CO_2 adsorption at high pressure–Manometric measurement method.

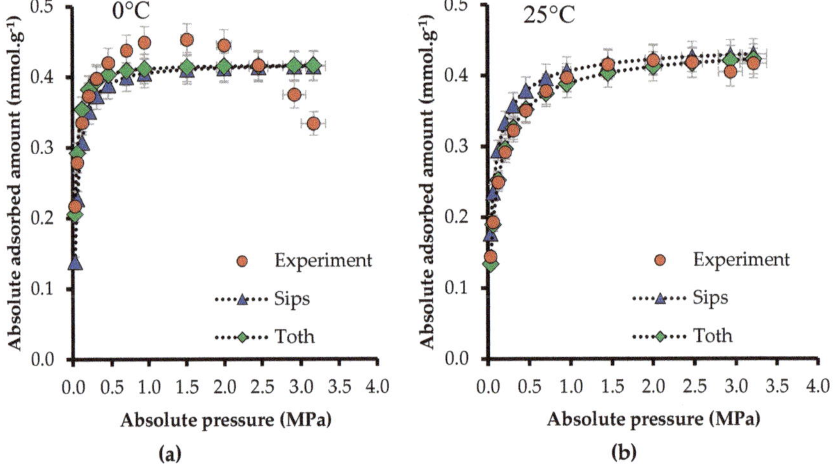

Figure A1. *Cont.*

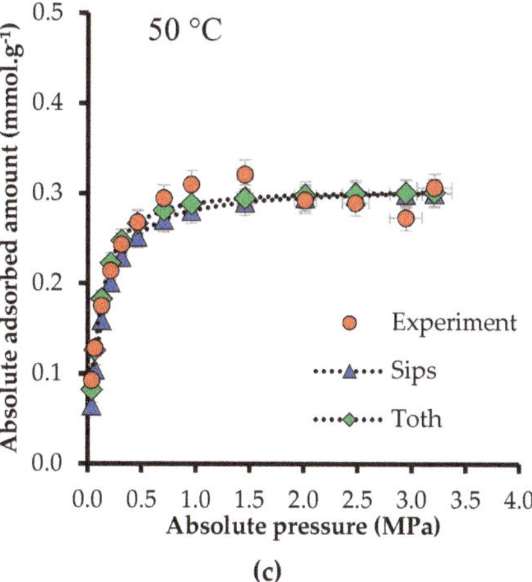

(c)

Figure A1. Adsorption isotherms fitted by the Sips and Toth models for sandstone impregnated with a mass fraction of 10% of carbon nanospheres synthesized with L-Lysine (CN.LYS2) at (**a**) 0° C, (**b**) 25 °C, and (**c**) 50 °C.

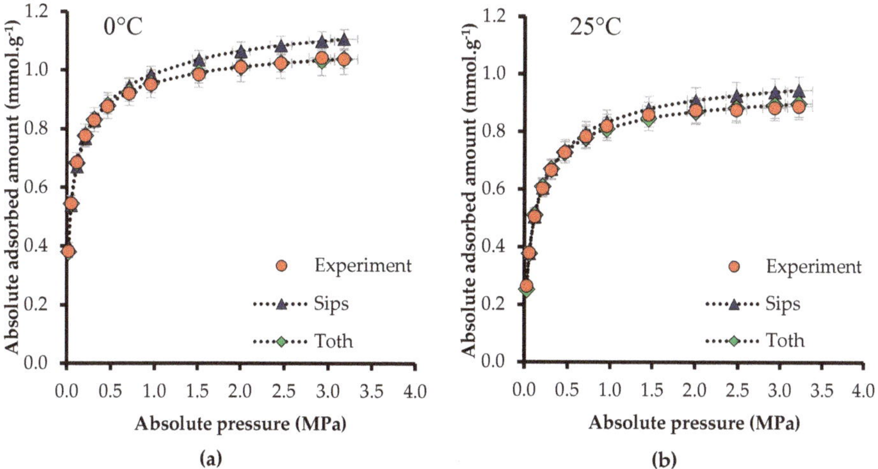

(a)   **(b)**

Figure A2. *Cont.*

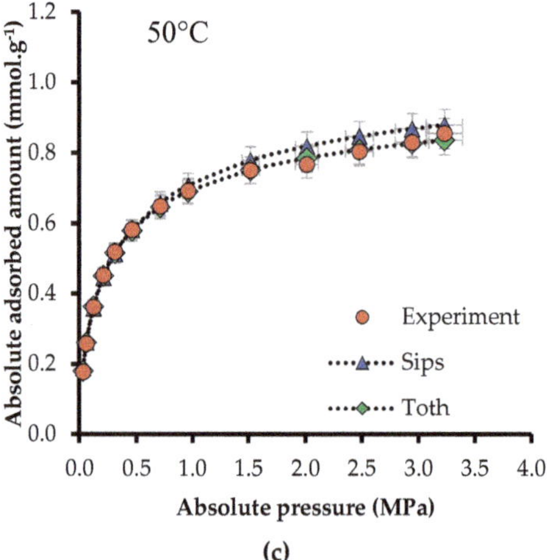

(c)

Figure A2. Adsorption isotherms fitted the Sips and Toth models for sandstone impregnated with a mass fraction of 20% of carbon nanospheres synthesized with L-Lysine (CN.LYS2) at (**a**) 0 °C, (**b**) 25 °C, and (**c**) 50 °C.

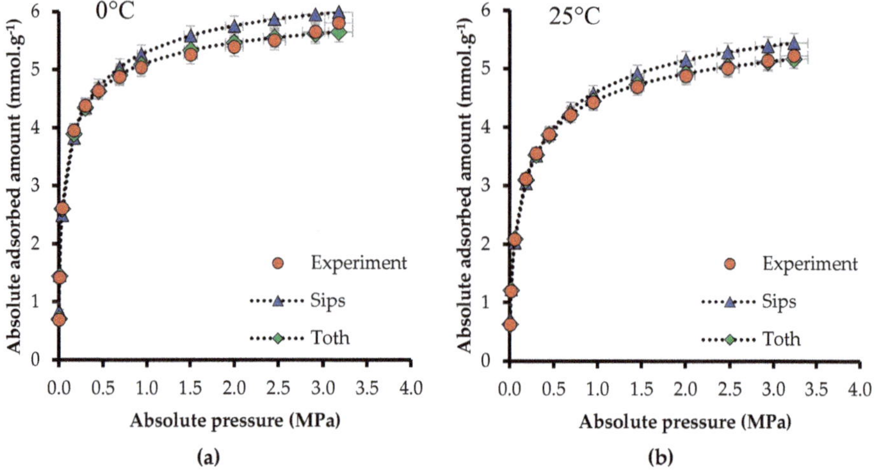

(a)   (b)

Figure A3. *Cont.*

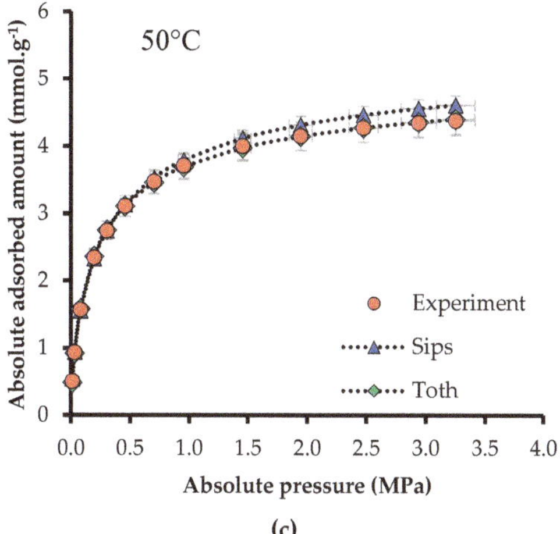

(c)

Figure A3. Adsorption isotherms fitted by the Sips and Toth models for carbon nanospheres synthesized with L-lysine (CN.LYS2) at (**a**) 0 °C, (**b**) 25 °C, and (**c**) 50 °C.

References

1. Root, T.L.; Price, J.T.; Hall, K.R.; Schneider, S.H.; Rosenzweig, C.; Pounds, J.A. Fingerprints of global warming on wild animals and plants. *Nature* **2003**, *421*, 57. [CrossRef]
2. Vitousek, P.M. Beyond global warming: Ecology and global change. *Ecology* **1994**, *75*, 1861–1876. [CrossRef]
3. McGlade, C.; Ekins, P. The geographical distribution of fossil fuels unused when limiting global warming to 2 C. *Nature* **2015**, *517*, 187. [CrossRef]
4. Harvey, L.D. *Global Warming*; Routledge: Abingdon, UK, 2018.
5. Baer, H.; Singer, M. *Global Warming and the Political Ecology of Health: Emerging Crises and Systemic Solutions*; Routledge: Abingdon, UK, 2016.
6. Lashof, D.A.; Ahuja, D.R. Relative contributions of greenhouse gas emissions to global warming. *Nature* **1990**, *344*, 529. [CrossRef]
7. Anderson, T.R.; Hawkins, E.; Jones, P.D. CO_2, the greenhouse effect and global warming: From the pioneering work of Arrhenius and Callendar to today's Earth System Models. *Endeavour* **2016**, *40*, 178–187. [CrossRef]
8. US-EPA, E.P.A. Global Greenhouse Gas Emissions Data. 2019. Available online: https://www.epa.gov/ghgemissions/global-greenhouse-gas-emissions-data (accessed on 15 May 2019).
9. Edenhofer, O.; Pichs-Madruga, R.; Sokona, Y.; Minx, J.C.; Farahani, E.; Kadner, S.; Seyboth, K.; Adler, A.; Baum, I.; Brunner, S.; et al. Climate Change 2014, Mitigation of Climate Change. In *Contribution of Working Group III to the Fifth Assessment Report of the Intergovernmental Panel on Climate Change*; IPCC, 2014: Summary for Policymakers; Cambridge University Press: Cambridge, UK, 2014.
10. Tan, Y.; Nookuea, W.; Li, H.; Thorin, E.; Yan, J. Property impacts on Carbon Capture and Storage (CCS) processes: A review. *Energy Convers. Manag.* **2016**, *118*, 204–222. [CrossRef]
11. Change, N.G.C. *Vital Signs of the Planet*; Earth Science Communications Team at NASA's Jet Propulsion Laboratory: Pasadena, CA, USA, 2018; p. 30.
12. NASA. Global Climate Change. Vital Signs of the Planet. 2019. Available online: https://climate.nasa.gov/vital-signs/carbon-dioxide/ (accessed on 30 March 2019).
13. Norby, R.J.; Luo, Y. Evaluating ecosystem responses to rising atmospheric CO_2 and global warming in a multi-factor world. *New Phytol.* **2004**, *162*, 281–293. [CrossRef]
14. Halmann, M.M. *Chemical Fixation of Carbon DioxideMethods for Recycling CO_2 into Useful Products*; CRC Press: Boca Raton, FL, USA, 2018.
15. Cox, P.M.; Betts, R.A.; Jones, C.D.; Spall, S.A.; Totterdell, I.J. erratum: Acceleration of global warming due to carbon-cycle feedbacks in a coupled climate model. *Nature* **2000**, *408*, 750. [CrossRef]

16. Wang, J.; Huang, L.; Yang, R.; Zhang, Z.; Wu, J.; Gao, Y.; Wang, Q.; O'Hare, D.; Zhong, Z. Recent advances in solid sorbents for CO_2 capture and new development trends. *Energy Environ. Sci.* **2014**, *7*, 3478–3518. [CrossRef]

17. Conti, J.; Holtberg, P.; Diefenderfer, J.; LaRose, A.; Turnure, J.T.; Westfall, L. *International Energy Outlook 2016 with Projections to 2040*; USDOE Energy Information Administration (EIA): Washington, DC, USA, 2016.

18. Metz, B.; Davidson, O.; de Coninck, H. *Carbon Dioxide Capture and Storage: Special Report of the Intergovernmental Panel on Climate Change*; Cambridge University Press: Cambridge, UK, 2005.

19. Kang, S.-P.; Lee, H. Recovery of CO_2 from flue gas using gas hydrate: Thermodynamic verification through phase equilibrium measurements. *Environ. Sci. Technol.* **2000**, *34*, 4397–4400. [CrossRef]

20. Yang, Q.; Xue, C.; Zhong, C.; Chen, J.F. Molecular simulation of separation of CO_2 from flue gases in CU-BTC metal-organic framework. *AIChE J.* **2007**, *53*, 2832–2840. [CrossRef]

21. Song, C.; Pan, W.; Srimat, S.T.; Zheng, J.; Li, Y.; Wang, Y.H.; Xu, B.O.; Zhu, Q.M. Tri-reforming of methane over Ni catalysts for CO_2 conversion to Syngas with desired H_2 CO ratios using flue gas of power plants without CO_2 separation. *Stud. Surf. Sci. Catal.* **2004**, *153*, 315–322.

22. Bui, M.; Adjiman, C.S.; Bardow, A.; Anthony, E.J.; Boston, A.; Brown, S.; Fennell, P.S.; Fuss, S.; Galindo, A.; Hackett, L.A.; et al. Carbon capture and storage (CCS): The way forward. *Energy Environ. Sci.* **2018**, *11*, 1062–1176. [CrossRef]

23. Knorr, W. Is the airborne fraction of anthropogenic CO_2 emissions increasing? *Geophys. Res. Lett.* **2009**, *36*. [CrossRef]

24. Cook, P.; Causebrook, R.; Gale, J.; Michel, K.; Watson, M. What have we learned from small-scale injection projects? *Energy Procedia* **2014**, *63*, 6129–6140. [CrossRef]

25. IEA, I. *World Energy Outlook 2011*; International Energy Agency: Paris, France, 2011; p. 666.

26. Balat, H.; Öz, C. Technical and Economic Aspects of Carbon Capture an Storage—A Review. *Energy Explor. Exploit.* **2007**, *25*, 357–392. [CrossRef]

27. Gough, C. State of the art in carbon dioxide capture and storage in the UK: An experts' review. *Int. J. Greenh. Gas Control* **2008**, *2*, 155–168. [CrossRef]

28. Gough, C. *Carbon Capture and Its Storage: An Integrated Assessment*; Routledge: Abingdon, UK, 2016.

29. Bailon-García, E.P.C.; Agustín, F.; Elizabeth, R.A.; Francisco, C.M. Nanoparticle Fabrication Methods. In *Formation Damage in Oil and Gas Reservoirs. Nanotechnology Applications for Its Inhibition/Remediation*; Franco, C.A.a.C.C., Farid, B., Eds.; Nova Science Publishers: Hauppauge, NY, USA, 2018; pp. 69–150.

30. Franco, C.A.C.C.; Farid, B. *Formation Damage in Oil and Gas Reservoirs. Nanotechnology Applications for Its Inhibition/Remediation*; Nova Science Publishers: Hauppauge, NY, USA, 2018.

31. Franco, C.A.; Zabala, R.; Cortés, F.B. Nanotechnology applied to the enhancement of oil and gas productivity and recovery of Colombian fields. *J. Pet. Sci. Eng.* **2017**, *157*, 39–55. [CrossRef]

32. Franco, C.A.; Nassar, N.N.; Ruiz, M.A.; Pereira-Almao, P.; Cortés, F.B. Nanoparticles for inhibition of asphaltenes damage: Adsorption study and displacement test on porous media. *Energy Fuels* **2013**, *27*, 2899–2907. [CrossRef]

33. Moncayo-Riascos, I.; Franco, C.A.; Cortés, F.B. Dynamic Molecular Modeling and Experimental Approach of Fluorocarbon Surfactant-Functionalized SiO_2 Nanoparticles for Gas-Wettability Alteration on Sandstones. *J. Chem. Eng. Data* **2019**, *64*, 1860–1872. [CrossRef]

34. Hurtado, Y.; Beltrán, C.; Zabala, R.D.; Lopera, S.H.; Franco, C.A.; Nassar, N.N.; Cortés, F.B. Effects of Surface Acidity and Polarity of SiO_2 Nanoparticles on the Foam Stabilization Applied to Natural Gas Flooding in Tight Gas-Condensate Reservoirs. *Energy Fuels* **2018**, *32*, 5824–5833. [CrossRef]

35. Cardona, L.; Arias-Madrid, D.; Cortés, F.; Lopera, S.; Franco, C. Heavy oil upgrading and enhanced recovery in a steam injection process assisted by NiO-and PdO-Functionalized SiO_2 nanoparticulated catalysts. *Catalysts* **2018**, *8*, 132. [CrossRef]

36. Yang, D.; Wang, S.; Zhang, Y. Analysis of CO_2 migration during nanofluid-based supercritical CO_2 geological storage in saline aquifers. *Aerosol Air Qual. Res.* **2014**, *14*, 1411–1417. [CrossRef]

37. Silvestre-Albero, J.; Reinoso, F.R. Nuevos materiales de carbón para la captura de CO_2. *Boletín del Grupo Español del Carbón* **2012**, *24*, 2–6.

38. Zhang, X.Q.; Li, W.C.; Lu, A.H. Designed porous carbon materials for efficient CO_2 adsorption and separation. *New Carbon Mater.* **2015**, *30*, 481–501. [CrossRef]

39. Bandosz, T.J.; Seredych, M.; Rodríguez-Castellón, E.; Cheng, Y.; Daemen, L.L.; Ramírez-Cuesta, A.J. Evidence for CO_2 reactive adsorption on nanoporous S-and N-doped carbon at ambient conditions. *Carbon* **2016**, *96*, 856–863. [CrossRef]

40. Lithoxoos, G.P.; Labropoulos, A.; Peristeras, L.D.; Kanellopoulos, N.; Samios, J.; Economou, I.G. Adsorption of N_2, CH_4, CO and CO_2 gases in single walled carbon nanotubes: A combined experimental and Monte Carlo molecular simulation study. *J. Supercrit. Fluids* **2010**, *55*, 510–523. [CrossRef]

41. Bikshapathi, M.; Sharma, A.; Sharma, A.; Verma, N. Preparation of carbon molecular sieves from carbon micro and nanofibers for sequestration of CO_2. *Chem. Eng. Res. Des.* **2011**, *89*, 1737–1746. [CrossRef]

42. Chowdhury, S.; Balasubramanian, R. Highly efficient, rapid and selective CO_2 capture by thermally treated graphene nanosheets. *J. CO2 Util.* **2016**, *13*, 50–60. [CrossRef]

43. Alonso, A.; Moral-Vico, J.; Markeb, A.A.; Busquets-Fité, M.; Komilis, D.; Puntes, V.; Sánchez, A.; Font, X. Critical review of existing nanomaterial adsorbents to capture carbon dioxide and methane. *Sci. Total Environ.* **2017**, *595*, 51–62. [CrossRef]

44. Ma, Y.; Wang, Z.; Xu, X.; Wang, J. Review on porous nanomaterials for adsorption and photocatalytic conversion of CO_2. *Chin. J. Catal.* **2017**, *38*, 1956–1969. [CrossRef]

45. Babu, D.J.; Bruns, M.; Schneider, R.; Gerthsen, D.; Schneider, J.J. Understanding the influence of N-doping on the CO_2 adsorption characteristics in carbon nanomaterials. *J. Phys. Chem. C* **2017**, *121*, 616–626. [CrossRef]

46. Chen, A.; Li, S.; Yu, Y.; Liu, L.; Li, Y.; Wang, Y.; Xia, K. Self-catalyzed strategy to form hollow carbon nanospheres for CO_2 capture. *Mater. Lett.* **2016**, *185*, 63–66. [CrossRef]

47. Heydari-Gorji, A.; Belmabkhout, Y.; Sayari, A. Degradation of amine-supported CO_2 adsorbents in the presence of oxygen-containing gases. *Microporous Mesoporous Mater.* **2011**, *145*, 146–149. [CrossRef]

48. Dong, Y.-R.; Nishiyama, N.; Egashira, Y.; Ueyama, K. Basic Amid Acid-Assisted Synthesis of Resorcinol−Formaldehyde Polymer and Carbon Nanospheres. *Ind. Eng. Chem. Res.* **2008**, *47*, 4712–4716. [CrossRef]

49. Bai, X.; Li, J.; Cao, C.; Hussain, S. Solvothermal synthesis of the special shape (deformable) hollow g-C_3N_4 nanospheres. *Mater. Lett.* **2011**, *65*, 1101–1104. [CrossRef]

50. Franco-Aguirre, M.; Zabala, R.D.; Lopera, S.H.; Franco, C.A.; Cortés, F.B. Interaction of anionic surfactant-nanoparticles for gas-Wettability alteration of sandstone in tight gas-condensate reservoirs. *J. Nat. Gas Sci. Eng.* **2018**, *51*, 53–64. [CrossRef]

51. Schaefer, S.; Fierro, V.; Izquierdo, M.T.; Celzard, A. Assessment of hydrogen storage in activated carbons produced from hydrothermally treated organic materials. *Int. J. Hydrog. Energy* **2016**, *41*, 12146–12156. [CrossRef]

52. Schaefer, S.; Fierro, V.; Szczurek, A.; Izquierdo, M.T.; Celzard, A. Physisorption, chemisorption and spill-over contributions to hydrogen storage. *Int. J. Hydrog. Energy* **2016**, *41*, 17442–17452. [CrossRef]

53. Tzabar, N.; Brake, H.T. Adsorption isotherms and Sips models of nitrogen, methane, ethane, and propane on commercial activated carbons and polyvinylidene chloride. *Adsorption* **2016**, *22*, 901–914. [CrossRef]

54. Álvarez-Gutiérrez, N.; Gil, M.V.; Rubiera, F.; Pevida, C. Adsorption performance indicators for the CO_2/CH_4 separation: Application to biomass-based activated carbons. *Fuel Process. Technol.* **2016**, *142*, 361–369. [CrossRef]

55. Abdeljaoued, A.; Querejeta, N.; Durán, I.; Álvarez-Gutiérrez, N.; Pevida, C.; Chahbani, M. Preparation and Evaluation of a Coconut Shell-Based Activated Carbon for CO_2/CH_4 Separation. *Energies* **2018**, *11*, 1748. [CrossRef]

56. Simon, C.M.; Smit, B.; Haranczyk, M. pyIAST: Ideal adsorbed solution theory (IAST) Python package. *Comput. Phys. Commun.* **2016**, *200*, 364–380. [CrossRef]

57. Cavenati, S.; Grande, C.A.; Rodrigues, A.E. Adsorption equilibrium of methane, carbon dioxide, and nitrogen on zeolite 13X at high pressures. *J. Chem. Eng. Data* **2004**, *49*, 1095–1101. [CrossRef]

58. Himeno, S.; Tomita, T.; Suzuki, K.; Yoshida, S. Characterization and selectivity for methane and carbon dioxide adsorption on the all-silica DD3R zeolite. *Microporous Mesoporous Mater.* **2007**, *98*, 62–69. [CrossRef]

59. Dunne, J.; Mariwala, R.; Rao, M.; Sircar, S.; Gorte, R.J.; Myers, A.L. Calorimetric heats of adsorption and adsorption isotherms. 1. O_2, N_2, Ar, CO_2, CH_4, C_2H_6, and SF6 on silicalite. *Langmuir* **1996**, *12*, 5888–5895. [CrossRef]

60. Siriwardane, R.V.; Shen, M.S.; Fisher, E.P.; Poston, J.A. Adsorption of CO_2 on molecular sieves and activated carbon. *Energy Fuels* **2001**, *15*, 279–284. [CrossRef]

Materials **2019**, *12*, 2088

61. Himeno, S.; Komatsu, T.; Fujita, S. High-pressure adsorption equilibria of methane and carbon dioxide on several activated carbons. *J. Chem. Eng. Data* **2005**, *50*, 369–376. [CrossRef]

Article

Electroanalytical Performance of Nitrogen-Doped Graphene Films Processed in One Step by Pulsed Laser Deposition Directly Coupled with Thermal Annealing

Florent Bourquard [1], Yannick Bleu [1], Anne-Sophie Loir [1], Borja Caja-Munoz [2], José Avila [2], Maria-Carmen Asensio [3], Gaëtan Raimondi [4], Maryam Shokouhi [4], Ilhem Rassas [4], Carole Farre [4], Carole Chaix [4], Vincent Barnier [5], Nicole Jaffrezic-Renault [4], Florence Garrelie [1] and Christophe Donnet [1,*]

[1] Laboratoire Hubert Curien, UMR 5516 CNRS, Université de Lyon, Université Jean Monnet, F-42000 Saint-Étienne, France; Florent.Bourquard@univ-st-etienne.fr (F.B.); yannick.bleu@univ-st-etienne.fr (Y.B.); Anne.Sophie.Loir@univ-st-etienne.fr (A.-S.L.); Florence.Garrelie@univ-st-etienne.fr (F.G.)

[2] Synchrotron SOLEIL, Université Paris-Saclay, Saint Aubin, F-91192 Gif sur Yvette, France; borja.caja-munoz@synchrotron-soleil.fr (B.C.-M.); jose.avila@synchrotron-soleil.fr (J.A.)

[3] Instituto de Ciencia de Materiales de Madrid, Sor Juana Inés de la Cruz, 328049 Madrid, Spain; mc.asensio@csic.es

[4] Institut des Sciences Analytiques, UMR 5280 CNRS, Université de Lyon, Université Claude Bernard Lyon 1, F-69100 Villeurbanne, France; gaetan.raimondi@isa-lyon.fr (G.R.); m.shokouhi89@gmail.com (M.S.); ilhemras@hotmail.fr (I.R.); carole.farre@univ-lyon1.fr (C.F.); carole.chaix-bauvais@univ-lyon1.fr (C.C.); Nicole.Jaffrezic@univ-lyon1.fr (N.J.-R.)

[5] Mines Saint-Etienne, Université de Lyon, UMR 5307 CNRS, Centre SMS, F-42023 Saint-Etienne, France; barnier@emse.fr

* Correspondence: christophe.donnet@univ-st-etienne.fr

Received: 29 January 2019; Accepted: 20 February 2019; Published: 23 February 2019

Abstract: Graphene-based materials are widely studied to enable significant improvements in electroanalytical devices requiring new generations of robust, sensitive and low-cost electrodes. In this paper, we present a direct one-step route to synthetize a functional nitrogen-doped graphene film onto a Ni-covered silicon electrode substrate heated at high temperature, by pulsed laser deposition of carbon in the presence of a surrounding nitrogen atmosphere, with no post-deposition transfer of the film. With the ferrocene methanol system, the functionalized electrode exhibits excellent reversibility, close to the theoretical value of 59 mV, and very high sensitivity to hydrogen peroxide oxidation. Our electroanalytical results were correlated with the composition and nanoarchitecture of the N-doped graphene film containing 1.75 at % of nitrogen and identified as a few-layer defected and textured graphene film containing a balanced mixture of graphitic-N and pyrrolic-N chemical functions. The absence of nitrogen dopant in the graphene film considerably degraded some electroanalytical performances. Heat treatment extended beyond the high temperature graphene synthesis did not significantly improve any of the performances. This work contributes to a better understanding of the electrochemical mechanisms of doped graphene-based electrodes obtained by a direct and controlled synthesis process.

Keywords: graphene; nitrogen-doped graphene; pulse laser deposition; electrochemical analysis; oxygen peroxide oxidation

---

1. Introduction

Graphene is considered to be a promising 2D material thanks to its unique versatile properties, in particular high thermal and electrical conductivity, with many potential applications as an electrode in chemistry and as an electrocatalyst in fuel cells, field emitters, batteries, supercapacitors, and sensors (for a review, see Reference [1]).

The structure and electronic properties of graphene can be tailored by heteroatom doping, in particular by incorporating nitrogen, thereby opening the band gap and transforming graphene into a semiconductor, as recently reviewed by Yadav et al. [2] and Xu et al. [3].

Some authors have studied the electrochemical properties of N-doped graphene (NG) in various experimental conditions. Wang et al. [4] obtained NG with a nitrogen content ranging between 0.11 and 1.35 at %, by nitrogen plasma exposure of graphene prepared by chemical reduction of graphene oxide. The NG films exhibited consistent electrocatalytic activity for the reduction of hydrogen peroxide (H_2O_2), as well as high sensitivity and selectivity for glucose biosensing. Shao et al. [5] compared the electroanalytical properties of NG electrode with pure graphene (G) and Pt/C electrode. Oxygen reduction reaction (ORR) overpotential is lower on NG than G, meaning that N doping significantly increases the electrocatalytic activity of graphene towards ORR. Although the NG electrode exhibited a lower initial electrocatalytic activity than Pt/C, it was much more stable and durable, suggesting that it may be possible to replace expensive Pt with low-cost NG. Moreover, unlike Pt, ORR on NG was not influenced by fuel molecules, making NG in direct liquid fuel cells very promising. The overpotential during electrocatalytic H_2O_2 reduction was significantly reduced, and a well-defined and enhanced H_2O_2 reduction peak was observed around -0.2 V, demonstrating better NG electrocatalytic activity compared to undoped graphene for H_2O_2 sensing. Ruiyi et al. [6] incorporated nitrogen in multiple graphene aerogel/gold nanostars (N-doped MGA/GNS) and reported that such a sensor was more sensitive than that of all reported DNA sensors to date. Saengsookwaow et al. [7] showed that, during cyclic voltammetry, a screen-printed carbon electrode (SPCE) functionalized by NG/Polyvinylpyrrolidone PVP/Gold nanoparticles (AuNPs) increased the anodic peak current by a factor of 10 compared to unmodified SCPE, due to a significant improvement in the interfacial charge transfer. This type of electrode showed higher electrochemical sensitivity and electrocatalytic activity toward hydrazine oxidation, leading to successful determination of hydrazine content in fruit and vegetable samples. Recently, Li et al. [8] also studied SPCE functionalized by NG sheets (NGS) obtained from graphene oxide and reported consistent sensitivity, selectivity and stability (with less overpotential required for oxidation) for nicotine detection, including when the molecule was in urine and tobacco samples.

However, most previous studies were performed with NG synthetized using rather complex chemical routes, with limited control of the nitrogen concentration incorporated in the graphene network. Thus, the investigation of electrochemical properties of NG films obtained by more simple routes is still of great interest for the next generations of electrodes. In particular, synthesis of NG film in one step from a solid carbon source directly onto silicon electrodes has been less explored than NG films obtained by other routes, including CVD processes and various reduction processes of GO by thermal annealing, plasma treatment, hydrothermal or solvothermal reactions in the presence of a nitrogen precursor. In a previous paper [9], we reported on the performance of a graphene electrode processed in one step by pulsed laser deposition directly coupled with in situ thermal annealing (PLD-TA). The electrochemical behavior of the NG film was studied in the presence of the redox probe, ferrocene methanol, which was shown to be the most suitable for quantifying electron transfer with graphene. Cyclic voltammetry revealed excellent electrochemical kinetic and quasi-reversibility performances. The attachment of ethynyl aryl groups on the surface of the electrode was robust, paving the way for the specific attachment of molecules bearing an azide function using the click reaction.

Moving on from there, in the present study we investigated the electroanalytic performance of a NG-functionalized electrode obtained using the one-step PLD-TA process previously successfully used for the electroanalytical investigation of pure graphene film. In a recent review [10], we mentioned that only three papers reported on the synthesis of NG films from an amorphous nitrogenated carbon film (a-C:N) obtained by PLD. Kumar et al. [11] reported in situ growth of n-type NG films (2 at % of nitrogen) by PLD performed at 973 K, with an increase in electrical conductivity with increased nitrogen partial pressure. Ren et al. [12] used the same approach and obtained NG films (1.7 to 3.2 at %) showing improved chemical enhancement for Raman analysis of absorbed Rhodamine 6G molecules, compared to pristine graphene. Recently, our group reported for the first time the synthesis of NG-doped few-layer graphene from a solid state nitrogen carbide (a-C:N film) synthetized by femtosecond pulse laser ablation [13]. We investigated the nanostructure and chemical composition of an NG film obtained by a vacuum thermal annealing at 780 °C of an a-C:N film previously deposited onto a SiO_2 substrate and further covered by a Ni catalytic film (150 nm). Here we optimize the previous protocol to form the NG film directly by high temperature condensation of the laser-induced carbon plasma plume in the presence of nitrogen atmosphere, onto the Si electrode previously covered by a Ni catalytic film. The structure and composition of the NG film, compared to undoped ones, were investigated by Raman spectroscopy, X-ray photoelectron spectroscopy (XPS) and scanning electron microscopy (FEG-SEM). To show the interest of using NG film as an electrode for biological applications, its performance in the detection of hydrogen peroxide (H_2O_2) was compared to that of graphene film. H_2O_2 is the product of the enzymatic detection of glucose in diabetes diagnosis and its electrochemical detection is implemented in a commercial glucometer. Moreover, H_2O_2 is involved in different signal transduction pathways and cell fate decisions. The "redox signaling" mechanism includes the H_2O_2-mediated reversible oxidation of redox sensitive cysteine residues in enzymes and transcription factors, thereby altering their activities. In comparison to normal cells, cancer cells are characterized by an increased H_2O_2 production rate and an impaired redox balance, thereby affecting the microenvironment as well as the antitumoral immune response [14]. There is consequently a strong demand for hydrogen peroxide detection in the cell environment.

2. Materials and Methods

2.1. Graphene Film Synthesis and Characterization

Figure 1 is a schematic diagram of NG film synthesis. Nickel film (150 nm thick) was deposited by thermal evaporation on the top of an n-doped silicon substrate, previously cleaned (in acetone then in ethanol and DI water baths) in a vacuum chamber pumped at a base pressure of 10^{-6} mbar. High purity (99.99%) Ni was molten thermally in a tungsten nacelle and evaporated towards the substrate. The deposition rate was set at 1.5 nm/min to minimize residual stress in the growing film, thereby limiting film delamination. The Ni/Si samples were introduced in the PLD chamber pumped at a base pressure of 10^{-6} mbar, annealed at 780 °C for 30 min to increase the Ni grain size. While maintaining the temperature of 780 °C, carbon was ablated from a high purity graphite (99.9995%) target using a femtosecond laser (wavelength = 800 nm; pulse width = 60 ns, repetition rate = 1 kHz, energy density = 5 J/cm^2) at a temperature of 780 °C. The Ni/Si substrates were mounted on a sample holder placed at a distance of 36 mm from the graphite target. Nitrogen gas (99.9995% purity) was introduced as a reactant gas in the vacuum chamber at a pressure of 10^{-1} mbar.

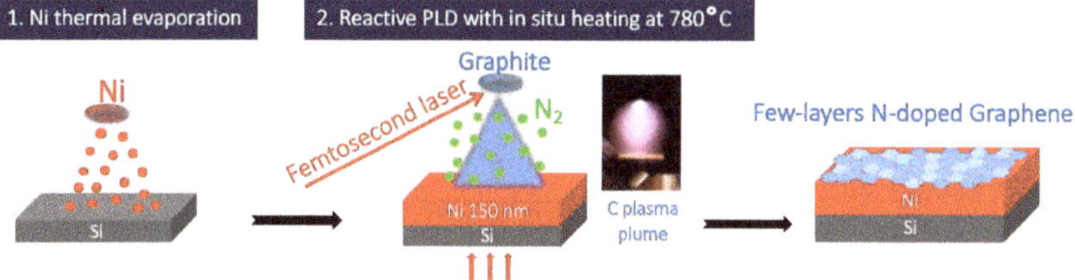

Figure 1. Synthesis of the N-doped graphene electrode.

Undoped graphene films were also synthetized using the same protocol, but without introducing nitrogen gas during carbon ablation. The ablation time was adjusted to keep both carbon and nitrogenated carbon film thicknesses equal to 40 nm. In some cases, the temperature of 780 °C was maintained for a defined period after deposition to observe the effect of longer heating periods. The procedure ended with natural cooling of the samples, before opening the vacuum chamber. In the present paper, we selected four deposition conditions to highlight the major effects due to nitrogen doping, compared to undoped films. The conditions are summarized in Table 1.

The morphology of the sample was observed using a field emission gun scanning electron microscope (FEG-SEM) Novananosem 200, (FEI, Hillsboro, OR, USA) operated at 15 kV. Micro-Raman analyses were performed using an Aramis spectrometer (Horiba Jobin Yvon, Gières, France), with 442 nm (2.81 eV) excitation laser focused through a ×100 objective with high aperture, ensuring the micrometric resolution of the analysis, and allowing for precise Raman mapping of the samples.

Table 1. Specific deposition parameters of the graphene films.

Graphene Films	N_2 Pressure during Deposition at 780 °C	Additional Period of Annealing at 780 °C after Deposition
NG-0	10^{-1} mbar	No additional annealing
NG-60	10^{-1} mbar	60 min
G-60	–	60 min
G-90	–	90 min

For each sample, $20 \times 20\ \mu m^2$ areas were scanned, recording a Raman spectrum every 2 μm (giving a total of 100 spectra per scanned area). The laser power was kept below 3 mW to avoid damaging the surface of the film. Raman components were associated with a Lorentzian fit, safe for the asymmetric G peak, which was fitted using a Breit-Wigner-Fano function. A custom fitting function was computed for all recorded spectra, and a simple linear function was added to eliminate the background. This enabled access to the exact values of the various peak positions, widths and maximum peak height intensity. XPS analysis was performed at the SOLEIL Synchrotron (Saclay, France) on the ANTARES beam line. The ring operating conditions were 2.5 GeV electron energy, with injection currents of 500 mA and "Top-up" mode. Radiation was monochromatized using a plane-grating monochromator (PGM), which is characterized by a slitless entrance and the use of two varied linear spacing (VLS) gratings with variable groove depth (VGD) along the grating lines. The diameter of the X-ray spot impinging the surface is 140 μm and the X-ray energy was fixed at 700 eV for analysis of the graphene films. The photoemission spectra were taken with incident photon energies of 700 and 350 eV, with 190 meV and 140 meV energy resolution, respectively.

2.2. Electrochemical Measurements

Electrochemical measurements were carried out in a conventional one compartment three electrode cell with an internal volume of 5 mL. This electrochemical cell was designed to maintain a fixed distance between the electrodes. It was manufactured with two inlets, one for positioning the reference electrode and the other for injecting H_2O_2. This feature prevented further manipulation or movements of the electrodes (fixing the geometry of the cell and also ensuring the reproducibility of measurements). For this work, a saturated calomel electrode from Hach Lange (Marne-la-Vallée, France) was used as the reference electrode, a planar platinum electrode ($0.59\ cm^2$) was used as the counter electrode and the nitrogen doped and undoped graphene samples were the working electrodes. The active surface of the working electrode, determined by a polyethylene terephthalate (PTFE) O-ring seal, was $0.07\ cm^2$. This three-electrode system was connected to a Bio-Logic potentiostat VMP2 (Bio-Logic Science Instruments, Seyssinet-Pariset, France). Results were recorded using EC-Lab software (v11.27) from Bio-Logic Science Instruments. In order to characterize the electron transfer rate for the different nitrogen doped and undoped graphene samples, cyclic voltammetry was performed in a 0.5 M 1,1′ ferrocene-dimethanol solution of 0.1 M $NaClO_4$. The scan rate was 100 mV/s. The detection of H_2O_2 in a non-deaerated 0.1 M phosphate buffer saline (PBS) solution (pH 7.4) was detected through linear sweep voltammetry in the anodic range from 0 to 1000 mV with a scan rate of 100 mV/s.

3. Results

Cyclic voltammetry measurements obtained with ferrocene methanol on pure graphene and NG films are presented in Figure 2, and the main results of the electrochemical measurements are listed in Table 2. The capacitive current appears to be higher with pure graphene film, due to the formation of more edge structures [15].

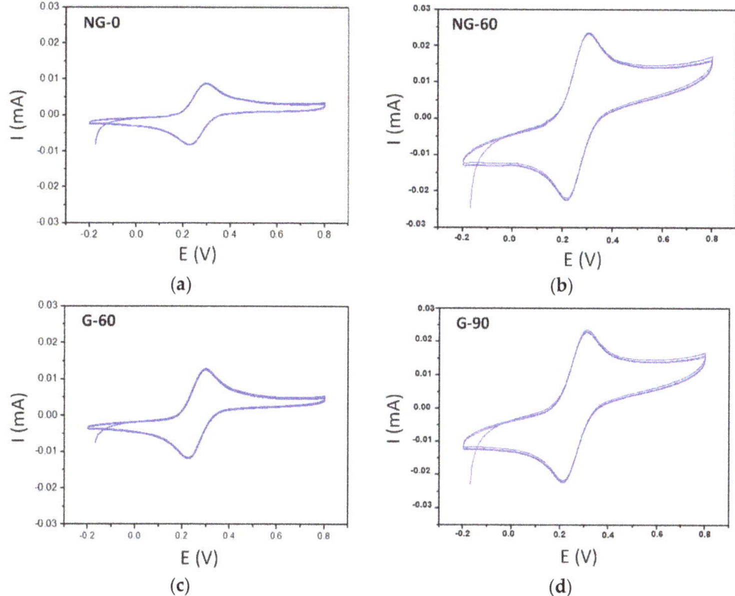

Figure 2. Cyclic voltammetry on (**a**) NG-0; (**b**) NG-60; (**c**) G-60 and (**d**) G-90 films, in a 0.5 M 1,1′ ferrocene-dimethanol solution of 0.1 M $NaClO_4$. The scan rate was 100 mV/s.

Table 2. Results of electrochemical measurements on NG and pure graphene films.

Graphene Films	Anodic Peak Intensity	ΔE between Anodic and Cathodic Peaks	Intensity for 500 mM H_2O_2
NG-0	4.0 μA	60 mV	1200 μA
NG-60	10 μA	78 mV	700 μA
G-60	5.7 μA	65 mV	5 μA
G-90	8.7 μA	82 mV	4 μA

The length of the annealing time following graphene growth had no significant effect on the intensity of the anodic peak of ferrocene methanol. The value of ΔE between anodic and cathodic peaks of ferrocene methanol increased with an increase in annealing time for both graphene and NG films. For the NG-0 film, in the absence of annealing following growth, the value of ΔE was close to the theoretical value of 59 mV, showing the high reversibility of the redox probe.

The oxidation of hydrogen peroxide began at a potential value of 600 mV for both types of films, as shown in Figure 3. The values of the oxidation intensities, reported in Table 2 for 500 mM of hydrogen peroxide, were measured at a potential value of 1 V. Our results show that the NG-0 film has excellent electrocatalytic properties without additional annealing after graphene growth, leading to the high reversibility of the ferrocene methanol redox probe and high sensitivity for hydrogen peroxide detection, with a detection limit of 1 mM and an oxidation intensity 240 times higher for 500 mM of H_2O_2 than undoped G-60 film.

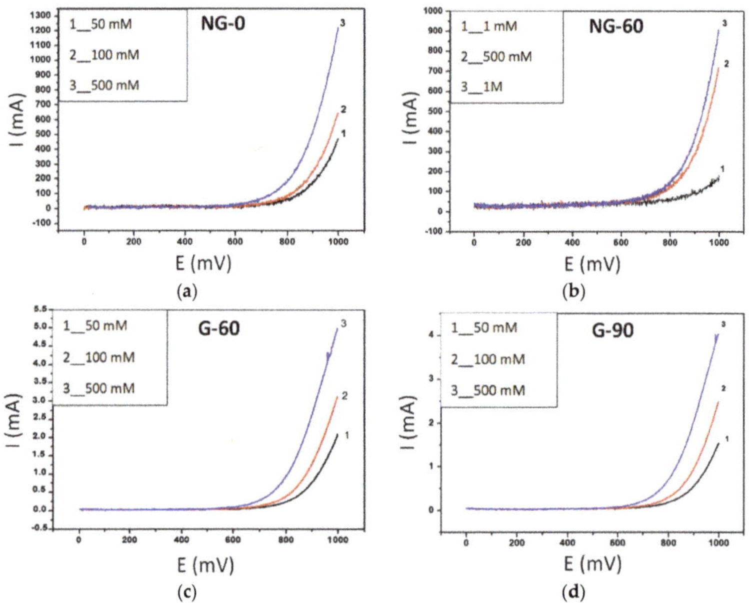

Figure 3. Linear sweep voltammetry of the (**a**) NG-0; (**b**) NG-60; (**c**) G-60 and (**d**) G-90 films, in the presence of different concentrations of H_2O_2 in 0.1 M PBS solution (pH 7.4).

The direct observation of the graphene-covered electrode by FEG-SEM (Figure 4) highlights a textured surface, with grain sizes in the range of 100 nm whatever the nature of the graphene film. Such a surface architecture is typical of graphene films synthetized on Ni films elaborated by thermal evaporation with subsequent Ni grain growth during the annealing process used to form the graphene films [9]. We do not observe any significant differences of morphology, with the FEG-SEM resolution at hand, between the four different graphene layers. Probably, the thermal heating during the PLD process is the crucial step inducing such a morphology, and the additional thermal annealing carried out on NG-60, G-60 and G-90 does not affect the surface morphology. Certainly, the nature and composition of the films, as deduced from XPS and Raman investigations, with such a significant difference between undoped and N-doped graphene films versus thermal annealing, is worth underlining. XPS was carried out on the NG-0 film obtained without post-annealing, given the significant electroanalytical result related to hydrogen peroxide oxidation. In Figure 5A, the XPS survey spectrum of the NG-0 film shows carbon located near 284 eV, nitrogen located near 400 eV, and some traces of oxygen near 533 eV.

The N/C intensity ratio deduced from the spectra was 0.01786, which is consistent with a nitrogen doping of 1.75 at %. Figure 5B shows deconvolution of the C1s into three components. The most intense component was centered at 284.4 eV and was attributed to sp^2 hybridized C atoms in graphene. The two other ones were located at 284.9 and 285.8 eV, respectively. The component at 284.9 eV was attributed to disordered carbon (C_B) and may be considered as an intermediate state between sp^2 and sp^3 hybridizations that can be found in nano-diamond and amorphous carbon films, in agreement with References [16–18].

The component at 285.8 eV was attributed to C–N or C–O bonds [19,20]. However, it is known that the peak of C–O is overlaid with C–N; therefore, it may be difficult to discriminate between the C–N and C–O oxygen group [4,9,12].

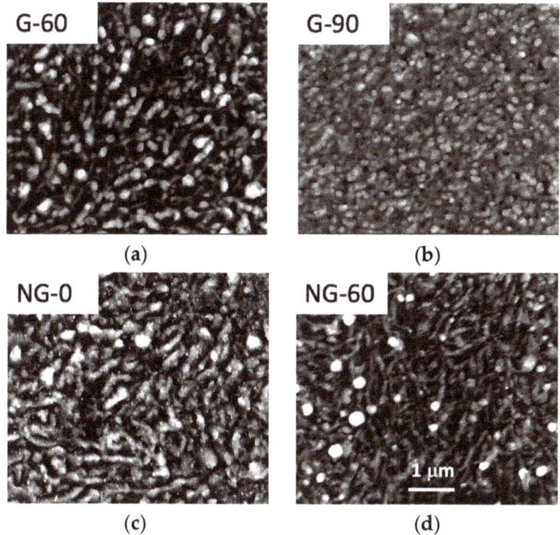

Figure 4. FEG-SEM images of (**a**) pure graphene with 60 mn post-deposition annealing; (**b**) pure graphene with 90 mn post-deposition annealing; (**c**) N-doped graphene with no post-deposition annealing; (**d**) N-doped graphene with 60 mn post-deposition annealing. The sub-micrometer texture was attributed to the texturing of the Ni catalyst film caused by thermal annealing. The four images are depicted with the same magnification 1 μm as noted on the image related to NG-60.

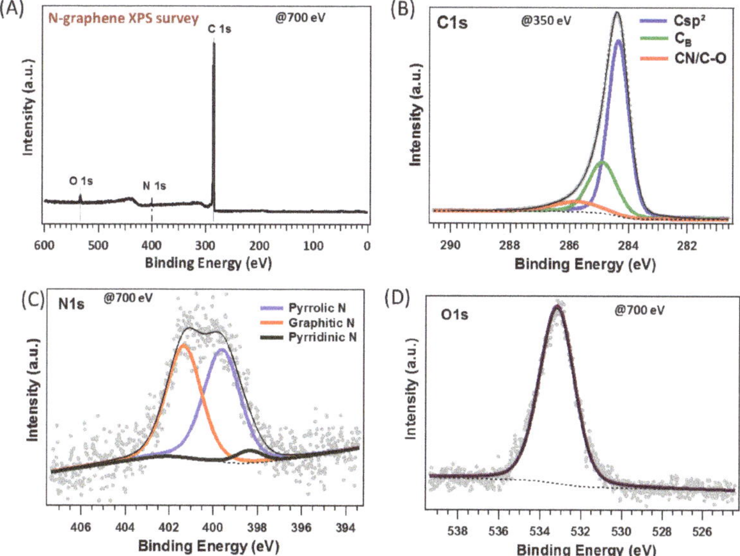

Figure 5. XPS spectra of the N-doped NG-0 graphene film; (**A**) XPS 700 eV overview spectrum; (**B**) XPS 350 eV C1*s* core level spectrum; (**C**) XPS 700 eV N1*s* core level spectrum; and (**D**) XPS 700 eV O1*s* core level spectrum.

Raman results related to the undoped and doped graphene films are shown in Figure 6. All the samples had the bands traditionally found in Raman spectrometry graphene materials; these results are in agreement with those reported in References [21,22]. For graphene films, the D, G and 2D bands are the most significant for the characterization of the thin film structure. The G band is associated with covalent C–C bonding vibrations in the graphite matrix and is present in every carbon material containing sp^2 bonding. The D band is associated with the pulsation of aromatic circles ("breathing mode") and appears only in the presence of defects and dislocations in the graphitic matrix. The intensity ratio between the D band and G band (D/G) is thus an indication of disorder in the carbon structure. The 2D band is associated with a double resonance Raman scattering process between two aromatic circles. It appears in both graphene and graphite, and the intensity ratio of the 2D band versus the G band (2D/G) is a good indicator of the graphene-like quality of a thin film, a ratio higher than 1 being indicative of monolayer graphene. In non-defective graphene, the study of the 2D peak position and width is also a good way to count the number of layers and to identify their stacking configuration [22,23]. Additional D + D″, D + D′ and 2D′ bands are also observed in Figure 6. The D + D″ and 2D′ peaks are, as the G and 2D peaks, usual features in most graphene samples [24]. They emerge, like the 2D peak, as a combination of two phonon mode individually associated with defects (D′ and D″) allowing so-called breathing of aromatic rings in carbon materials. The combination of those resonances can appear without defects as the two phonons can verify momentum conservation provided they have opposite wavevectors. In the case of the D + D′ band, also sometimes labelled as D + G, the comprehension of excitation mechanisms remains rather unclear, but whether it is a combination of D and G phonons or D and D′ does not change the fact that respectively one or both of the phonons need a defect to arise. Thus, the D + D′ band mostly appears in defective graphene-like material [25], which explains its presence in our materials.

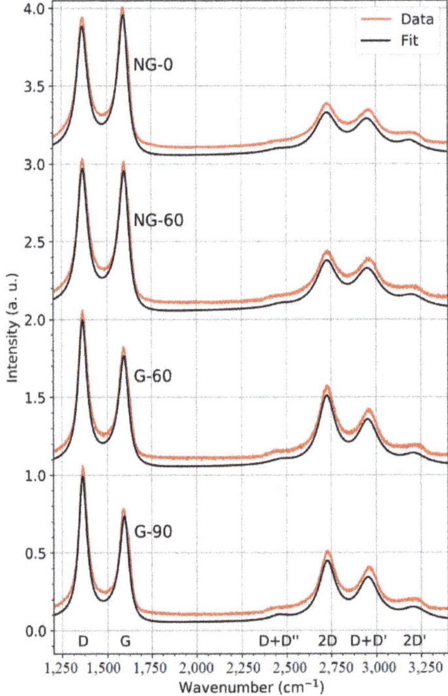

Figure 6. Typical Raman spectra of undoped and N-doped graphene films. The temperature during PLD graphene synthesis was 780 °C in all cases. Post-annealing times (min) are indicated in parentheses.

All the Raman spectra shown in Figure 6 present a higher D band compared to the G band, implying a very defective nature of the thin films. The 2D peak maximum was always lower than the G peak maximum, which may be associated with the multilayer nature of the graphene. A reduction in the 2D band in N-doped samples should also be noted; this is expected when nitrogen is introduced in the graphene matrix [13], although one would also expect an increase in the D/G ratio [26], which was not the case here. Typical maps of peak ratios, like the one in Figure 7 related to the multilayer pure graphene sample obtained with 90 mn post-annealing, enable evaluation of the uniformity of the synthesized thin films. In the case of pure graphene, a relative lack of uniformity appears in both the 2D/G and D/G ratios, with respective variations from 0.5 to 0.9 and from 1.0 to 1.3. The lack of homogeneity at the micrometric scale is consistent with the defective nature of the film. A slight correlation between areas with low D/G ratio and low 2D/G ratio can be observed, consistent with a slightly higher number of defects in areas where there are fewer graphene layers.

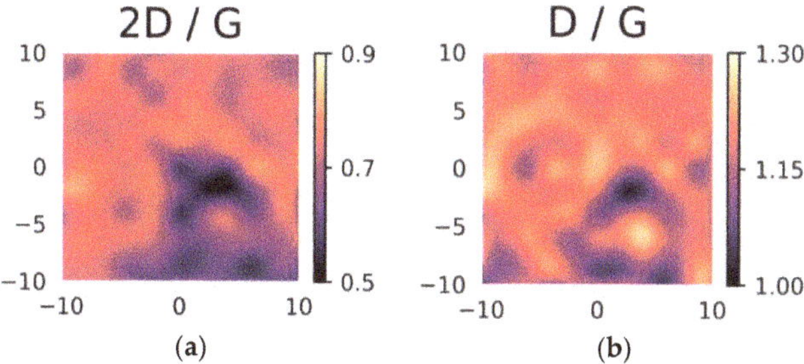

Figure 7. Maps of the Raman (**a**) 2D/G and (**b**) D/G intensity ratios recorded at the 442 nm excitation wavelength on a 20 × 20 μm^2 area of the pure graphene film annealed at 780 °C for 90 min (G-90). *X-Y* scales in μm.

Figure 8 shows the impact of incorporating nitrogen in the sample synthesized at 780 °C with no post-annealing. The intensity ratio of the 2D and G peaks exhibit considerably less variation than in Figure 7, whereas the D/G intensity ratios still show high variability (note that in Figures 7 and 8 the scales of the color bars are not the same) but are generally lower. This may mean that incorporating nitrogen helps to stabilize the multilayer graphene on the substrate, giving it a more organized structure.

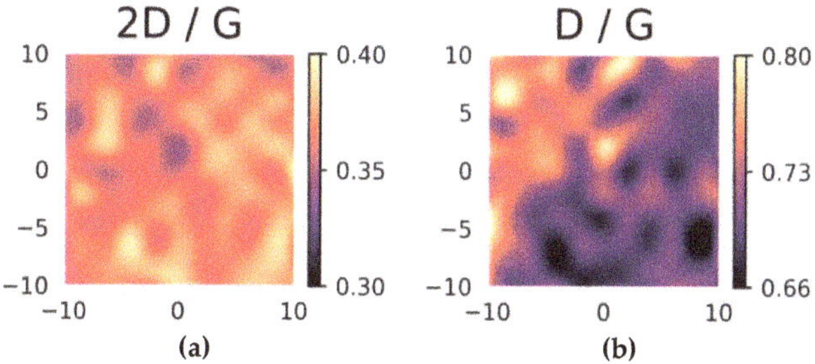

Figure 8. Maps of the Raman (**a**) 2D/G and (**b**) D/G intensity ratios recorded at an excitation wavelength of 442 nm on a 20 × 20 μm^2 area of the N-doped graphene film with no post-deposition annealing (NG-0). *X-Y* scales in μm.

The Raman findings are summarized in Table 3 where the average fit parameters obtained for scanned areas relate to the four samples. As mentioned above, we focused on the D, G and 2D peak parameters. The 2D/G and D/G intensity ratios further confirm the information provided by Figure 6.

Pure graphene samples exhibited higher ratios of both the 2D/G and D/G bands than the N-doped graphene samples.

Table 3. Average Raman fit parameters and parameter ratios for four undoped and doped multilayer graphene samples with different post-annealing durations. The standard deviation for each parameter is in parentheses.

Samples	Intensity Ratio (Standard Deviation)		Peak Position (cm^{-1}) (Standard Deviation (cm^{-1}))			Peak Full Width Half Maximum (cm^{-1}) (Standard Deviation (cm^{-1}))		
	2D/G	D/G	D	G	2D	D	G	2D
NG-0	0.369 (0.011)	0.722 (0.032)	1365.6 (0.7)	1589.8 (1.7)	2721.7 (1.9)	78.2 (2.0)	73.9 (0.8)	139 (3.2)
NG-60	0.386 (0.011)	0.903 (0.026)	1366.8 (0.6)	1595.0 (1.00)	2724.4 (1.9)	69.5 (2.1)	71.0 (1.4)	129.0 (5.0)
G-60	0.659 (0.036)	1.242 (0.029)	1365.6 (0.8)	1594.5 (1.0)	2727.0 (1.7)	59.5 (1.4)	68.1 (2.1)	111.1 (2.6)
G-90	0.712 (0.060)	1.183 (0.042)	1365.9 (0.8)	1591.9 (1.7)	2726.2 (2.0)	60.4 (5.2)	66.8 (3.1)	108.4 (4.6)

It should be noted that, in the course of this study, further post-annealing durations were tested in addition to those presented here. It was impossible to obtain any kind of pure graphene material without at least 30 min of post-deposition annealing. However, it was possible to synthesize an N-doped graphene film without post-annealing, i.e., the substrate was left to cool naturally immediately after the a-C:N pulsed-laser deposition. We would like to underline that this possibility implies an effect of the nitrogen environment on the catalysis of graphene growth by Ni, which opens the way for a more rapid fabrication of N-doped graphene. This also implies that, despite the high temperature used for graphene synthesis using annealing on a nickel catalyst, the as-deposited carbon structure may have a strong influence on the catalytic process. Additionally, it appears that lengthening the post-annealing period is an advantage with pure few-layer graphene, as the peak D/G ratio decreased and the 2D/G ratio increased when post-annealing time was increased from 60 to 90 min. This is in contrast to the fact that for N-doped graphene films, post-annealing only appeared to increase the D/G ratio, pointing to a higher number of defects.

All the observed peaks appear to be more intense than those of monocrystalline non-defective graphite or graphene. The full width at half maximum (FWHM) of the G peak is generally around 15 cm^{-1} [22] while here it was 70 cm^{-1}. The width of this peak may be associated with the distortion of the C–C sp^2 bonding angle. This is to be expected due to the nanotexturing of our samples, as observed by FEG-SEM shown in Figure 4. The position of the peak was also always upshifted here compared to graphite or monolayers graphene, i.e., between 1590 and 1595 cm^{-1} compared to 1582 cm^{-1} [22].

These characteristics, combined with the high D versus G peak intensities ratio, are clear indicators of highly nano-crystallized graphene-like layers. The broadening of all peaks due to the nanotexturing of the substrate during annealing makes it impossible to draw clear conclusions concerning the precise number of graphene layers by studying the 2D band Full Width Half Maximum (FWHM).

4. Discussion

The objective of this section is to discuss the significant improvement of electroanalytical oxidation of H_2O_2 by the N-doped graphene films compared to undoped ones in more detail. What is the main effect among the defective nature of the N-doped graphene films, the dopant concentration or the nature/proportion of the nitrogen chemical functions in the graphene network?

According to the previous section, the morphologies of all films (doped and undoped) appear to be textured, probably due to the texturing of the nickel catalyst surface caused by thermal annealing. We previously showed that the formation of nickel silicide contributed to texturing when graphene was grown on a silicon substrate covered by a Ni thin film [27].

Likewise, all films have a defective structure, as shown by Raman investigations, and nitrogen doping at a concentration as low as 1.6 at % does not significantly influence the defective structure of graphene, which is probably inherent to the synthesis process, as already observed by Schiros et al., who compared CVD synthesis of graphene and N-doped graphene films [28]. Moreover, when extended beyond the N-doped graphene synthesis, heat treatments do not cause significant changes to the nanoarchitecture of graphene film in terms of texture and number of graphene layers. Based on the comparison of the 2D/G and D/G ratio maps, the textured films are heterogeneous and the concentration of defects was slightly higher in areas where there are fewer layers of graphene.

However, a huge difference in electroanalytical oxidation of H_2O_2 was observed between undoped and N-doped graphene films, and the difference was more pronounced when the heat treatment was limited during the PLD film growth in the presence of nitrogen gas. The in situ nitrogen doping process during PLD graphene growth described in the previous section, led to the formation of a few-layer graphene film containing 1.75 at % of nitrogen, with a similar proportion of pyrrolic-N and graphitic-N, and a negligible amount of pyridinic-N. According to the literature, in particular, in Wang et al. [4] and in Shao et al. [5], the presence of incorporated nitrogen induces a change in the Fermi level, which is responsible for the doping effect and opens the band gap of the graphene structure, thus enhancing electrochemical reactivity. The high level of electronic state density and the efficient quantity of free electrons in N-doped graphene facilitates H_2O_2 oxidation. In particular, carbon atoms adjacent to nitrogen dopants may have a substantially positive charge density to counterbalance the higher electronic affinity of N atoms, consistent with the increased adsorption of H_2O_2 involved in the oxidation reaction. Such a mechanism has also been shown by density functional theory (DFT) simulating the physisorption process of H_2O_2 onto graphene-based surfaces [29]. Additionally, the structural defects resulting from N doping increased the amount of unsaturated carbon atoms located at graphene edge sites, which appeared to be very active in reacting with oxygen containing groups. During H_2O_2 electrocatalytic oxido-reduction, the O–O bond in H_2O_2 was more easily broken at the surface of N-doped graphene because N doping induced the charge delocalization of graphene. With our NG-0 film, in the absence of annealing following growth, the oxidation of H_2O_2 began at a potential value of 600 mV. This value is rather high compared to that reported by Wang et al. (200 mV), who observed a four times higher H_2O_2 oxidation signal with the NG films compared to the undoped ones [4]. Our results show that the NG-0 film presents excellent electrocatalytic properties leading to the high reversibility of the ferrocene methanol redox probe and high sensitivity for hydrogen peroxide detection, with a detection limit of 1 mM. A 240 times higher H_2O_2 oxidation signal was observed for the NG-0 film compared to the average value obtained with the undoped graphene films.

Such high electroanalytical reactivity is generally attributed both to chemical functionalization (both pyrrolic, pyridinic and graphitic nitrogen-carbon forms are generally reported in the literature) and to structural defects caused by nitrogen atoms. However, based on our experiments, the nature and proportions of the various N–C chemical functions certainly play a more significant role in the electroanalytical oxidation of H_2O_2 than the number of defects, which is similar with or without N doping. Such an influence was already highlighted by Xu et al. [3]. According to these authors, the relationship between the nature of N functions and the properties of N-doped graphene needs to be clarified, so that more desirable properties for specific applications can be selected. Based on experimental and theoretical (DFT calculations) considerations [28,30], the nature and level of doping in N-doped graphene depends on the proportion of the three main functional groups: graphitic-N is responsible for an n-doping effect and pyridinic-N and pyrrolic-N are responsible for a p-doping effect. The balance between the three chemical functions may strongly affect the electroanalytical performance of the N-doped graphene. More precisely, nitrogen in the graphitic-N configuration has 4 of the 5 electrons filling in the σ- and π-orbitals, leaving one extra electron. About 50% of the additional charge is localized on the N dopant coupled with its nearest carbon neighbors, whereas the remaining 50% is distributed in the local network of carbon π-states, thus inducing n-doping and preserving high electron mobility. Pyridinic-N and pyrrolic-N dopants have the opposite electronic effect, as they withdraw charge from their carbon neighbors. In the case of pyridinic-N, two electrons fill σ bonds with carbon neighbors, and two electrons form a lone pair in the graphene plane. The remaining electron occupies the nitrogen π-state. As a consequence, pyridinic-N is the equivalent of a nominal carbon in graphene, but an π electron is missing due to the vacant site, hence p-doping the graphene. Our NG-0 film contained 49% of graphitic-N and 47% of pyrrolic-N, whereas pyridinic-N contents were as low as 4%. With a 1.35 at % N content close to our doping concentration (1.75 at %), Wang et al. [4] obtained a nitrogen-doped graphene film with significantly higher proportions of pyrrolic-N and pyridinic-N and a lower proportion of graphitic-N than we obtained. We conclude that our nitrogen doping process induced a rather higher proportion of mixed graphitic-N and pyrrolic-N than other N-doped graphene films reported in the literature, and this may be responsible for the very high electroanalytical H_2O_2 oxidation performance. However, our results do not follow the reactivity scale simulated by DFT by Wu et al. [29] who concluded that H_2O_2 oxidation reactivity occurred in the following order: pyridinic-N > Pyrrolic-N > Graphitic-N > Pristine graphene. However, it is difficult to compare DFT calculations performed with only one nitrogen-based function, even if true, with an experimental system comprised of a mixture of nitrogen-based functions embedded in a textured few-layer graphene material. Further studies are thus recommended to optimize N-doped graphene films with better control of the N-based chemical functions, in particular with distinct proportions of the various N-based functions, in order to confirm their effect as key factors for electrochemical applications of N-doped graphene films.

5. Conclusion

Here we report on experimental work on the electroanalytical performance of N-doped graphene silicon-based electrodes, obtained in one step by pulse laser deposition of carbon performed in a vacuum at high temperature, in the presence of a surrounding nitrogen atmosphere. The main conclusions are the following:

- The electrode is covered by a few-layer defective and textured N-doped graphene film, containing 1.75 at % of nitrogen distributed in both graphitic-N and pyrrolic-N chemical functions at similar proportions.

- With the ferrocene methanol system, the electrode displays excellent reversibility, 60 mV, close to the theoretical value of 59 mV, and very high sensitivity for hydrogen peroxide oxidation characterized by an intensity 240 times higher than that obtained with undoped graphene synthetized using the same process, and a detection limit of 1 mM of hydrogen peroxide.

- These significant electroanalytical results are correlated with the amount of N doping and with the proportion of both graphitic-N and pyrrolic-N chemical functions incorporated into the defective and textured few-layer graphene film.

- Additional heat treatment following the deposition process does not significantly modify the nanoarchitecture of the N-doped graphene films and slightly decreases the electroanalytical performance in terms of reversibility and hydrogen peroxide oxidation performance, compared to the N-doped films obtained with no additional heating.

Further works are recommended to achieve better control of the different N-based chemical functions embedded in the graphene network and to quantify their effects on the electroanalytical performances of N-doped graphene electrodes.

Author Contributions: Conceptualization, C.D., N.J.-R., C.C., M.-C.A. and F.G.; Methodology, F.B., A.-S.L. and C.F.; Investigation, Y.B., G.R., B.C.-M., J.A., M.S. and I.R.; Resources, V.B.; Writing–Original Draft Preparation, C.D., F.B., Y.B. and N.J.-R.; Writing–Review & Editing, C.D.; Visualization, C.D.; Supervision, C.D., F.G. and N.J.-R..; Project Administration, F.B. and C.D.

Funding: This research was funded by the program PEPS "Risques et Environnement 2016" of Université de Lyon–CNRS ("PEPS 3D-GraPS"), and by the LABEX MANUTECH-SISE (ANR-10-LABX-0075) of Université de Lyon, within the program "Investissements d'Avenir" (ANR-11-IDEX-0007) operated by the French National Research Agency (ANR). This research was partly funded by CAMPUS FRANCE, through PHC Maghreb

#39382RE. Maryam Shokouhi thanks the ministry of Science, Research and Technology, Islamic Republic of Iran-Department of Scholarships and Overseas Students' Affairs for the grant.

Conflicts of Interest: The authors declare no conflict of interest.

References

1. Lu, W.; Soukiassian, P.; Boeckl, J. Graphene: Fundamentals and functionalities. *MRS Bull.* **2012**, *37*, 1119–1124. [CrossRef]
2. Yadav, R.; Dixit, C.K. Synthesis, characterization and prospective applications of nitrogen-doped graphene: A short review. *J. Sci. Adv. Mater. Dev.* **2017**, *2*, 141–149. [CrossRef]
3. Xu, H.; Ma, L.; Jin, Z. Nitrogen-doped graphene: Synthesis, characterizations and energy applications. *J. Energy Chem.* **2018**, *27*, 146–160. [CrossRef]
4. Wang, Y.; Shao, Y.; Matson, D.W.; Li, J.; Lin, Y. Nitrogen-doped graphene and its application in electrochemical biosensing. *ACS Nano* **2010**, *4*, 1790–1798. [CrossRef] [PubMed]
5. Shao, Y.; Zhang, S.; Engelhard, M.H.; Li, G.; Shao, G.; Wang, Y.; Liu, J.; Aksay, I.A.; Lin, Y. Nitrogen-doped graphene and its electrochemical application. *J. Mater. Chem.* **2010**, *20*, 7491–7496. [CrossRef]
6. Li, R.; Liu, L.; Bei, H.; Li, Z. Nitrogen-doped multiple graphene aerogel/gold nanostar as the electrochemical sensing platform for ultrasensitive detection of circulating free DNA in human serum. *Biosens. Bioelectron.* **2016**, *79*, 457–466.
7. Saengsookwaow, C.; Rangkupan, R.; Chailapakul, O.; Rodthongkum, N. Nitrogen-doped multiple graphene-polyvinylpyrrolidone/gold nanoparticles modified electrode as a novel hydrazine sensor. *Sens. Actuators* **2016**, *227*, 524–532. [CrossRef]
8. Li, X.; Zhao, H.; Shi, L.; Zhu, X.; Lan, M.; Zhang, Q.; Fan, Z.H. Electrochemical sensing of nicotine using screen-printed carbon electrodes modified with nitrogen-doped graphene sheets. *J. Electroanal. Chem.* **2017**, *784*, 77–84. [CrossRef]
9. Fortgang, P.; Tite, T.; Barnier, V.; Zehani, N.; Maddi, C.; Lagarde, F.; Loir, A.-S.; Jaffrezic-Renault, N.; Donnet, C.; Garrelie, F.; et al. Robust electrografting on self-organized 3D graphene electrodes. *ACS Appl. Mater. Interfaces* **2015**, *8*, 1424–1433. [CrossRef] [PubMed]
10. Bleu, Y.; Bourquard, F.; Tite, T.; Loir, A.-S.; Maddi, C.; Donnet, C.; Garrelie, F. Review of graphene growth from a solid carbon source by pulsed laser deposition (PLD). *Front. Chem.* **2018**, *6*, 574. [CrossRef] [PubMed]
11. Kumar, S.R.; Nayak, P.K.; Hedhili, M.N.; Khan, M.A.; Alshareef, H.N. In situ growth of p and n-type graphene thin films and diodes by pulsed laser deposition. *Appl. Phys. Lett.* **2013**, *103*, 192109. [CrossRef]
12. Ren, P.; Pu, E.; Liu, D.; Wang, Y.; Xiang, B.; Ren, X. Fabrication of nitrogen-doped graphenes by pulsed laser deposition and improved chemical enhancement for Raman spectroscopy. *Mater. Lett.* **2017**, *204*, 65–68. [CrossRef]
13. Maddi, C.; Bourquard, F.; Barnier, V.; Avila, J.; Asensio, M.C.; Tite, T.; Donnet, C.; Garrelie, F. Nano-architecture of nitrogen-doped graphene films synthesized from a solid CN source. *Sci. Rep.* **2018**, *8*, 3247. [CrossRef] [PubMed]
14. Szatrowski, T.P.; Nathan, C.F. Production of large amounts of hydrogen peroxide by human cells. *Cancer Res.* **1991**, *51*, 794–798. [PubMed]
15. Zribi, B.; Castro-Arias, J.M.; Decanini, D.; Gogneau, N.; Dragoe, D.; Cattoni, A.; Ouerghi, A.; Korri-Youssoufi, H.; Haghiri-Gosnet, A.M. Large area graphene nanomesh: An artificial platform for edge-electrochemical biosensing at the sub-attomolar level. *Nanoscale* **2016**, *8*, 15479–15485. [CrossRef] [PubMed]
16. Blume, R.; Rosenthal, D.; Tessonnier, J.-P.; Li, H.; Knop-Gericke, A.; Schlög, R. Characterizing graphitic carbon with X-ray photoelectron spectroscopy: A step-by-step approach. *ChemCatChem* **2015**, *7*, 2871–2881. [CrossRef]
17. Weathertrup, R.S.; Bayer, B.C.; Blume, R.; Ducati, C.; Baehtz, C.; Schlögl, R.; Hofmann, S. In situ characterization of alloy catalysts for low-temperature graphene growth. *Nano Lett.* **2011**, *11*, 4154–4160. [CrossRef] [PubMed]
18. Ma, K.; Tang, J.; Zou, Y.; Ye, Q.; Zhang, W.; Lee, S. Photoemission spectroscopic study of nitrogen-incorporated nanocrystalline diamond films. *Appl. Phys. Lett.* **2007**, *90*, 92105. [CrossRef]

19. Malitesta, C.; Losito, I.; Sabbatini, L.; Zambonin, P.G. New Findings on Polypyrrole Chemical Structure by XPS Coupled to Chemical Derivatization Labelling. *J. Electron Spectrosc. Relat. Phenom.* **1995**, *76*, 629–634. [CrossRef]

20. Ganguly, A.; Sharma, S.; Papakonstantinou, P.; Hamilton, J. Probing the thermal deoxygenation of graphene oxide using high-resolution in situ X-ray-based spectroscopies. *J. Phys. Chem. C* **2011**, *115*, 17009–17019. [CrossRef]

21. Ferrari, A.C.; Meyer, J.C.; Scardaci, V.; Casiraghi, C.; Lazzeri, M.; Mauri, F.; Piscanec, S.; Jiang, D.; Novoselov, K.S.; Roth, S.; et al. Raman spectrum of graphene and graphene layers. *Phys. Rev. Lett.* **2006**, *97*, 187401. [CrossRef] [PubMed]

22. Malard, L.M.; Pimenta, M.A.; Dresselhaus, G.; Dresselhaus, M.S. Raman spectroscopy of graphene. *Phys. Rep.* **2009**, *473*, 51–87. [CrossRef]

23. Park, J.S.; Reina, A.; Saito, R.; Kong, J.; Dresselhaus, G.; Dresselhaus, M.S. G′ band Raman spectra of single, double and triple layer graphene. *Carbon* **2009**, *47*, 1303–1310. [CrossRef]

24. Ferrari, A.C.; Basko, D.M. Raman spectroscopy as a versatile tool for studying the properties of graphene. *Nat. Nanotechnol.* **2013**, *8*, 235–246. [CrossRef] [PubMed]

25. Iqbal, M.W.; Singh, A.K.; Iqbal, M.Z.; Eom, J. Raman fingerprint of doping due to metal adsorbates on graphene. *J. Phys. Condens. Matter* **2012**, *24*, 335301. [CrossRef] [PubMed]

26. Zhang, C.; Fu, L.; Liu, N.; Liu, M.; Wang, Y.; Liu, Z. Synthesis of nitrogen-doped graphene using embedded carbon and nitrogen sources. *Adv. Mater.* **2011**, *23*, 1020–1024. [CrossRef] [PubMed]

27. Tite, T.; Barnier, V.; Donnet, C.; Loir, A.-S.; Reynaud, S.; Michalon, J.-Y.; Vocanson, F.; Garrelie, F. Surface enhanced Raman spectroscopy platform based on graphene with one-year stability. *Thin Solid Films* **2016**, *604*, 74–80. [CrossRef]

28. Schiros, T.; Nordlund, D.; Pálová, L.; Prezzi, D.; Zhao, L.; Kim, K.S.; Wurstbauer, U.; Gutiérrez, C.; Delongchamp, D.; Jaye, C.; et al. Connecting dopant bond type with electronic structure in N-doped graphene. *Nano Lett.* **2012**, *12*, 4025–4031. [CrossRef] [PubMed]

29. Wu, O.; Du, P.; Zhang, H.; Cai, C. Microscopic effects of the bonding configuration of nitrogen-doped graphene and its reactivity toward hydrogen peroxide reduction reaction. *Phys. Chem. Chem. Phys.* **2013**, *15*, 6920–6928. [CrossRef] [PubMed]

30. Marsden, A.J.; Brommer, P.; Mudd, J.J.; Dyson, M.A.; Cook, R.; Asensio, M.-C.; Avila, J.; Levy, A.; Sloan, J.; Quigley, D.; et al. Effect of oxygen and nitrogen functionalization on the physical and electronic structure of graphene. *Nano Res.* **2015**, *8*, 2620–2635. [CrossRef]

Article

Effect of Reaction Temperature on Structure, Appearance and Bonding Type of Functionalized Graphene Oxide Modified *P*-Phenylene Diamine

Hong-Juan Sun [1,*,†], Bo Liu [2,*,†], Tong-Jiang Peng [1] and Xiao-Long Zhao [1]

[1]   Key Laboratory of Ministry of Education for Solid Waste Treatment and Resource Recycle, Southwest University of Science and Technology, Mianyang 621010, China; tjpeng@swust.edu.cn (T.-J.P.); skygraphene@163.com (X.-L.Z.)

[2]   School of National Defense Science and Technology, Southwest University of Science and Technology, Mianyang 621010, China

*   Correspondence: sunhongjuan@swust.edu.cn (H.-J.S.); liuboswust@126.com (B.L.)

†   These authors contributed equally to this work.

Received: 23 March 2018; Accepted: 19 April 2018; Published: 23 April 2018

Abstract: In this study, graphene oxides with different functionalization degrees were prepared by a facile one-step hydrothermal reflux method at various reaction temperatures using graphene oxide (GO) as starting material and *p*-phenylenediamine (PPD) as the modifier. The effects of reaction temperature on structure, appearance and bonding type of the obtained materials were investigated by X-ray diffraction (XRD), Fourier transform infrared spectroscopy (FT-IR), X-ray photoelectron spectroscopy (XPS), and scanning electron microscopy (SEM). The results showed that when the reaction temperature was 10–70 °C, the GO reacted with PPD through non-covalent ionic bonds ($-COO^-H_3{}^+N-R$) and hydrogen bonds (C–OH ... H_2N-X). When the reaction temperature reached 90 °C, the GO was functionalized with PPD through covalent bonds of C–N. The crystal structure of products became more ordered and regular, and the interlayer spacing (*d* value) and surface roughness increased as the temperature increased. Furthermore, the results suggested that PPD was grafted on the surface of GO through covalent bonding by first attacking the carboxyl groups and then the epoxy groups of GO.

Keywords: graphene oxide; *p*-phenylene diamine; functionalized graphene oxide; cross-link bond type; bonding type

---

1. Introduction

Graphene is attracting increasing attention in physics, chemistry and material research due to its unique laminar crystal structure [1], high electrical conductivity [2], high thermal conductivity [3], excellent flexibility, and mechanical properties [4–7]. The chemical reduction of graphene oxide has been regarded as an effective way to achieve large-scale preparation of grapheme [8–10]. However, the addition of only hydrazine hydrates without other materials will likely result in aggregated graphene [11,12]. Therefore, the formation of monolithic and high-performance graphene will require further treatments besides the necessary reduction reaction.

Compared to graphene, graphene oxide (GO) possesses a large number of carboxyl groups (–COOH) near the edges, with many epoxy groups (C–O–C) and hydroxyl groups (C–OH) on the surface. These oxygen-containing groups increase the reaction activity of GO. Therefore, functionalization of GO can be achieved using various approaches, such as through hydrogen, ionic, and covalent bonding [13–15]. Hence, the structure of graphene can be regulated in order to modify its light, electrical, and magnetic properties [15,16].

In recent years, various amine based chemicals have been used to modify GO. These include ammonia [17], cetylamine [18], octadecylamine [19], ethidenediamine [20], poly-o-phenylenediamine [21], and amino acids [22]. For instance, Shanmugharaj et al. [23] used alkylamines with different chain lengths as modifiers and mixed them with GO under ultrasonic conditions at normal temperature, followed by suction filtration during the first reaction stage. Next, they subjected the suspension to vacuum drying treatment to yield functionally modified GO. They detected the formation of ionic, hydrogen, and covalent bonds between aliphatic amine and GO. The surface roughness degree of GO increased after modification, and roughness degree rose asalkylamine chain length was extended. Matsuo et al. [24] intercalated alkylamines with different chain lengths into GO, and identified three types of interactions between alkylamines in GO. In addition, the arrangement patterns of alkylamines with different chain lengths were different, yielding various interlamellar spacing. Hung et al. [25] used three amine monomers (ethidenediamine, butanediamine, and p-phenylenediamine) as cross-linking agents to prepare GO skeleton sheets with different interlamellar spacing using pressure-assisted self-assembling technology. During the modification process, amine monomer was combined with GO through chemical bonding to yield GO with significantly changed hydrophobicity. Furthermore, the cross-linking mode changed from non-aromatic cyclamine cross-linking to aromatic cyclamine cross-linking, and the composite film-water contact angle varied from 24.4° to 80.6°.

According to previous literature [26,27], GO was functionalized with amine by reacting at a temperature over 90 °C. However, the reason for selecting this temperature and the effects of reaction temperature on structure, appearance, and interaction mechanisms of FGO have not been reported so far. In this paper, FGO composites were prepared using a facile one-step hydrothermal reflux method under different reaction temperatures using GO as starting material and PPD as the modifier. The structures, functional groups and compositions of the prepared composites were investigated by XRD, Raman, SEM, FT-IR, and XPS analyses. The effects of reaction temperature on structure and appearance of modified composites were also examined. Finally, the mechanism of interaction between PPD and GO at different reaction temperatures was explored.

2. Experimental

2.1. Reagents

The following reagents were used in this study: natural flake graphite (Tangseng Gou, Xinghe County, Inner Mongolia, China, with carbon content $\geq$90%, screened by 200 mesh), potassium permanganate and concentrated sulfuric acid (Sinopharm, Shanghai, China), H_2O_2 solution (5%) and 0.05 mol·L^{-1} HCl solution (Chengdu Jinshan Chemical Reagent, Chengdu, China), p-phenylene diamine(PPD), and methyl alcohol ($\geq$99.5%, Chengdu Kelong Chemical Reagent Factory, Chengdu, China). All reagents were of analytical grade. Ultrapure water with a resistivity of >18.25 MΩ·cm was used for the experiments.

2.2. Preparation of Graphene Oxide (GO)

GO was prepared from natural graphite powder using the improved Hummers method. The detailed preparation method is described in our previous report [28].

2.3. Preparation of FGO

First, 0.2 g graphene oxide powder was added to 250 mL ultrapure water and stirred for 120 min under ultrasonic dispersion. This yielded a GO dispersion with concentration of 0.8 mg·mL^{-1}. Secondly, 0.4 g PPD was added to the GO dispersion, followed by 10 min of ultrasound mixing. Next, the solution was poured into a 500 mL 3-neck boiling flask and magnetically stirred in a water bath with refluxing at 10 °C for 24 h. Subsequently, the solution was filtered off using polypropylene (PP) thin film with mean pore size of 0.2 µm, and washed five times with ethanol and ultrapure water.

Finally, the PP thin film was dried at 80 °C for 24 h to obtain the product. The above processes were then repeated by changing the reaction temperature to 30, 50, 70, and 90 °C, respectively. The composite thin films obtained at different temperatures were denoted as, FGO-T (T = 10, 30, 50, 70, and 90), where T represents the temperature. All samples were immersed in methanol solution for 10 h and then dried at 80 °C for 2 h, resulting in samples denoted as FGOS-T (T = 10, 30, 50, 70, and 90).

2.4. Characterization

Nicolet-5700 infrared spectrometer (FT-IR, Thermo Nicolet Corporation, Madison, WI, USA) with a scanning range of 4000~500 cm^{-1} was used for bonding characterization using the KBrpellet method. XSAM800 multifunctional electron spectrometer was employed for surface analysis (XPS, Kratos Company, Manchester, UK), with Al target (1486.6 eV) and X-ray gun (12 KV × 15 mA, in FAT mode). The data were corrected using carbon contamination C1s, X'pert MPD Pro X-ray diffractometer (XRD, PANalytical B.V., Almelo, The Netherlands), with Cu target, DS: (1/2)°, SS: 0.04 rad, AAS: 5.5 mm, at the scan range from 5° to 45°. Raman spectroscopy was performed on a Renishaw InVia spectrograph (Wharton Andech, UK), with Ar$^+$ excitation source, wavelength of 514.5 nm, and scanning range from 400 to 4000 cm^{-1}. Microstructure analysis was conducted using Ultra 55 field emission scanning electron microscope (FE-SEM, Zeiss Instruments, Stuttgart, Germany).

3. Results and Discussion

3.1. Structural Changes of FGO at Different Reaction Temperatures

Figure 1 shows the XRD patterns and interlamellar spacing *d* values of GO and FGO-T. The *d* value of GO was determined as 0.86 nm while the *d* values of FGO-T (T = 10, 30, 50, 70, and 90) were 0.94, 0.94, 1.03, 1.04, and 1.07 nm, respectively. Compared to GO, the *d* value of FGO-T was larger and increased with the reaction temperature, showing different connection modes between PPD and GO at various temperatures. According to the results of Lu et al. [29], the XRD data and spacing values indicated that the PPD molecules were grafted into the layers of GO.

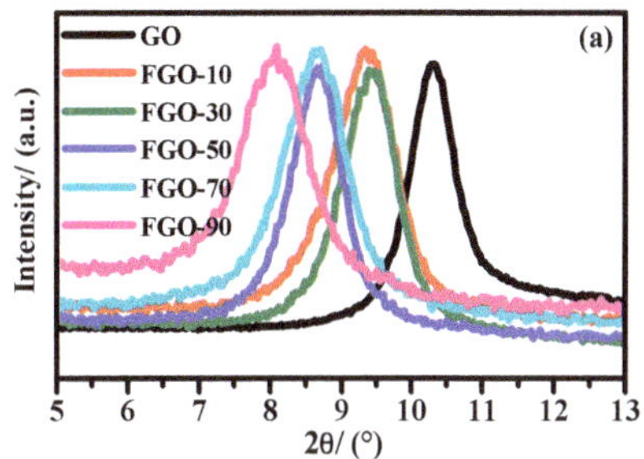
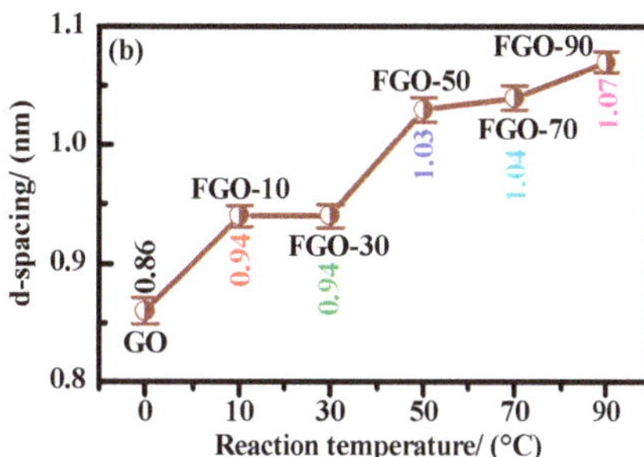

Figure 1. XRD patterns (**a**) and d-spacing (**b**) of GO and FGO-T samples.

Figure 2 depicts the Raman spectra of GO and FGO-T. In the first order Raman spectral region, both samples showed two major characteristic peaks at 1353 and 1599 cm^{-1}, corresponding to the D and G bands of graphene oxide structures, respectively. The D peak was attributed to the double resonance Raman scattering process near critical point K at graphene Brillouin zone, indicating the presence of structural defects [30], while the G peak was caused by E$_{2g}$ eigen vibration of sp^2 carbon domains [31]. As shown in Figure 2a, the G peaks of FGO-T were shifted from 1585 to 1599 cm^{-1}, suggesting that oxidation and PPD caused changes in the graphene structure. The ratio of the integral intensity between the D and G peaks (I_D/I_G) determines the disorder level in crystal structure [32]. With the increase in reaction temperature, the value of I_D/I_G basically presented a decreasing trend, indicating that the structural order in samples was partially restored.

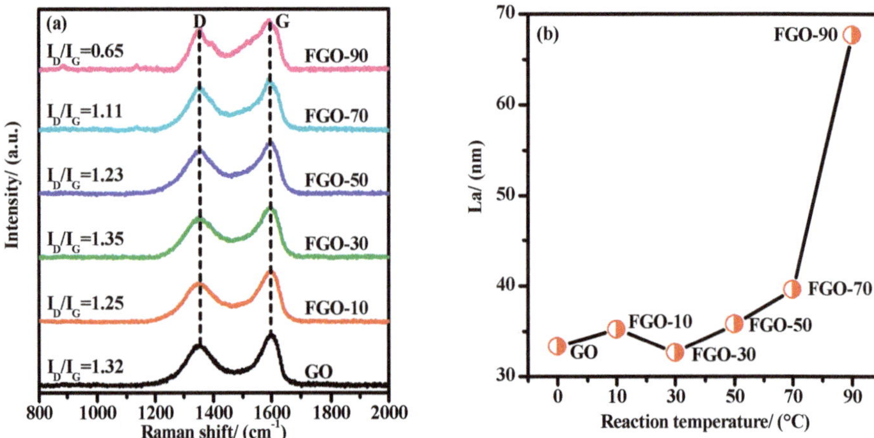

Figure 2. Raman spectra (**a**) and L_a values (**b**) of GO and FGO-T samples.

Furthermore, it has been empirically found that the disorder decreases the sp^2 plane domain (L_a) in product structures [33].The L_a value can be obtained with the formula of $L_a = (2.4 \times 10^{-10}) \times \lambda^4_{laser} (I_D/I_G)^{-1}$, where λ_{laser} is the excitation wavelength [34]. As shown in Figure 2b, compared to GO, the L_a value of FGO-T generally increased, indicating that addition of PPD molecules increased the order and the sp^2 plane in GO was restored gradually. Moreover, from 10~70 °C, the change in L_a values was relatively small, while over 70 °C, the change in L_a was the largest. This indicated that elevated temperature could promote the amidation reactions, and the optimum temperature for amidation reaction should be above 70 °C [35].

3.2. Effect of Reaction Temperature on FGO Appearance

The morphologies of GO and FGO-T (T = 10, 50, 70, and 90) are shown in Figure 3. The most significant difference between GO and FGO-T is their surface roughness. The GO surface appeared flat and smooth (Figure 3a,b). However, as shown in Figure 3c–f, when the GO interacted with PPD, the surface of FGO-T films became rough. PPD is a rigid structure and acts as a nanospace barrier, blocking the stacking of GO sheets. Therefore, the surface of FGO became rough and showed more cracks. The surface roughness was similar to that reported by Shanmugharaj et al. [23] for composites which were prepared by the reaction between GO and alkylamines of varying chain lengths. However, the effect of temperature on surface roughness was less than that of chain lengths. As the temperature increased, the surface roughness of FGO-T films became smaller. This can be attributed to the fact that the elevated temperature promoted amidation reactions instead of adsorption.

Figure 3g,h presents the TEM images of the FGO-50 sample. It can be seen clearly that some PPD particles were wrapped within or on the surface of graphene sheets. With the increase in reaction temperature (Figure 3i,j), the adsorbed PPD particles disappeared. The FGO-90 sample presented a wrinkled and transparent nanosheet with a lateral dimension of about several micrometers. The TEM results are consistent with the SEM data.

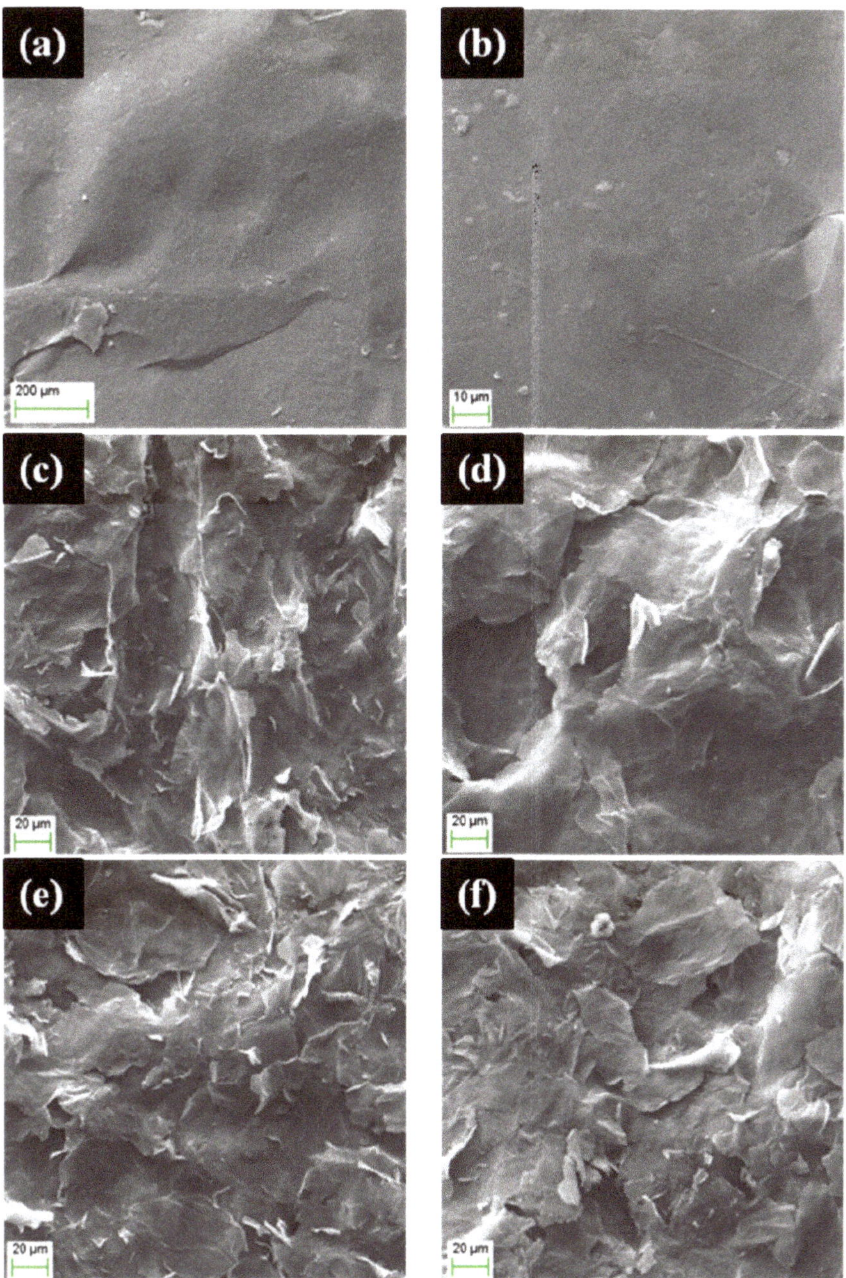

Figure 3. *Cont.*

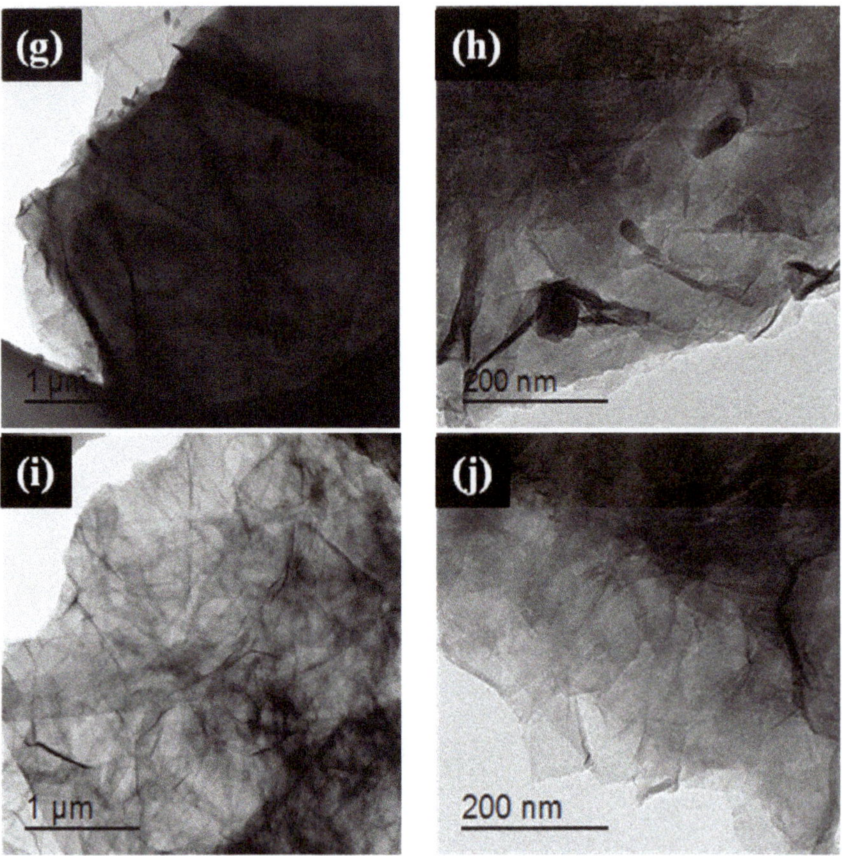

Figure 3. SEM images of (**a**,**b**) GO; (**c**) FGO-10; (**d**) FGO-50; (**e**) FGO-70; and (**f**) FGO-90; TEM images of (**g**;**h**) FGO-50 and (**i**,**j**) FGO-90.

In solution, the COOH at the edge of GO nanosheet will be deprotonated, making its lamellae negatively charged [36]. Due to the electrostatic repulsion between adjacent sheets and a large number of hydrophilic oxygen-containing functional groups, the GO can be dispersed steadily in water. This feature can be expressed by Equation (1):

$$R\text{-COOH} \rightleftharpoons R\text{-COO}^- + H^+ \tag{1}$$

After addition of PPD molecules, the activity of NH_2 in PPD molecule decreased at low temperatures. The attachment of PPD molecules between GO slice layers increased due to physical adsorption. On the other hand, in solution, some PPD molecules were protonated according to Equation (2):

$$R\text{-NH}_2 + H_2O \rightleftharpoons R\text{-NH}_3^+ + OH^- \tag{2}$$

Equations (1) and (2) indicate that when H^+ combines with OH^-, water molecule will be produced and the forward reaction will be favoured. Therefore, most GO slices will be negatively charged while protonated PPD molecule will be positively charged. Then, both will be connected through ionic bonding. In addition, the non-protonated NH_2 will connect with some oxygen-containing groups though non-covalent hydrogen bonding. With increase in reaction temperature, the activity of NH_2 in

PPD molecule will grow, and the ionic bonding between PPD and GO will strengthen. This would result in stacked and compact GO slices.

At the reaction temperature of 90 °C, the sample surface contained numerous pits but less scattered pieces. Compared to FGO-T at low temperature, the GO slice surface appeared more flat and smooth. This was probably due to the higher activity of –NH$_2$ in PPD molecule, which caused large amounts of PPD molecules to covalently react with the oxygen-containing groups in GO. This lowered the space barrier role of PPD for GO slices. However, some GO slices were cross-linked with PPD monomer, making the connection between slices more tight and smooth, hence increasing the order of the structure.

3.3. Influence of Temperature on Oxygen-Containing Functional Groups and Types of Interactions

Figure 4 shows the FT-IR spectra of GO and FGO-T before and after soaking in methyl alcohol. Numerous oxygen-containing groups were present in GO (Figure 4a). The absorption peaks at 3431, 2922, 2845, 1731, 1625, 1400, 1096 and 1038 cm^{-1} were attributed to the stretching vibration of hydroxyl (OH), CH$_2$ anti-symmetry and symmetry in benzene ring framework, carbonyl group (C=O), benzene ring skeleton (C=C), carboxyl group (O–C=O), epoxy group (C–O–C) and alkoxy group (C–O), respectively [37,38].

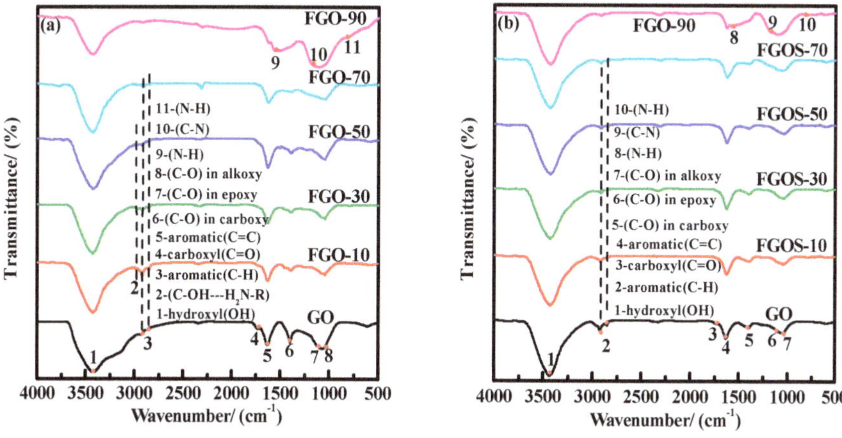

Figure 4. FT-IR spectra of GO and FGO-T samples: (**a**) without and (**b**) with methanol soaking.

By contrast, FGO-T (T = 10, 30, 50, and 70 °C) showed new absorption peaks at 2982 cm^{-1} and 833 cm^{-1}, which can be ascribed to the hydrogen-bond interaction between –NH$_2$ and oxygen-containing group in GO, and N–H bending vibration in PPD, respectively [39]. The new absorption peaks of FGO-90 at 1583 and 1180 cm^{-1} were ascribed to N–H stretching vibration and C–N stretching vibration, respectively [27,36].These results indicate the occurrence of covalent reaction between PPD molecule and GO at 90 °C.

Therefore, according to SEM analyses, the type of bond that would allow the PPD monomer to attach onto GO at temperatures from 10~70 °C can only be based on physical adsorption: (I) hydrogen-bond (C–OH ... H$_2$N–X) between oxygen-containing group of GO and NH$_2$ group of PPD molecule, and (II) ionic bond (–COO$^-$H$_3^+$N–R) between protonated PPD and weakly acidic GO. However, above 90 °C, NH$_2$ in PPD reacted with the oxygen-containing group of GO, resulting in C–N bond (Figure 5).

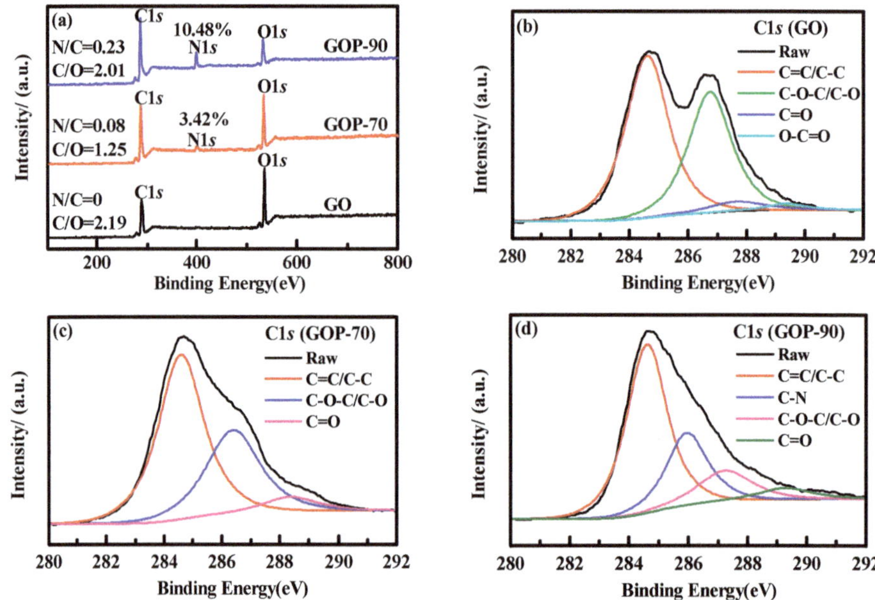

Figure 5. XPS data of GO and GOP samples: (**a**)contents of C, N and O, (**b**) C1s spectrum of GO, (**c**) C1s spectrum of GOP-70, and (**d**) C1s spectrum of GOP-90.

To further confirm the above hypothesis, FT-IR analysis of FGOS-T was performed and the results are presented in Figure 4b. Compared to Figure 4a, the absorption peaks of FGO-T (T = 10, 30, 50, 70) at 833 and 2982 cm^{-1} vanished. These peaks were ascribed to N–H bending vibration with hydrogen bonding between –NH$_2$ in PPD and the oxygen-containing group of GO, respectively. The absorption peaks of FGOS-90 at 1583, 1176 and 834 cm^{-1} attributed to N–H stretching vibration, C–N stretching vibration and N–H bending vibration, respectively, were consistent with those observed for FGO-90. In addition, compared to GO, the absorption peaks caused by O–C=O and C–O–C stretching vibrations vanished from the spectra. Therefore, it can be concluded that from 10~70 °C, PPD monomer interacted with GO through non-covalent hydrogen and ionic bonds. After soaking in methyl alcohol, the physical adsorptions in PPD were removed. At 90 °C, PPD monomer covalently reacted with GO, resulting in a C–N covalent bond that was still intact after being soaked in methyl alcohol.

To further verify the proposed hypothesis, XPS was performed on GO, FGO-70, and FGO-90 (Figure 5a). GO showed two spectral peaks at 286.0 and 535.0 eV, corresponding to C1s and O1s peaks, respectively [25,28]. By contrast, FGO-70 and FGO-90 displayed not only C1s and O1s peaks but also a new N1s peak near 401 eV, indicating the presence of N atom [27,36]. In addition, the relative percentage contents of N atom were estimated to be 3.42% and 10.48%, suggesting that N content in FGO-90 was higher than that in FGO-70. Figure 5b–d shows the peak-differentiating of GO, FGO-70, and FGO-90. GO revealed characteristic peaks at 284.6, 286.7, 287.9 and 289.4 eV, corresponding to C=C/C–C, C–O–C/C–O, O–C=O and C=O, respectively. FGO-70 displayed characteristic peaks of C=C/C–C, C–O–C/C–O, and C=O. Finally, FGO-90 showed not only the above characteristic peaks but also a new peak at 285.9 eV, ascribed to C–N bond [28,36].

The above analyses confirmed the absence of covalent C–N bond in FGO-70. In other words, from 10~70 °C, only non-covalent bonding existed between PPD and GO. However, covalent C–N bond formed at 90 °C. FT-IR and XPS data showed that the carboxyl (O–C=O) group in GO was removed by epoxy group (C–O–C) during the reaction process. This indicated that PPD preferred to react first with the carboxyl group and then with the epoxy group. The results of this study were slightly different from a previous report on amidation reactions. Xue et al. [40] found that the grafting of ethylenediamine on the surface of GO can be carried out under very mild conditions, even at 273 K, and the amine was grafted on the surface of GO mainly by a nucleophilic ring opening reaction between the amine and the epoxy group of GO. This may be explained as follows. Previous studies have suggested that benzene ring in PPD and its amine lone pair possess high resonance stability [25,28,36], reducing the reactivity of amine with epoxy groups. Moreover, the PPD's aromatic ring has large steric hindrance, which would not allow it to either attack the carbon of GO nanosheet or undergo a ring-opening reaction with epoxy at low temperatures. At 90 °C, the amine of PPD showed a higher activity. The NH_2 in PPD underwent amidation reaction with COOH at edge of GO, and nucleophilic substitution with the surface C–O–C groups. The types of interactions at different reaction temperatures based on SEM, FT-IR and XPS results are summarized in Figure 6.

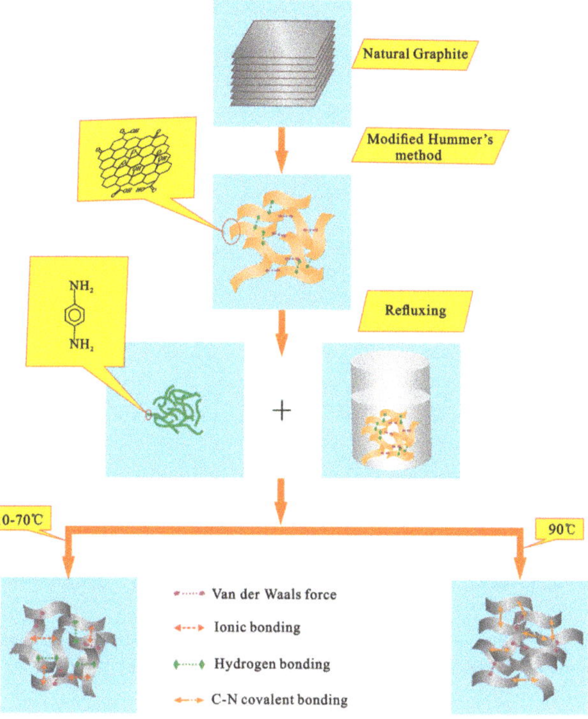

Figure 6. Schematic representation of reaction between GO and PPD.

4. Conclusions

In this work, FGO was synthesized via the reaction between GO and PPD through a simple one-step hydrothermal reflux method. The structures, functional groups and compositions of the products were investigated by XRD, Raman, SEM, FT-IR, and XPS analyses. The effects of reaction temperature on the structure of modified composites were also examined. It was found that the surface roughness and the d value of the product increased with the rise in reaction temperature. When the reaction temperature was low (10~70 °C), most of the PPD monomer reacted with GO through non-covalent ionic and hydrogen bonds. As the reaction temperature increased, the order of the crystal structure gradually improved. When the reaction temperature was high (>70 °C), grafting of PPD was the primary reaction. PPD was grafted on the surface of GO through covalent C–N bonding by first attacking the carboxyl groups and then the epoxy groups of GO.

Acknowledgments: This work was financially supported by the National Science Foundation of China (NSFC) (Grant Nos. 41772036 and U1630132) and the scientific research plan for Longshan academic talents supported by Southwest University of Science and Technology (Grant No. 17LZXT11).

Author Contributions: Hong-Juan Sun and Bo Liu contributed equally to this article. In this work, Hong-Juan Sun and Bo Liu designed the experiments, produced the nanomaterials, performed the SEM, TEM, and XPS characterizations and wrote the manuscript. Tong-Jiang Peng and Xiao-Long Zhao performed XRD, FTIR, and analyzed the results. The manuscript was corrected and improved by all authors.

Conflicts of Interest: The authors declare no conflict of interest.

References

1. Hu, T.; Gerber, I.C. Theoretical study of the interaction of electron donor and acceptor molecules with graphene. *J. Phys. Chem. C* **2017**, *117*, 2411–2420. [CrossRef]
2. Reina, A.; Jia, X.; Ho, J.; Nezich, D.; Son, H.; Bulovic, V.; Dresselhaus, M.S.; Kong, J. Layer area, few-layer graphene films on arbitrary substrates by chemical vapor deposition. *Nano Lett.* **2009**, *9*, 3087. [CrossRef]
3. Balandin, A.A.; Ghosh, S.; Bao, W.; Calizo, I.; Teweldebrhan, D.; Miao, F.; Lau, C.N. Superior thermal conductivity of single-layer graphene. *Nano Lett.* **2008**, *8*, 902–907. [CrossRef] [PubMed]
4. El-Kady, M.F.; Shao, Y.; Kaner, R.B. Graphene for batteries, supercapacitors and beyond. *Nat. Rev. Mater.* **2016**, *1*, 16033. [CrossRef]
5. Gonçalves, J.A.; Nascimento, R.; Matos, M.J.S.; de Oliveira, A.B.; Chacham, H.; Batista, R.J.C. Edge-reconstructed, few-layered graphene nanoribbons: Stability and electronic properties. *J. Phys. Chem. C* **2017**, *121*, 5836–5840. [CrossRef]
6. Bottari, G.; Herranz, M.A.; Wibmer, L.; Volland, M.; Rodriguez-Perez, L.; Guldi, D.M.; Hirsch, A.; Martin, N.; D'Souza, F.; Torres, T. Chemical functionalization and characterization of graphene-based materials. *Chem. Soc. Rev.* **2017**, *46*, 4464–4500. [CrossRef] [PubMed]
7. Suresh, S.; Wu, Z.P.; Bartolucci, S.F.; Basu, S.; Mukherjee, R.; Gupta, T.; Hundekar, P.; Shi, Y.; Lu, T.-M.; Koratkar, N. Protecting silicon film anodes in lithium-ion batteries using an atomically thin graphene drape. *ACS Nano* **2017**, *11*, 5051–5061. [CrossRef] [PubMed]
8. Zhu, Y.; Ji, H.; Cheng, H.-M.; Ruoff, R.S. Mass production and industrial applications of graphene materials. *Natl. Sci. Rev.* **2017**, *0*, 1–12. [CrossRef]
9. Emiru, T.F.; Ayele, D.W. Controlled synthesis, characterization and reduction of graphene oxide: A convenient method for large scale production. *Egypt. J. Basic Appl. Sci.* **2017**, *4*, 74–79. [CrossRef]
10. Phiri, J.; Gane, P.; Maloney, T.C. General overview of graphene: Production, properties and application in polymer composites. *Mater. Sci. Eng. B* **2017**, *215*, 9–28. [CrossRef]
11. Guo, H.; Li, X.; Li, B.; Wang, J.; Wang, S. Thermal conductivity of graphene/poly(vinylidene fluoride) nanocomposite membrane. *Mater. Des.* **2017**, *114*, 355–363. [CrossRef]
12. Zhou, Z.; Zhang, X.; Wu, X.; Lu, C. Self-stabilized polyaniline@graphene aqueous colloids for the construction of assembled conductive network in rubber matrix and its chemical sensing application. *Compos. Sci. Technol.* **2016**, *125*, 1–8. [CrossRef]
13. Georgakilas, V.; Tiwari, J.N.; Kemp, K.C.; Perman, J.A.; Bourlinos, A.B.; Kim, K.S.; Zboril, R. Noncovalent functionalization of graphene and graphene oxide for energy materials, biosensing, catalytic, and biomedical applications. *Chem. Rev.* **2016**, *116*, 5464–5519. [CrossRef] [PubMed]
14. Parhizkar, N.; Shahrabi, T.; Ramezanzadeh, B. A new approach for enhancement of the corrosion protection properties and interfacial adhesion bonds between the epoxy coating and steel substrate through surface treatment by covalently modified amino functionalized graphene oxide film. *Corros. Sci.* **2017**, *123*, 55–75. [CrossRef]
15. Zhang, Y.; Gong, S.; Zhang, Q.; Ming, P.; Wan, S.; Peng, J.; Jiang, L.; Cheng, Q. Graphene-based artificial nacre nanocomposites. *Chem. Soc. Rev.* **2016**, *45*, 2378–2395. [CrossRef] [PubMed]
16. Cheng, H.; Huang, Y.; Shi, G.; Jiang, L.; Qu, L. Graphene-based functional architectures: Sheets regulation and macrostructure construction toward actuators and power generators. *Acc. Chem. Res.* **2017**, *50*, 1663–1671. [CrossRef] [PubMed]
17. Park, M.S.; Lee, S.; Lee, Y.S. Mechanical properties of epoxy composites reinforced with ammonia-treated graphene oxides. *Carbon Lett.* **2017**, *21*, 1–7. [CrossRef]

18. Manna, R.; Srivastava, S.K. Fabrication of functionalized graphene filled carboxylated nitrile rubber nanocomposites as flexible dielectric materials. *Mater. Chem. Front.* **2017**, *1*, 780–788. [CrossRef]

19. Zahirifar, J.; Karimi-Sabet, J.; Moosavian, S.M.A.; Hadi, A.; Khadiv-Parsi, P. Fabrication of a novel octadecylamine functionalized graphene oxide/pvdf dual-layer flat sheet membrane for desalination via air gap membrane distillation. *Desalination* **2018**, *428*, 227–239. [CrossRef]

20. Liu, S.; Li, D.; Wang, L.; Yang, H.; Han, X.; Liu, B. Ethylenediamine-functionalized graphene oxide incorporated acid-base ion exchange membranes for vanadium redox flow battery. *Electrochim. Acta* **2017**, *230*, 204–211. [CrossRef]

21. Deng, W.; Zhang, Y.; Tan, Y.; Ma, M. Three-dimensional nitrogen-doped graphene derived from poly-o-phenylenediamine for high-performance supercapacitors. *J. Electroanal. Chem.* **2017**, *787*, 103–109. [CrossRef]

22. Pandit, S.; De, M. Interaction of amino acids and graphene oxide: Trends in thermodynamic properties. *J. Phys. Chem. C* **2017**, *121*, 600–608. [CrossRef]

23. Shanmugharaj, A.M.; Yoon, J.H.; Yang, W.J.; Ryu, S.H. Synthesis, characterization, and surface wettability properties of amine functionalized graphene oxide films with varying amine chain lengths. *J. Colloid Interface Sci.* **2013**, *401*, 148–154. [CrossRef] [PubMed]

24. Matsuo, Y.; Miyabe, T.; Fukutsuka, T.; Sugie, Y. Preparation and characterization of alkylamine-intercalated graphite oxides. *Carbon* **2007**, *45*, 1005–1012. [CrossRef]

25. Hung, W.-S.; Tsou, C.-H.; De Guzman, M.; An, Q.-F.; Liu, Y.-L.; Zhang, Y.-M.; Hu, C.-C.; Lee, K.-R.; Lai, J.-Y. Cross-linking with diamine monomers to prepare composite graphene oxide-framework membranes with varying d-spacing. *Chem. Mater.* **2014**, *26*, 2983–2990. [CrossRef]

26. Eng, A.Y.S.; Chua, C.K.; Pumera, M. Facile labelling of graphene oxide for superior capacitive energy storage and fluorescence applications. *Phys. Chem. Chem. Phys.* **2016**, *18*, 9673–9681. [CrossRef] [PubMed]

27. Wang, T.; Wang, L.; Wu, D.; Xia, W.; Zhao, H.; Jia, D. Hydrothermal synthesis of nitrogen-doped graphene hydrogels using amino acids with different acidities as doping agents. *J. Mater. Chem. A* **2014**, *2*, 8352–8361. [CrossRef]

28. Peng, T.; Sun, H.; Peng, T.; Liu, B.; Zhao, X. Structural regulation and electroconductivity change of nitrogen-doping reduced graphene oxide prepared using p-phenylene diamine as modifier. *Nanomaterials* **2017**, *7*, 292. [CrossRef] [PubMed]

29. Lu, X.; Li, L.; Song, B.; Moon, K.-S.; Hu, N.; Liao, G.; Shi, T.; Wong, C. Mechanistic investigation of the graphene functionalization using p-phenylenediamine and its application for supercapacitors. *Nano Energy* **2015**, *17*, 160–170. [CrossRef]

30. Hu, N.; Wang, Y.; Chai, J.; Gao, R.; Yang, Z.; Kong, E.S.-W.; Zhang, Y. Gas sensor based on p-phenylenediamine reduced graphene oxide. *Sens. Actuators B Chem.* **2012**, *163*, 107–114. [CrossRef]

31. Liu, B.; Sun, H.J.; Peng, T.J. Factor group analysis of molecular vibrational modes of graphene and density functional calculations. *Acta Phys. Chim. Sin.* **2012**, *28*, 799–804.

32. Eigler, S.; Dotzer, C.; Hirsch, A. Visualization of defect densities in reduced graphene oxide. *Carbon* **2012**, *50*, 3666–3673. [CrossRef]

33. Tomita, S.; Sakurai, T.; Ohta, H.; Fujii, M.; Hayashi, S. Structure and electronic properties of carbon onions. *J. Chem. Phys.* **2001**, *114*, 7477–7482. [CrossRef]

34. Roghani-Mamaqani, H.; Haddadi-Asl, V. In-plane functionalizing graphene nanolayers with polystyrene by atom transfer radical polymerization: Grafting from hydroxyl groups. *Polym. Compos.* **2014**, *35*, 386–395. [CrossRef]

35. Liu, Z.; Zhou, H.; Huang, Z.; Wang, W.; Zeng, F.; Kuang, Y. Graphene covalently functionalized with poly(p-phenylenediamine) as high performance electrode material for supercapacitors. *J. Mater. Chem. A* **2013**, *1*, 3454–3462. [CrossRef]

36. Han, Z.; Tang, Z.; Li, P.; Yang, G.; Zheng, Q.; Yang, J. Ammonia solution strengthened three-dimensional macro-porous graphene aerogel. *Nanoscale* **2013**, *5*, 5462–5467. [CrossRef] [PubMed]

37. Park, S.; Dikin, D.A.; Nguyen, S.B.T.; Ruoff, R.S. Graphene oxide sheets chemically cross-linked by polyallylamine. *J. Phys. Chem. C* **2016**, *113*, 15801–15804. [CrossRef]

38. Zhang, D.D.; Zu, S.Z.; Han, B.H. Inorganic–organic hybrid porous materials based on graphite oxide sheets. *Carbon* **2009**, *47*, 2993–3000. [CrossRef]

39. Ma, H.-L.; Zhang, H.-B.; Hu, Q.-H.; Li, W.-J.; Jiang, Z.-G.; Yu, Z.-Z.; Dasari, A. Functionalization and reduction of graphene oxide with p-phenylene diamine for electrically conductive and thermally stable polystyrene composites. *ACS Appl. Mater. Interfaces* **2012**, *4*, 1948–1953. [CrossRef] [PubMed]

40. Xue, B.; Zhu, J.; Liu, N.; Li, Y. Facile functionalization of graphene oxide with ethylenediamine as a solid base catalyst for knoevenagel condensation reaction. *Catal. Commun.* **2015**, *64*, 105–109. [CrossRef]

Article

Universal Effectiveness of Inducing Magnetic Moments in Graphene by Amino-Type sp^3-Defects

Tao Tang [1,*], Liting Wu [1], Shengqing Gao [1], Fang He [1], Ming Li [1], Jianfeng Wen [1], Xinyu Li [1] and Fuchi Liu [2,*]

[1] College of Science & Key Laboratory of Nonferrous Materials and New Processing Technology, Guilin University of Technology, Guilin 541004, China; tingabce@163.com (L.W.); gaoshengqing@hust.edu.cn (S.G.); hefang7132@163.com (F.H.); liming928@163.com (M.L.); wjfculater@163.com (J.W.); lixinyu5260@163.com (X.L.)

[2] College of Physics and Technology, Guangxi Normal University, Guilin 541004, China

[*] Correspondence: tangtao@glut.edu.cn (T.T.); liufuchi@gxnu.edu.cn (F.L.); Tel./Fax: +86-773-5897053 (T.T.)

Received: 25 March 2018; Accepted: 12 April 2018; Published: 17 April 2018

Abstract: Inducing magnetic moments in graphene is very important for its potential application in spintronics. Introducing sp^3-defects on the graphene basal plane is deemed as the most promising approach to produce magnetic graphene. However, its universal validity has not been very well verified experimentally. By functionalization of approximately pure amino groups on graphene basal plane, a spin-generalization efficiency of ~1 μ_B/100 NH$_2$ was obtained for the first time, thus providing substantial evidence for the validity of inducing magnetic moments by sp^3-defects. As well, amino groups provide another potential sp^3-type candidate to prepare magnetic graphene.

Keywords: graphene; sp^3-defect; amino group; magnetic moment

---

1. Introduction

The introduction of magnetic moments in graphene is a long-standing hot topic [1]. Generally speaking, the net spins in graphene come from unpaired electrons; however, all the electrons in the intrinsic graphene are compensated for owning to the π-symmetry system. Thus, breaking the symmetric structure of graphene is a feasible approach to make graphene magnetic. These approaches to introduce magnetic moments in graphene can be divided two ways [2]: (i) creating sp^3-defects on the basal plane of graphene sheets via atoms or functional groups chemisorbed on carbon networks to form covalent sp^3-type bonds, typically H [3], F [4] or hydroxyl group [2,5,6]; and (ii) producing edge-type defects at the edge sites via bombarding graphene sheets to introduce vacancies [4,7], cutting graphene into quantum dots [8], nanoribbons [9] or nanomeshes [10], or substituting vacancy-site carbon atoms by nitrogen atoms [11]. From the perspective of the spintronics application of a 2-dimensional film, the former is superior to the latter, since it does not need to damage the graphene sheet and can maintain the integrity of the film. However, a theoretical prediction has not been experimentally verified yet: is it indeed universal to introduce magnetic moments in graphene by covalent sp^3-type defects [12]?

In fact, atomic-scale control of the distribution of H atoms to paint magnetism on graphene has been achieved by scanning tunneling microscopy (STM) [13]. Graphene has also been proved to turn from diamagnetic to paramagnetic by fluorination [4]. It is undoubted that both H and F are effective at inducing magnetism in graphene and beyond that, only the effectiveness of hydroxyl-functionalized sp^3-type defects has been reported [2]. However, the existing reports indicate that hydroxyl-functionalization of graphene can only be achieved by further processing of graphene oxide (GO), which needs a strong oxidant such as potassium permanganate in Hummers' method [2] or potassium dichromate in Brodie's method [14] because graphene is inert and hard to be chemically functionalized. That is to say, magnetic metal impurities such as manganese or chromium will have to be brought into hydroxyl-functionalization. To make sure the magnetic signals are intrinsic to graphene while not from the magnetic pollutants, repetitious washing of hydroxyl-functionalized graphene or GO with acid and deionized water must be done. Even so, it is still hard to guarantee all the 3-D contaminants are completely disposed of. Namely, hydroxyl sp^3-defect may be not a good choice to experimentally confirm the universal validity for inducing magnetism in graphene. Amino group (NH_2) provides a better choice than the hydroxyl group.

In fact, there are a lot of ways to introduce highly atomic N in graphene [15,16]; however, the N types obtained are generally in-plane and there is no evidence they can generate amino sp^3-type defects on the graphene basal plane. It has been reported that illuminating graphene under white light in ammonia atmosphere is a feasible way to form amino-type sp^3-defects on the graphene basal plane [17,18]. By such a method, we can obtain a sufficient amount of sp^3-type graphene suitable for measurement on a superconducting quantum interference device (SQUID) without importing original 3-D metals, as long as graphene is not originated from GO and can be massively produced—Parvez et al. provided a good way to solve such a problem by electrolytic exfoliation of graphite [19]. Graphene material is also prospective in high-precision low magnetic measurements using new switching sensing devices, which has high sensitivity, and compensate temperature drift [20,21].

In this study, we illuminated electrolyzed graphene (EG) with a decent few-layer ratio in ammonia to successfully obtain sp^3-type N-doped graphene (sp^3-NG). Our results demonstrate that almost all the N atoms are bonded to graphene basal-plane carbon atoms in the form of amino groups, and these amino-type sp^3-defects can effectively introduce magnetic moments in graphene with an efficiency of $\sim$1 μ_B/100 NH_2. We firstly experimentally proved the universal validity of inducing magnetic moments by amino-type sp^3-defects and provide another potential candidate to prepare magnetic graphene, which is regarded as significantly crucial in the application of graphene spintronics. Furthermore, by using such sp^3-type magnetic graphene to introduce localized magnetic moments on graphene, controlling the spin scattering to control the magnetoresistance as dilute F-doped sp^3-functionalized graphene [22] is hopeful, and a potential alternative of light-element magnet [23] can be expected as well.

2. Experimental Section

2.1. Preparation

The graphene sheets were obtained through electrolytic exfoliation [19] of commercial graphite rods (99.999% purity, Beijing Gaochun, Beijing, China). Both electrodes were adopted as graphite rods and the electrolyte was ammonium sulfate solution with the concentration of 0.1 M. The distance between the two rods was 2 cm. The electrolytic voltage between the two poles was kept as 10 V and the initial output power was set as 10 W. Next, the exfoliated graphene sheets were collected with a PTFE membrane filter (0.2-μm pore size) by vacuum filtration and then washed by deionized water for 10 times. The graphene sheets were further exfoliated through ultrasonication in alcohol for 10 min at low power, and after 24 h of standing, only the supernatant was kept to render the sheets with a high few-layer ratio. To avoid any possible magnetic contaminants, the as-prepared graphene was washed by dilute nitric acid once and then by deionized water three times. Finally, after drying it in an oven at 60 °C, the original EG sample was successfully prepared.

Amino-functionalized sp^3-NG was obtained by irradiating EG for 30 min with light from a 500 W high-pressure Hg lamp in NH_3 (99%) atmosphere at a rate of 80 sccm [17]. For comparison, we also heat EG for 1 h at 200 °C under Ar atmosphere to get thermally treated graphene (TG) and under NH_3 atmosphere to get nitrogen-doped graphene (NG), respectively.

2.2. Instrumentation

The morphologies of the samples were characterized using transmission electron microscopy (TEM, JEM-2100F, JEOL, Tokyo, Japan), and the X-ray photoelectron spectroscopy (XPS) measurements were performed on ESCALAB 250Xi (TMAG, Waltham, MA, USA) using an Al Kα radiation. Raman spectra were performed on Renishaw inVia (Wotton-under-Edge, UK) using a laser excitation of 532 nm. The magnetic properties of the samples were measured using SQUID magnetometer with a sensitivity less than 10^{-8} emu (Quantum Design MPMS-XL, San Diego, CA, USA), and all data was corrected for the diamagnetic contribution by subtracting the corresponding linear diamagnetic background at room temperature. The 3-D impurity elements of all the samples are measured by inductively coupled plasma (ICP) spectrometry (Jarrell-Ash, Waltham, MA, USA).

3. Results and Discussion

Shown in Figure 1 are the typical TEM images of EG. It is easily found that the graphene sheets obtained by electrolytic exfoliation maintain two-dimensional ultrathin flexible structure and μm scale integrity, and are quite different from GO sheets with many ripples [2], the EG sheets are much flatter, maybe because there is no violent oxidation process during the preparation as GO has. From the curved edge of EG (see the red rectangle in Figure 1b), one can find a three- or four-layered graphene sheet, which means that by such kind of electrolytic exfoliation, few-layered graphene can be successfully obtained. Here we have to point out that, in the previous report [19], Parvez et al. got high-ratio mono-layered graphene through graphite electrolysis; however, it seems impossible to separate the mono-layered graphene sheets with sufficient quantity suitable for SQUID measurement, because we need at least several milligrams for each measurement to ensure the magnetic signals of the graphene samples are not completely flooded by the background signals.

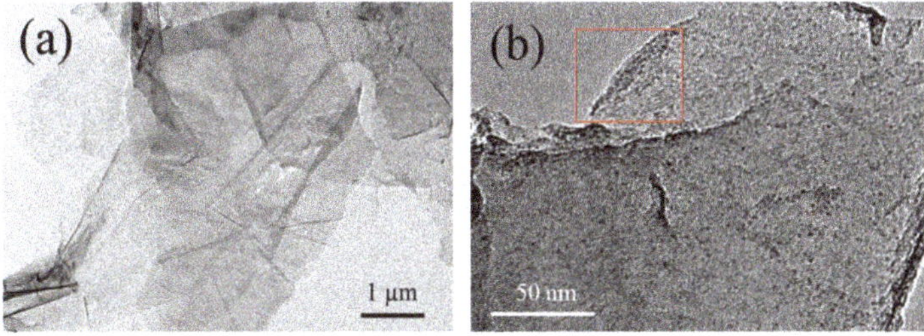

Figure 1. Typical TEM images of EG. The scale bar is (**a**) 1 μm and (**b**) 50 nm. In the red rectangular zone, the curved edge indicates the graphene sheet is few-layered.

The main features in the Raman spectra (Figure 2a) of carbon-based materials are the D, G and 2-D peaks that center at around 1350, 1580 and 2700 cm^{-1}, respectively. The shape and height of 2-D peak is similar to the previous report [19], typically characteristic of electrolytic few-layered graphene sheets [24]. Obviously, EG sheet cannot maintain the pristine sp^2-carbon-based structure of graphite or graphene since the D peak is prominent ($I_D/I_G = 0.44$), which demonstrates that during the electrolysis a lot of defects had been induced on the graphene sheet. By the determination of XPS measurement (Table 1), we found that EG has an oxygen atomic ratio of 12.1 at %, implying some oxygen groups are physically or chemically adsorbed on [19]. After thermally heating, TG has a distinctly lower D peak ($I_D/I_G = 0.32$) and oxygen content (5.5 at %), which means a lot of oxygen groups were removed and the pristine sp^2-aromatic structure was restored to a certain extent. From Figure 2b, one can find that the G peak of TG is located at ~1585 cm^{-1}, that is, with respect to 1580 cm^{-1} of pristine graphene, such blue shift means a lot of non-crystalline fractions still exist after heating [25]. It also can be seen in Figure 2b, by doping N through either thermal treatment (NG) or light treatment (sp^3-NG), the defect ratios are dramatically increased (I_D/I_G is 0.83 for NG and 0.90 for sp^3-NG, respectively). Note although the nitrogen contents of NG and sp^3-NG are very close (Table 1), the D peak of sp^3-NG is more prominent and the slight red-shift (~2 cm^{-1}) of G peak of NG imply they may have different N-doping types. Since in-plane N atoms implanting into graphene will lead to the red shift of the G peak, we may guess the N-doping type of sp^3-NG is out of plane, that is, amino sp^3-type [17]. Moreover, the Raman spectrum of sp^3-NG is very alike to lightly sp^3-functionalized F-doping few-layered graphene [26]. The following XPS measurements confirmed our guess.

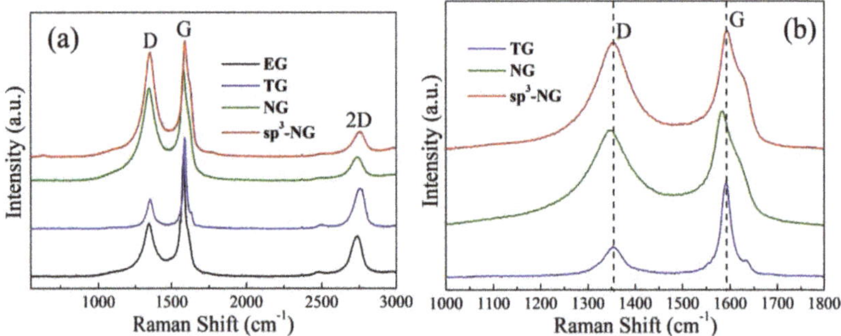

Figure 2. (a) Raman spectra of EG, TG, NG, and sp^3-NG; (b) The D and G peaks of TG, NG, and sp^3-NG.

Table 1. The ratios of different types of N and elemental contents of the typical samples.

Samples (at %)	Pyridinic-N	Pyrrolic-N	Graphite-N	Amino-N	N	O	C
EG	-	-	-	-	0.8	12.1	87.1
TG	-	-	-	-	0.7	5.5	93.8
NG	1.0	1.1	0.5	0	2.6	4.6	92.8
sp^3-NG	0	0	0.3	2.6	2.9	8.5	88.6

From the XPS spectra in Figure 3a, one can find that nitrogen peaks are almost invisible in EG and TG while evident in NG and sp^3-NG, so we know that by thermal or light treatment EG in ammonia, the nitrogen atoms can be successfully inserted into the carbon skeleton of graphene. To identify the difference of the N-bonding environments of these two kinds of nitrogen doping, we carefully deconvolute the fine-scanned N 1s spectra of NG and sp^3-NG (Figure 3b). Fairly interestingly, the sp^3-NG sample shows nearly only a single strong peak which located at ~399.4 eV, typically manifested as amino-type N bonding to carbon (NH$_2$–C) [17], accompanied with a very weak graphite-N subpeak located at ~401.7 eV. Unlike sp^3-NG, aside from a similar weak graphite-N subpeak, NG presents another two subpeaks which sit at ~398.3 and 400 eV, typically identified as pyridinic- and pyrrolic-N, respectively. As is known, pyridinic-, pyrrolic- and graphite-N are all in-plane in graphene sheet while amino-N is out of plane to form sp^3-defect, so we name the illuminated N-doped graphene as sp^3-NG. As shown in Table 1, according to the deconvoluted subpeak area, we calculated the different N-type ratios and found that NG has a total N content of 2.6 at % with a proportion of pyridinic-N:pyrrolic-N:graphite-N $\approx$ 2:2:1, and sp^3-NG is almost purely amino-N functionalized with a total N content of 2.9 at %. Naturally, the different types of N bonding will lead to different physical properties of N-doped graphene.

All the samples were compressed into a diamagnetic plastic bag for magnetism measurement. To make the results as accurate as possible, we used at least 20 mg samples for measurement of each run. Figure 4a shows the relationship of the mass magnetization (M) and the different applied magnetic field (H) of the typical samples under the temperature of 2 K. EG, TG, NG and sp^3-NG are all diamagnetic, which means the intrinsic diamagnetism of graphene is predominant in these samples. However, the diamagnetic degrees are different. The diamagnetic measurements of TG (blue dots in

Figure [4]a) can be perfectly fitted with a line (blue line in Figure [4]a), while EG, NG and sp^3-NG have paramagnetic signals mixed in the diamagnetic signals. Based on the Experimental procedure and XPS measurements, it is easily known that TG is obtained by thermal treatment of EG and has lower oxygen content (see Table [1]), indicating after the heating a lot of oxygen groups physiosorbed or chemisorbed on EG had been removed, so TG can be seen as a purer graphene sample than EG. Namely, the magnetic signals of TG can be taken as the pristine signals of diamagnetic graphene. Therefore, we can assess the magnetic moments induced by heteroatoms in EG, NG and sp^3-NG by performing the subtraction of the magnetization of TG (ΔM). We plotted ΔM under different H in Figure [4]b. Apparently, due to the positive ΔM of EG, NG and sp^3-NG, both O and N heteroatoms have successfully introduced magnetic moments in graphene. To fit the ΔM–H curves with Brillouin function

$$\Delta M = \Delta M_s \left[\frac{2S+1}{2S} Coth\left(\frac{2S+1}{2S}x\right) - \frac{1}{2S}Coth\left(\frac{x}{2S}\right)\right] \tag{1}$$

where saturated magnetization $\Delta M_s = NgS\mu_B$, $x = gS\mu_B H/(k_B T)$, k_B is the Boltzmann constant, N is the number of present magnetic moments, S is the spin angular momentum number, and g is the Landau factor assumed to be 2, we found that, all EG, NG and sp^3-NG can be well fitted by using $S = 1/2$, exhibiting the typical spin-1/2 paramagnetic behaviors of single point defects [4]. Correspondingly, ΔM_S of EG, NG and sp^3-NG are 0.045, 0.032 and 0.082 emu/g, respectively.

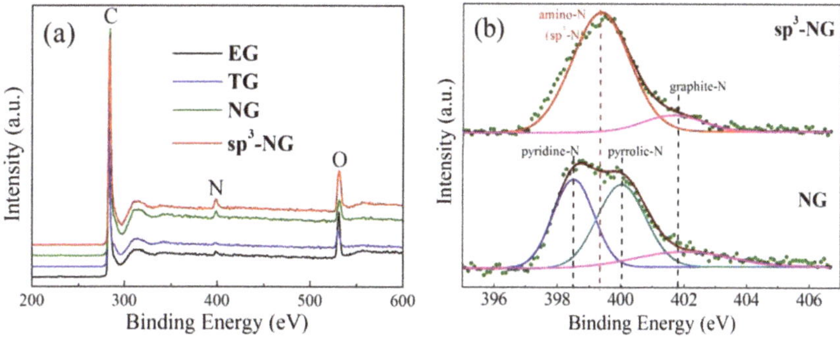

Figure 3. (**a**) XPS spectra of EG, TG, NG, and sp^3-NG; (**b**) Typical fine-scanned XPS spectra of N 1 s of NG and sp^3-NG. Blue, green, magenta, and red subpeaks are ascribed to pyridinic-, pyrrolic-, graphite- and amino-type (sp^3-type) N–C bonding, respectively.

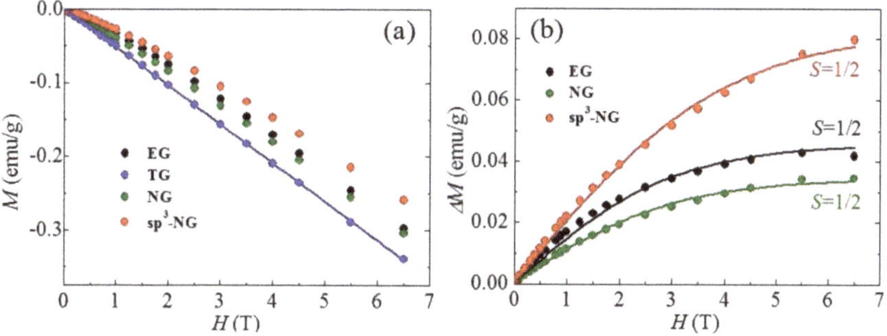

Figure 4. (**a**) Mass magnetization dependences on the applied magnetic field (*M*–*H*) of EG, TG, NG, and sp^3-NG. The dots are the measurements and the blue solid line is linearly fit to the blue dots; (**b**) The dependences of mass magnetization of EG, NG and sp^3-NG by subtracting which of TG on applied magnetic field (ΔM–*H*). The solid lines are fit to Brillouin function with $g = 2$. The measurement temperature is 2 K.

To analyze the magnetic sources of these samples, first, we must make certain the magnetic moments are not originated from 3-D contaminants. Through ICP measurement, the possible impurity contents are listed in Table 2, and one can find that their contents are trivial enough to be ignored. For instance, the highest Fe content in EG is 0.61 ppm, and it can generate the maximal magnetization of $\sim 4 \times 10^{-4}$ emu/g, which is far less than the magnetization we got (10^{-2} order of magnitude). Note that we burned at least 30 mg sample each time for ICP measurement, which avoided the influence from the uneven distribution of 3-D metals. Thus, we can further study the magnetic sources without considering possible extrinsic factors. Previous reports [2,5,27,28] indicate that adsorbed heteroatoms on the graphene basal plane can produce magnetic moments, for the p_z orbitals of carbon atoms are partly occupied to change the symmetric structure of π electron system. That is, uncompensated electron spins are generated. During the electrolysis, due to the decomposition of water, many oxygen atoms are physically adsorbed on or even form hydroxyl groups to bond to the graphene basal plane [19], thus inducing magnetic moments in EG. After heat treatment, some oxygen atoms were released, or unstable oxygen groups were decomposed, the bipartite honeycomb lattice of graphene was partly recovered and hence, some magnetic structures were converted back into non-magnetic states. As a result, TG is more diamagnetic than EG. What we are more interested is the magnetic moments induced by N atoms. However, the magnetization of NG is even lower than EG, implying the N atoms in NG did not introduce remarkable magnetic moments in graphene. Different from NG, sp^3-NG has an improvement of magnetization to about twofold of EG, indicating the N atoms in sp^3-NG are effective to induce magnetic moments. Clearly, the different types of N atoms resided on NG and sp^3-NG bring forth such different effects.

Table 2. The contents of the typical 3-D metal impurities of the samples. The unit is 'ppm'. 'ND' denotes 'not found'.

Impurities	Fe	Co	Ni	Cr	Mn	Al
EG	0.61	ND	0.07	0.05	0.08	0.25
TG	0.32	ND	0.03	0.04	0.09	0.16
NG	0.31	ND	0.06	0.05	0.10	0.20
sp^3-NG	0.50	ND	0.03	0.04	0.09	0.12

According to the XPS deconvolution results (Figure 3b), possible N types of NG and sp^3-NG were schematically represented in Figure 5. It is known that all the in-plane pyridinic-, pyrrolic- and graphite-N atoms can produce magnetic moments in carbon materials due to the extra electrons of N both experimentally and theoretically [11,28–30], but they face two big problems: (i) pyridinic- and pyrrolic-N can only exist at the vacancy- or edge-site of graphene to form highly active dangling bonds, therefore they are easily bonded to other atoms and the extra electrons are consequently covalently paired to lose the magnetic moments [31]; and (ii) the contents of graphite-N are generally extremely low in N-doped graphene. Up till now it has been hard to experimentally testify its effectiveness of inducing magnetic moments in graphene. For this reason, our NG sample is weakly paramagnetic (after subtraction of pristine diamagnetic signals of TG, see Figure 4b), and it is even weaker than EG—in other words, the magnetic moments induced by pyridinic-, pyrrolic- and graphite-N is even less than adsorbed oxygen groups. We also implemented thermally doping N at a higher temperature (500 °C)—generally it's seen as the most effective temperature to dope more N and results in higher magnetization—and found that the magnetization of NG is even lower than that which was thermally N-doped at 200 °C. To sum up, in-plane N atoms did not bring in any noteworthy effect of inducing magnetic moments in our graphene samples.

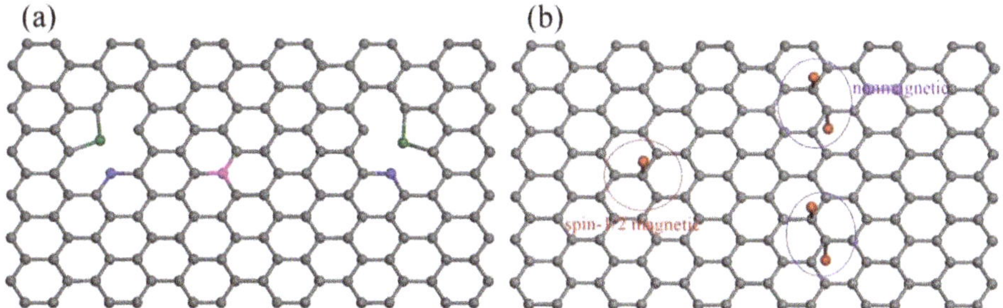

Figure 5. Schematic representation of (**a**) NG and (**b**) sp^3-NG. Carbon atoms are grey. Pyridinic-, pyrrolic- and graphite-N atoms are blue, green, and magenta, respectively. The red balls denote the amino groups (-NH$_2$) covalently bonded to the basal-plane carbon atoms. In the red oval region there is an isolated amino group and in blue ones two amino groups chemisorbed on different sublattices to form AB dimers.

As for an isolated amino-N, it creates a sp^3-type point defect on graphene basal-plane to introduce the magnetic moment of 1 μ_B (see red oval region in Figure 5b), without considering forming dangling bonds. Only when the nearest neighbor site is simultaneously occupied by another amino group (see blue oval region in Figure 5b) or when two adjacent sp^3-defects sit on the different sublattices (AB dimer) [12,13], will it be non-magnetic. Moreover, several near sp^3-defects sitting on the same sublattice can contribute large spin clusters (e.g., AA dimer and AAA trimer contribute 2 and 3 μ_B, respectively) [13]. However, the coupled spin clusters were not found in our sp^3-NG samples since its ΔM exhibit good spin-1/2 paramagnetic behavior (Figure 4b), so we can speculate the amino groups on sp^3-NG are all isolated or AB dual. The saturated ΔM is 0.082 emu/g, which means the efficiency of inducing magnetic moments in graphene by amino-type sp^3 defects is ~1 μ_B/3000 C or 1 μ_B/100 NH$_2$. Such an efficiency is close to fluorine-doped sp^3-defects (2–20 μ_B/1000 F) [4]. We tried to alter the amino coverage on the graphene basal-plane to tune the magnetism by changing the illumination time, but it cannot work because in a very short time (~1 min) the photochemical N-doping will nearly saturate and lose the ability of enhancing amino coverage. Anyway, as a sp^3-type defect, the amino group exhibits the universal effectiveness of inducing magnetic moments in graphene.

4. Conclusions

In summary, we have prepared almost purely amino-functionalized graphene by white-light illumination of EG. As a sp^3-type defect, the amino group can introduce magnetic moments in graphene with an efficiency of 1 μ_B/100 NH$_2$. As is known, the existence of localized magnetic moments is a prerequisite for magnetic coupling to induce ferromagnetism in graphene, which is deemed promising to design a spin field-effect transistor (SFET). Although the as-prepared amino-functionalized graphene is still diamagnetic, the validity of inducing magnetic moments of amino groups provides the imagination space to achieve ferromagnetic graphene. Moreover, unlike F atoms producing lots of holes on graphene basal plane to damage the integrity of graphene film when generating sp^3-defects [32], the photochemical process to dope amino groups on graphene is facile and keeps the completeness of 2-dimensional film, and thus, magnetic amino-functionalized graphene is more advantageous in SFET when film integrity is required. In short, the universal validity of spin generalization in graphene by sp^3-type defect was verified experimentally, laying a solid foundation for graphene magnetism theory, and paving the way for its potential applications in spintronics.

Acknowledgments: This work was financially supported by NSFC (Nos. 11604061, 51662004, 11664003, 11764011 and 11664007), Guangxi Natural Science Foundation (Nos. 2015GXNSFBA139002 and 2015GXNSFAA139015), the Scientific Research and Technology Development Program of Guilin (No. 2016012002), and Foundation of Guilin University of Technology (No. GUTQDJJ2002023), China.

Author Contributions: Tao Tang proposed the project and prepared the manuscript; Liting Wu, Shengqing Gao and Fang He performed the syntheses and measurements of the samples; Liting Wu, Shengqing Gao and Tao Tang analyzed the data. Ming Li, Jianfeng Wen, Xinyu Li and Fuchi Liu had valuable discussions and edited the manuscript.

Conflicts of Interest: The authors declare no conflict of interest.

References

1. Hollen, S.M.; Gupta, J.A. Painting magnetism on a canvas of graphene. *Science* **2016**, *352*, 415–416. [CrossRef] [PubMed]
2. Tang, T.; Tang, N.J.; Zheng, Y.P.; Wan, X.G.; Liu, Y.; Liu, F.C.; Xu, Q.H.; Du, Y.W. Robust magnetic moments on the basal plane of the graphene sheet effectively induced by OH groups. *Sci. Rep.* **2015**, *5*, 8448. [CrossRef] [PubMed]
3. Xie, L.; Wang, X.; Lu, J.; Ni, Z.; Luo, Z.; Mao, H.; Wang, R.; Wang, Y.; Huang, H.; Qi, D.; et al. Room temperature ferromagnetism in partially hydrogenated epitaxial graphene. *Appl. Phys. Lett.* **2011**, *98*, 193113. [CrossRef]
4. Nair, R.R.; Sepioni, M.; Tsai, I.L.; Lehtinen, O.; Keinonen, J.; Krasheninnikov, A.V.; Thomson, T.; Geim, A.K.; Grigorieva, I.V. Spin-half paramagnetism in graphene induced by point defects. *Nat. Phys.* **2012**, *8*, 199–202. [CrossRef]
5. Tang, T.; Liu, F.C.; Liu, Y.; Li, X.Y.; Xu, Q.H.; Feng, Q.; Tang, N.J.; Du, Y.W. Identifying the magnetic properties of graphene oxide. *Appl. Phys. Lett.* **2014**, *104*, 123104. [CrossRef]
6. Chen, J.; Zhang, W.L.; Sun, Y.Y.; Zheng, Y.P.; Tang, N.J.; Du, Y.W. Creation of localized spins in graphene by ring-opening of epoxy derived hydroxyl. *Sci. Rep.* **2016**, *6*, 26862. [CrossRef] [PubMed]
7. Ney, A.; Papakonstantinou, P.; Kumar, A.; Shang, N.-G.; Peng, N. Irradiation enhanced paramagnetism on graphene nanoflakes. *Appl. Phys. Lett.* **2011**, *99*, 102504. [CrossRef]
8. Sun, Y.Y.; Zheng, Y.P.; Chen, J.; Zhang, W.L.; Tang, N.J.; Du, Y.W. Intrinsic magnetism of monolayer graphene oxide quantum dots. *Appl. Phys. Lett.* **2016**, *108*, 033105. [CrossRef]
9. Rao, S.S.; Jammalamadaka, S.N.; Stesmans, A.; Moshchalkov, V.V.; van Tol, J.; Kosynkin, D.V.; Higginbotham-Duque, A.; Tour, J.M. Ferromagnetism in graphene nanoribbons: Split versus oxidative unzipped ribbons. *Nano Lett.* **2012**, *12*, 1210–1217. [CrossRef] [PubMed]
10. Yang, H.X.; Chshiev, M.; Boukhvalov, D.W.; Waintal, X.; Roche, S. Inducing and optimizing magnetism in graphene nanomeshes. *Phys. Rev. B* **2011**, *84*, 214404. [CrossRef]
11. Liu, Y.; Feng, Q.; Tang, N.J.; Wan, X.G.; Liu, F.C.; Lv, L.Y.; Du, Y.W. Increased magnetization of reduced graphene oxide by nitrogen-doping. *Carbon* **2013**, *60*, 549–551. [CrossRef]
12. Santos, E.J.G.; Ayuela, A.; Sánchez-Portal, D. Universal magnetic properties of sp^3-type defects in covalently functionalized graphene. *New J. Phys.* **2012**, *14*, 043022. [CrossRef]
13. Gonzalez-Herrero, H.; Gomez-Rodriguez, J.M.; Mallet, P.; Moaied, M.; Palacios, J.J.; Salgado, C.; Ugeda, M.M.; Veuillen, J.Y.; Yndurain, F.; Brihuega, I. Atomic-scale control of graphene magnetism by using hydrogen atoms. *Science* **2016**, *352*, 437–441. [CrossRef] [PubMed]
14. Khurana, G.; Kumar, N.; Kotnala, R.K.; Nautiyal, T.; Katiyar, R.S. Temperature tuned defect induced magnetism in reduced graphene oxide. *Nanoscale* **2013**, *5*, 3346–3351. [CrossRef] [PubMed]
15. Lin, Z.Y.; Waller, G.; Liu, Y.; Liu, M.L.; Wong, C.P. Facile synthesis of nitrogen-doped graphene via pyrolysis of graphene oxide and urea, and its electrocatalytic activity toward the oxygen-reduction reaction. *Adv. Energy Mater.* **2012**, *2*, 884–888. [CrossRef]
16. Samad, Y.A.; Li, Y.Q.; Schiffer, A.; Alhassan, S.M.; Liao, K. Graphene foam developed with a novel two-step technique for low and high strains and pressure-sensing applications. *Small* **2015**, *11*, 2380–2385. [CrossRef] [PubMed]
17. Liu, F.C.; Tang, N.J.; Tang, T.; Liu, Y.; Feng, Q.; Zhong, W.; Du, Y.W. Photochemical doping of graphene oxide with nitrogen for photoluminescence enhancement. *Appl. Phys. Lett.* **2013**, *103*, 123108. [CrossRef]
18. Xu, X.F.; Gao, F.H.; Bai, X.H.; Liu, F.C.; Kong, W.J.; Li, M. Tuning the photoluminescence of graphene quantum dots by photochemical doping with nitrogen. *Materials* **2017**, *10*, 1328. [CrossRef] [PubMed]
19. Parvez, K.; Wu, Z.S.; Li, R.J.; Liu, X.J.; Graf, R.; Feng, X.L.; Mullen, K. Exfoliation of graphite into graphene in aqueous solutions of inorganic salts. *J. Am. Chem. Soc.* **2014**, *136*, 6083–6091. [CrossRef] [PubMed]

20. Matko, V.; Jezernik, K. Greatly improved small inductance measurement using quartz crystal parasitic capacitance compensation. *Sensors* **2010**, *10*, 3954–3960. [CrossRef] [PubMed]
21. Matko, V. Next generation at-cut quartz crystal sensing devices. *Sensors* **2011**, *11*, 4474–4482. [CrossRef] [PubMed]
22. Hong, X.; Cheng, S.H.; Herding, C.; Zhu, J. Colossal negative magnetoresistance in dilute fluorinated graphene. *Phys. Rev. B* **2011**, *83*, 085410. [CrossRef]
23. Yazyev, O.V. Emergence of magnetism in graphene materials and nanostructures. *Rep. Prog. Phys.* **2010**, *73*, 056501. [CrossRef]
24. Ferrari, A.C.; Meyer, J.C.; Scardaci, V.; Casiraghi, C.; Lazzeri, M.; Mauri, F.; Piscanec, S.; Jiang, D.; Novoselov, K.S.; Roth, S.; et al. Raman spectrum of graphene and graphene layers. *Phys. Rev. Lett.* **2006**, *97*, 187401. [CrossRef] [PubMed]
25. Ferrari, A.C.; Robertson, J. Interpretation of Raman spectra of disordered and amorphous carbon. *Phys. Rev. B* **2000**, *61*, 14095–14107. [CrossRef]
26. Nair, R.R.; Ren, W.C.; Jalil, R.; Riaz, I.; Kravets, V.G.; Britnell, L.; Blake, P.; Schedin, F.; Mayorov, A.S.; Yuan, S.J.; et al. Fluorographene: A two-dimensional counterpart of Teflon. *Small* **2010**, *6*, 2877–2884. [CrossRef] [PubMed]
27. Lehtinen, P.O.; Foster, A.S.; Ayuela, A.; Krasheninnikov, A.; Nordlund, K.; Nieminen, R.M. Magnetic properties and diffusion of adatoms on a graphene sheet. *Phys. Rev. Lett.* **2003**, *91*, 017202. [CrossRef] [PubMed]
28. Dai, J.Y.; Yuan, J.M. Adsorption of molecular oxygen on doped graphene: Atomic, electronic, and magnetic properties. *Phys. Rev. B* **2010**, *81*, 165414. [CrossRef]
29. Ma, Y.C.; Foster, A.S.; Krasheninnikov, A.V.; Nieminen, R.M. Nitrogen in graphite and carbon nanotubes: Magnetism and mobility. *Phys. Rev. B* **2005**, *72*, 205416. [CrossRef]
30. Liu, Y.; Tang, N.J.; Wan, X.G.; Feng, Q.; Li, M.; Xu, Q.H.; Liu, F.C.; Du, Y.W. Realization of ferromagnetic graphene oxide with high magnetization by doping graphene oxide with nitrogen. *Sci. Rep.* **2013**, *3*, 2566. [CrossRef] [PubMed]
31. Nair, R.R.; Tsai, I.L.; Sepioni, M.; Lehtinen, O.; Keinonen, J.; Krasheninnikov, A.V.; Neto, A.H.C.; Katsnelson, M.I.; Geim, A.K.; Grigorieva, I.V. Dual origin of defect magnetism in graphene and its reversible switching by molecular doping. *Nat. Commun.* **2013**, *4*, 3010. [CrossRef] [PubMed]
32. Kashtiban, R.J.; Dyson, M.A.; Nair, R.R.; Zan, R.; Wong, S.L.; Ramasse, Q.; Geim, A.K.; Bangert, U.; Sloan, J. Atomically resolved imaging of highly ordered alternating fluorinated raphene. *Nat. Commun.* **2014**, *5*, 5902.

MDPI

St. Alban-Anlage 66

4052 Basel

Switzerland

Tel. +41 61 683 77 34

Fax +41 61 302 89 18

www.mdpi.com

Materials Editorial Office

E-mail: materials@mdpi.com

www.mdpi.com/journal/materials